Construction Technology for Tall Buildings

(2nd Edition)

Construction Technology for Tall Buildings

(2nd Edition)

CHEW Yit Lin, Michael

Department of Building
National University of Singapore

Published by

Singapore University Press
Yusof Ishak House, National University of Singapore
31 Lower Kent Ridge Road, Singapore 119078

and

World Scientific Publishing Co. Pte. Ltd.
P O Box 128, Farrer Road, Singapore 912805
USA office: Suite 1B, 1060 Main Street, River Edge, NJ 07661
UK office: 57 Shelton Street, Covent Garden, London WC2H 9HE

British Library Cataloguing-in-Publication Data
A catalogue record for this book is available from the British Library.

CONSTRUCTION TECHNOLOGY FOR TALL BUILDINGS (2nd Edition)

ISBN 981-02-4338-3 (pbk)

Printed in Singapore by FuIsland Offset Printing

To my father,
Chew Meng Ching

To my father

ACKNOWLEDGEMENT

I would like to express my sincere gratitude to my mother Madam Wong Shi Chin. Her courage and fortitude has inspired me to take on the task of writing this book. The work would not have been completed without her support and encouragement.

PREFACE

This new edition incorporates comprehensive revision and expansion to the previous edition with new topics and numerous new figures. The text introduces the latest construction practices and processes for tall buildings from foundation to roof. It attempts to acquaint readers with the methods, materials, equipment and systems used for the construction of tall buildings.

The target readers are practitioners and students in the related professions including architecture, engineering, building, real estate, project and property management, quantity and land surveying.

The text progresses through the stages of site investigation, excavation and foundations, basement construction, structural systems for the superstructure, site and material handling, wall and floor construction, cladding and roof construction. The construction sequence, merits and limitations of the various proprietary systems commonly used in these respective stages are discussed.

Chew Yit Lin, Michael
National University of Singapore

CONTENTS

CHAPTER 1
ASSEMBLY OF BUILDING

1.1. Development of Tall Buildings

The term "tall building" is not defined in specific terms related to height or the number of storeys. A building is considered tall when its structural analysis and design are in some way affected by the lateral loads, particularly sway caused by such loads [1].

It has always been a human aspiration to create taller and taller structures. Ancient structures such as the Tower of Babel, Colossus of Rhodes, the pyramids of Egypt, Mayan temples of Mexico, the Kutub Minar of India and many more were apparently built as symbols of power. They were monumental, protected and were infrequently used. Today, the determining factors for buildings to become higher are mainly the economic and social factors, although human ego and competition are still playing a role.

The history of the development of tall buildings can be broadly classified into three periods. The first period saw the erection of buildings such as the Reliance Building (Chicago, 1894, Figure 1.1), the Guaranty Building (Buffalo, 1895, Figure 1.2), and the Carson Pirie Scott Department Store (Chicago, 1904, Figure 1.3). Most of these buildings were masonry wall bearing structures with thick and messy walls. The horizontal and lateral loads of these structures were mainly resisted solely by the load bearing masonry walls. The 17-storey Manadnock Building (Chicago, 1891, Figure 1.4) for example, was built with 2.13 m thick masonry walls at the ground level. The area occupied by the walls of this building at the ground level is 15% of the gross floor area. In addition to reduced floor area, lightings and ventilations are major problems associated with thick wall construction.

Figure 1.1. Reliance Building, Chicago, 1894.

Figure 1.2. Guaranty Building, Buffalo, 1895.

Figure 1.3. Carson Pirie Scott Department Store, Chicago, 1904.

Figure 1.4. Manadnock Building, Chicago, 1891.

In the second period, with the evolution of steel structures, and sophisticated services such as mechanical lifts and ventilation, limitations on the height of buildings were removed. The demand for tall buildings increased in this period as corporations recognised the advertising and publicity advantages of connecting their names with imposing high-rise office buildings. It was also seen as sound financial investment as it could generate high rental income. The race for tallness commenced with a focus on Chicago and New York. Among the more famous buildings evolved during the period were the Woolworth Building (New York, 1930, Figure 1.5) and the Chrysler Building (New York, 1930, Figure 1.6). The race ended with the construction of the Empire State Building (New York, 1931, Figure 1.7) which, measuring 381 m with the television antenna, was the highest structure of the nineteenth century [2–6].

Reinforced concrete established its own identity in the 1950's into the third period which is now regarded as modernism in construction history. In contrast to the previous periods, where architectural emphasis was on external dressing and historical style, the third period placed emphasis on (a) reasons (b) functional and (c) technological facts [7, 8]. This new generation of buildings evolved from World Trade Centre (New York, 1972, Figure 1.8), Sears Tower (Chicago, 1974, Figure 1.9), to the recent Twin Towers (Kuala Lumpur, 1996, Figure 1.10).

The amount of materials needed in a tall building for the resistance of gravity load is almost linear with its height. However, the same materials needed for the resistance of lateral load (mainly wind load) increases as the square of the wind speed. The Sears Tower (Figure 1.9) which is about twice as tall as the Woolworth Building (Figure 1.5) has to resist wind effects four times as large as those on the Woolworth Building [9]. The third period of tall buildings saw the transition of structural systems from rigid frame to more efficient structural systems [9, 10]. The concept of channelling the gravity and wind loads using two or more separate structural systems, giving rise to buildings with flexible exterior frames and an inner core of stiff wind-bracing frames, reduces the building weight significantly. For taller buildings (in excess of roughly 60 storeys), the slender interior core and the planar frames are no longer sufficient to effectively resist the

Figure 1.5. Woolworth Building, New York, 1930.

Figure 1.6. Chrysler Building, New York, 1930.

Figure 1.7. Empire State Building, New York, 1931.

lateral force. The concept of an outer core, the perimeter structure of the building, must be activated to undertake this task by acting as a huge cantilever tube (e.g. World Trade Centre, Figure 1.8). Tubes are three-dimensional hollow structures internally braced by rigid floor diaphragms, with the cantilever out of the ground, such that overturning is resisted by the entire spatial structure as a unit and not as separate elements. To improve the shear stiffness of the framed perimeter tube, an inner braced steel or concrete tube may be added — tube in tube. The interaction causes the outer tube to primarily resist the rotation while the interior tube resists the shear (see Chapter 7 for more details). The Sears Tower (Figure 1.9) with nine square tubes of different heights "bundled" together is regarded as the most notable refinement of the tube concept.

In Singapore in the early fifties, only two buildings stood out as "high-rise", i.e. the old Cathay Building and the Asia Insurance building. Most other commercial and residential buildings were low-rise. The rapid economic growth over the last three decades in Singapore has seen an unprecedented explosion in residential, commercial and industrial development. Land scarce Singapore exhibited the necessity for the inevitable high-rise skyline, one which is now taken for granted.

Extensive coastal reclamation works have been made at Jurong, Tuas, Changi, the East Coast, stretching from Bedok to Tanjong Rhu and Marina South. Land at Jurong and Tuas were reclaimed from the swamp. Land was reclaimed from deeper waters, with marine clay substrata, at Changi, from Bedok to Tanjong Rhu and Marina South. Due to the urgent need for buildings, it has often been necessary to construct on these reclaimed areas almost immediately. The problem with excessive building settlement or large negative skin friction becomes critical.

1.2. Building Performance and System Integration

Performance is the measurement of achievement against intention. Building system integration is the act of creating a whole functioning building containing and including building systems in various combinations [11]. The various criteria including energy conservation, functional appropriateness,

Figure 1.8. World Trade Centre, New York, 1972.

Figure 1.9. Sears Tower, Chicago, 1974.

Figure 1.10. Twin Towers, Kuala Lumpur, 1996.

strength and stability, durability, fire safety, weathertightness, visual/acoustical comfort and economic efficacy, are only delivered when the entire building performs as an integrated whole [12]. Understanding the combination effect of the various systems on the delivery of each performance is thus important.

With buildings getting taller, more sophisticated and intelligent, integration between various aspects in physiology, psychology, sociology, economics, as well as the available technology is needed. A building needs to perform the functions of building enclosure against environmental degradation through moisture, temperature, air movement, radiation, chemical and biological attack or environmental disasters such as fire or flood. It also needs to provide interior occupancy requirements and the elemental parameters of comfort. Hartkopf [12] classified performance into six mandates: (1) spatial performance, (2) thermal performance, (3) indoor air quality, (4) acoustical performance, (5) visual performance and (6) building integrity. To achieve these performances requires good integration among all participants involved in the building process, from developer, designers, building professionals, fabricator to workmen on the site.

Building professionals nowadays are often required to participate right from the planning and design stage. In the case of design and build contract for instance, the contractor is responsible for both the design and construction of a project. It is thus important that a building professional understands the implications of these performances to the design, construction as well as maintenance of a building. Among others, consideration should be given to the following:

Structural Problems of Construction:

- Loadbearing: stability develops during construction (except for considerations of temporally works e.g. formwork).
- Frame construction: temporary provisions for stability, rigid joints, bracing, shear walls.
- Special structure: bridges, space structures, provision for erection stresses, sensitivity to construction or uneven loading.

Safety Margins, Construction Instability:

— Safety margins: reduced by accurate design.
— Failure characteristics: e.g. *in situ* versus prestressed.
— Construction instability: appropriate temporary support, e.g. shell construction, air-stabilised construction.

Settlement of Structures:

— Ground conditions: flexibility of structure where settlement occurs.
— Effect on building design and detailing.

Services Installation:

— Relationship to building use and structure.
— Provision of services
  - horizontal, vertical, ducting, ceiling spaces, penetration of slabs, beams etc.

— Integration of installation.

Table 1.1 illustrates the various performances required of a building.

Table 1.1. Building Performance Mandates [2].

Building Integrity	**Thermal Comfort**	**Acoustic Comfort**	**Visual Comfort**	**Air Quality**	**Spatial Comfort**
moisture temperature air infiltration radiation & light chemical attack biological attack fire protection wind, settlement	air & radiant temperature humidity air speed	reverberation & absorption vibration	artificial light & daylight	ventilation, natural & artificial mass pollution energy pollution	flexibility hidden services circulation

1.3. Cost, Quality and Time

The triangle involving cost, quality and time is well known with priority between the three relying on the client's objectives. Buildings related to commerce such as shopping complexes may require that time be the top priority so as to have the building commence operating before certain festive seasons, or in certain cases, to reduce financing bills etc. With a limited budget, cost may be the top priority. Quality may be emphasised in cases where the building itself is monumental or reputable in terms of height, architecture, appearance and background.

It is important that economic buildings do not necessarily mean unsafe buildings. Through proper design, management and execution, an economic building can provide the required standard at the lowest cost.

The basic resources for a building are: (1) money, (2) labour, (3) materials and (4) machinery. Labour must be employed and paid, materials must be purchased and machinery must be bought or hired. The manner in which materials are incorporated in the fabrication and structure of a building at the design stage and in which materials are handled and equipment deployed on the site or in a factory all affect the degree of expenditure of money and the overall economy of a building project [13, 14].

1.4. Building Regulations and Control

Building regulations are documents laying down the minimum requirements and standards that a building must comply with to ensure that the safety, hygiene, stability and level of amenity are compatible with environmental and social requirements at the time of construction and throughout the lifetime of the building [15–19].

The history of modern building legislation goes back to 1845 in the U.K. when the first Public Health Act was passed. Housing was then the primary concern and the defects to be fixed were mainly damp, structural instability, poor sanitation, fire risk and lack of light and ventilation [15]. In the U.K. in 1877, the first model bylaws were produced as a guide for local authorities on whom lay the responsibility for setting and enforcing minimum standards.

These applied only to new buildings, but in 1936 new legislation was enacted covering all buildings and requiring all local authorities to make and enforce "building bylaws". At this time, although guidance from central government was given, local authorities still held direct responsibility for building standards and issued their own bylaws which, although generally similar, had individual differences. The breakthrough came with the issue of the 1952 Model Bylaws, series IV (amended 1953) in which a different technique of control was used whereby standards of performance were stated, and these formed the mandatory part of the bylaws. A description of the actual structural minimal, which previously had been mandatory, were now contained in the so-called "deemed to satisfy provisions", leaving the way open for other newer methods and materials to be used, provided their performance could be established. This system has enabled increasing reference to be made to advisory publications such as British Standard Specifications and Codes of Practice.

In Singapore, the Building and Construction Authority (BCA) of the Ministry of National Development plays an important role in building control, especially after the collapse of Hotel New World in 1986. The Building Control Act empowers the Minister for Law and National Development to make such regulations for carrying out the purpose and provisions of the aforesaid Act in the interest of public health and safety. It sets out the scope of the regulations, namely, the submission of plans and specifications of works, the authorisation of persons qualified to submit the same and their duties and responsibilities, the construction, alteration and demolition of buildings with special emphasis on frontage, airspace, lighting, air conditioning, ventilation, height, approaches, entrances and exits, damp proofing, building materials, structural stability, drainage, sanitation, fire precautions and provision of car parking facilities.

The Fire Safety Bureau, a division of the Singapore Civil Defence Force, is another important authority in building control. Clearance from the Fire Safety Bureau is required for both the submission of plans as well as the application for the temporary occupation permit (TOP). Based on the Code of Practice for Fire Precautions in Building [20] and others, the authority scrutinises the design in terms of the escape exits, structural fire precautions, site planning and external fire fighting provisions, electrical power supplies, fire fighting systems, mechanical ventilation and smoke control systems.

There are also other subsidiary legislations and regulations imposed by other ministries. Examples are the "Environment Public Health Act" by the Ministry of Environment, and the "Factories (Amendment) Act" and "Factories Regulations" by the Ministry of Manpower.

1.5. Constraints and Resources

In Singapore, the combination of constraints and resources are probably unique compared with elsewhere. For this reason techniques of design and construction may be used successfully here where they have been considered unsuccessful elsewhere and vice versa.

The basic constraints on the construction process are control mechanism, construction resources, locational constraints, client requirements and restrictions, and design.

The basic resources are finance, time, technology & information, administrative & managerial skill, materials, labour and plant.

In Singapore, the government has been monitoring these constraints and resources through the then Construction Industry Development Board (CIDB) formed in 1984. In April, 1999, CIDB merged with Building Control Division of the Public Works Department (PWD) to form the Building and Construction Authority (BCA). The primary role of BCA is to develop and regulate Singapore's building and construction industry. The four key thrusts of BCA are:

- improving quality and productivity through high standards of excellence and the use of innovative construction technology,
- raising skills through training and testing to develop a professional construction workforce,
- ensuring building works are designed to comply with regulations and built to high safety standards,
- supporting industry growth through resource and information management.

1.6. Environmental Requirements

The exterior and interior environmental issues covering energy, resources and materials, transport, pollution, noise, landscape and ecology, waste management, etc. have been much discussed [21–26].

The effect of cyclic temperature and moisture, radiation such as ultraviolet light on building materials can be significant:

Temperature: The temperature range in the tropics is small compared with temperate countries with summer and winter as the two extreme seasons. However, joints are still very important on large areas and exposed faces and in some cases the requirements may be higher than those required in overseas design due to factors such as thermal shock. Thermal shock may also act to cause cracking of surfaces through the release of stresses built into materials during manufacture, e.g. bricks.

Sunlight: Ultraviolet light is considered significant in the deterioration of mastics, sealants, plastics and paints, etc. More details on the effect of ultraviolet on facade and roof coverings are discussed in Chapters 8 and 9.

Moisture: Since freezing may be ignored except for special applications, some design requirements used overseas are not significant here. Cyclic absorption and evaporation together with high temperature and humidity lead to problems with micro-organisms and fungi, etc. Detailing to minimise staining is important. Rising damp from the ground is common [27].

With tall buildings getting more sophisticated, so too the building users. They no longer tolerate thermal discomfort, glaring illumination, poor ventilation, energy wastage, poor acoustics and excessive noise, etc. Much improvement in environmental control has been made since the late eighteenth century with the advancement in building services. Rapid improvements are also seen in other areas e.g. activating the building fabric to provide the most favourable interior environment by controlling the utilisation of solar energy.

1.7. Industrialisation

Industrialisation in building is often related to prefabricating in a plant the maximum number of building works with the appropriate equipment and efficient technological and managerial methods. The greater the number of prefabricated components that are produced in the plant, the fewer the onsite works required. This will significantly reduce the dependence on skilled labour, the weather, the site and various other constraints [28–30].

The immediate benefits that can be derived from industrialisation are obvious:

- Saving in manual labour onsite especially in skilled trades such as formwork, masonry, plastering, painting, carpentry, tiling and M&E.
- Faster construction process.
- Higher quality of components attainable through careful choice of materials, equipment and quality control.

The application of industrialisation in Singapore was first attempted on 1963. It was implemented in a large scale in the 1980's. Figure 1.11 shows the basic sequence of erecting the main component of a building based on the system adopted by White Industries. In this system, horizontal units and volumetric units which include hollow core slabs, staircases, balcony spandrels, refuse chutes, bathroom units and service ducts are erected first. The floor slabs are lifted by vacuum suction lifters attached to the crawler crane and are placed horizontally onto the ground frames or preceding floor. They are then linked together by a tie system which includes internal, transverse and peripheral ties. Components like staircases, balcony spandrels, refuse chutes, bathroom box units and service ducts are hoisted by cables and placed in position. They are aligned and secured by dowel joints and reinforcement ties. Vertical panels like walls and frames, and internal partitions are lifted and aligned into the correct position on the floor slab by dowel bars. Packing mortar is applied to the joint where the panel is located and is held vertically by temporary props. The voids of the joints are cemented and the perimeter of the exposed joints are laid with grout seals to achieve watertightness. Structural connections in the form of tie beams are formed between vertical to vertical, horizontal to horizontal and vertical to horizontal components. When the grout to the joints or connections has hardened, the temporary props are removed. A 50 mm reinforced screed is then cast over the floor slab to ensure continuity. To complete the whole

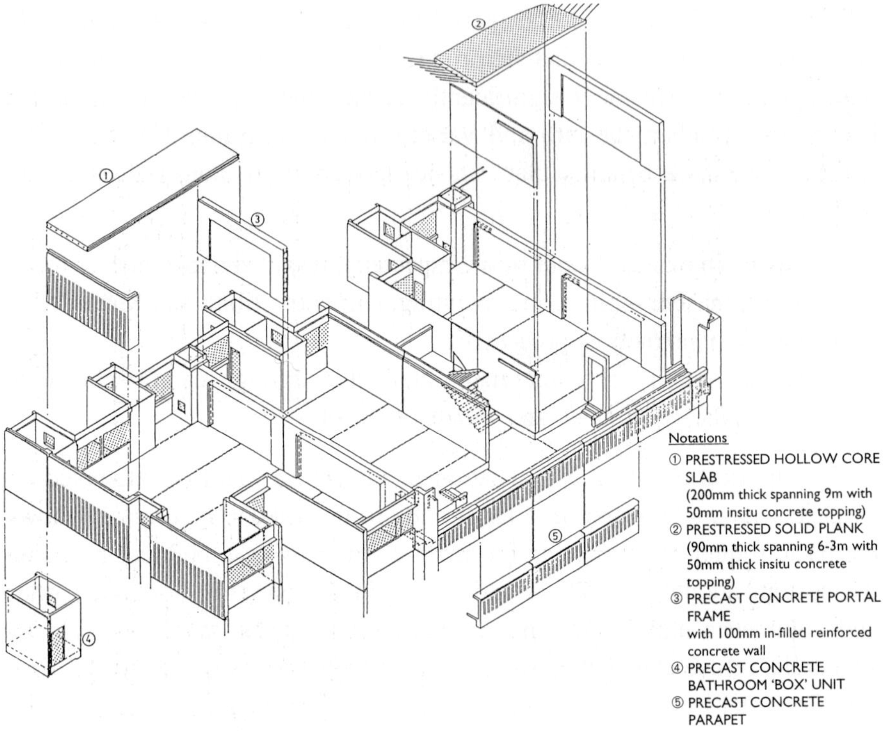

Notations
① PRESTRESSED HOLLOW CORE SLAB
(200mm thick spanning 9m with 50mm insitu concrete topping)
② PRESTRESSED SOLID PLANK
(90mm thick spanning 6-3m with 50mm thick insitu concrete topping)
③ PRECAST CONCRETE PORTAL FRAME
with 100mm in-filled reinforced concrete wall
④ PRECAST CONCRETE BATHROOM 'BOX' UNIT
⑤ PRECAST CONCRETE PARAPET

Figure 1.11. Isometric view of structural system by White Industries Pte Ltd.

building, the roof structure is erected with roof panels, parapet walls, roof water tank beams and panels.

1.8. Robotics in Construction

The casting, erection, jointing, connection and finishing of building components require a high level of skilled manual work onsite. The problem with the shortage of skilled personnel and the need to increase productivity in the industry has prompted research and development into robotics in the construction industry.

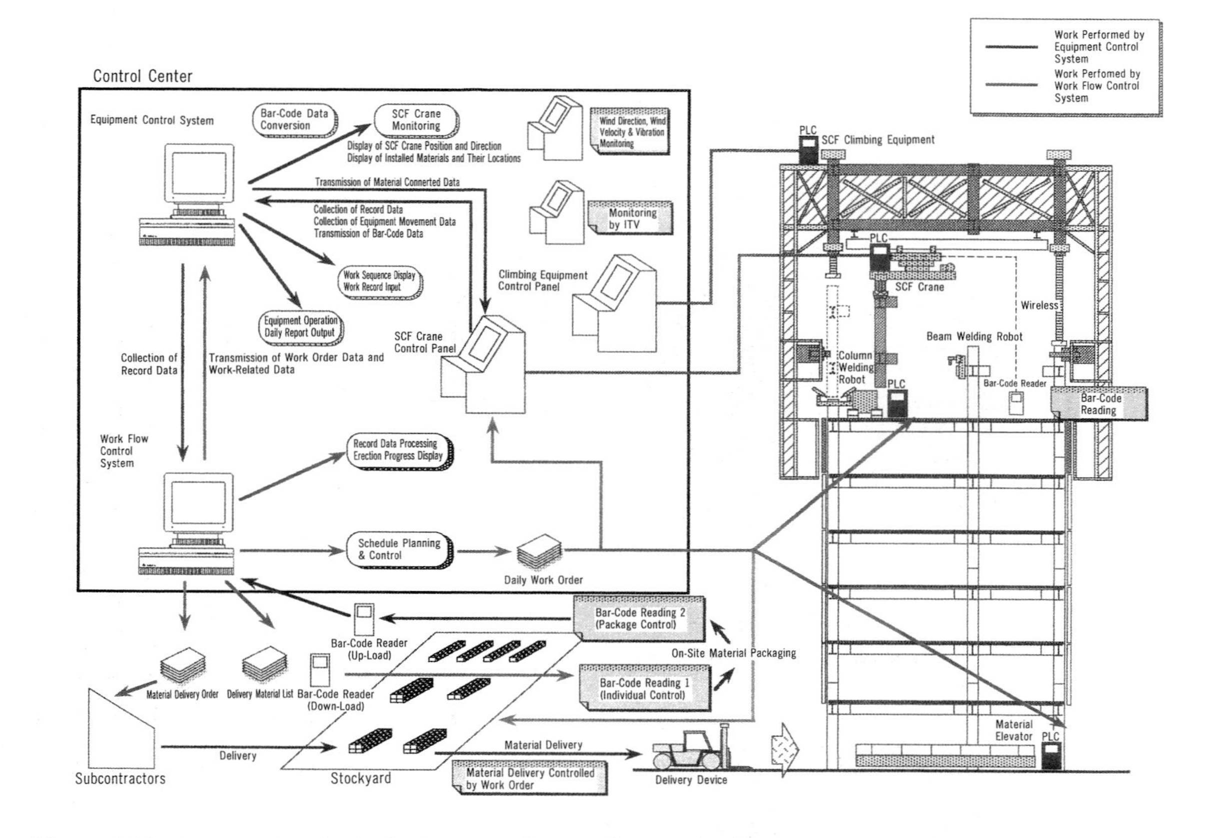

Figure 1.12. An example of robotised construction — Construction Factory (courtesy: Obayashi Corporation).

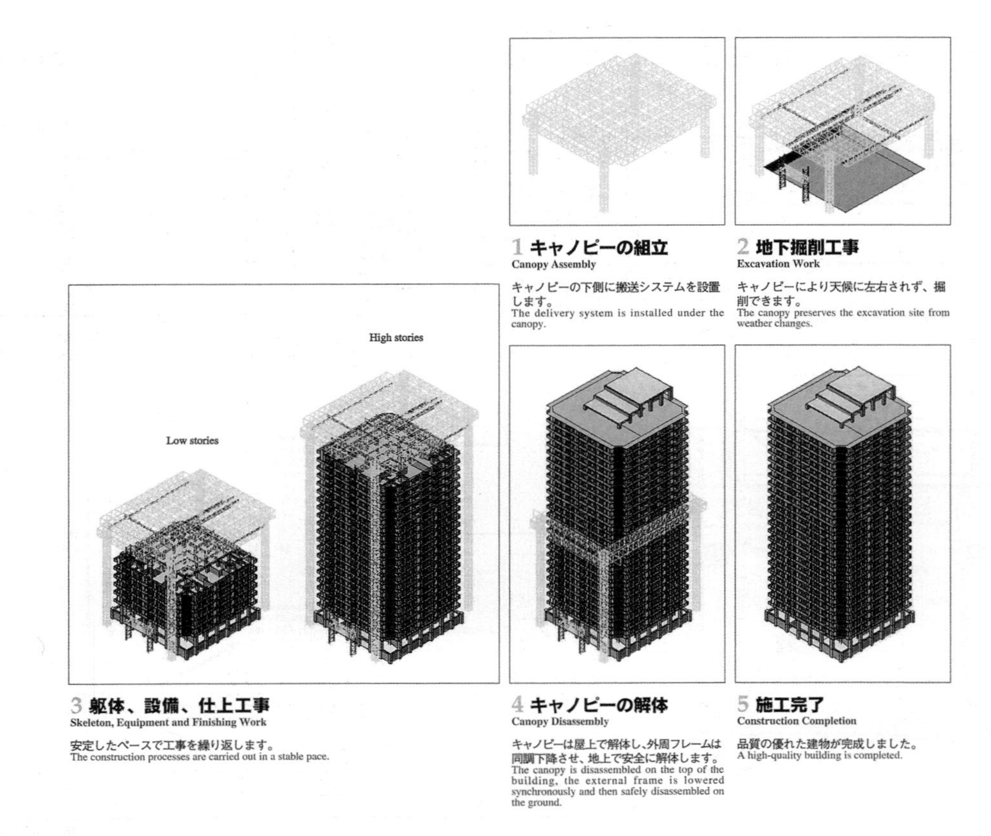

Figure 1.13. An example of robotised construction — Big Canopy (courtesy: Obayashi Corporation).

Recent progress in robot technology enables robots to perform sequences of tasks onsite, by interaction with its environment through electronic sensors. An example is Obayashi's "Super Construction Factory" which integrates the concepts of factory automation into the building site for steel structures (Figure 1.12). Building components and materials are delivered to the floor under construction through elevators and are lifted to the exact location of the floor by cranes. Welding and fastening are then carried out by robots. Upon completion of one floor, the factory is jacked up through an internal climbing system to commence work on the next floor.

Another system for reinforced concrete building named "Big Canopy" integrates technologies of climbing canopy, prefabricated components, automated assembly and computerised management systems (Figure 1.13). The canopy provides protection for the floor under construction from unfavourable weather and environmental conditions. Independent tower crane posts are used as four columns supporting the canopy. The rise of the canopy is performed by the climbing equipment of tower crane. Vertical movement of materials to and from the working story is by the use of lifts and horizontal movement by hoists. The movement of the hoists is entirely automated to improve work efficiency.

Mobile robots have also been used in isolated applications such as wall tile inspection [31], paint/concrete spraying, high pressure water jetting, concrete floor surface finishing, reinforcement laying etc. A survey of the use of construction robotics in Japan shows that the best returns are from tunnelling applications because of the hostile working environment and the shortage of skilled workers [32].

References

[1] B. S. Taranath, *Structural Analysis and Design of Tall Buildings*, McGraw-Hill, 1988.

[2] J. S. Foster, *Structure and Fabric*, Part 1, 5th Edition, Longman Scientific & Technical, 1994.

[3] C. Jencks, *Skyscrapers — Skycities*, Rizzoli, 1980.

[4] P. Goldberger, *The Skyscraper*, Allen Lane, 1981.

[5] L. S. Beedle, editor-in-chief, "Second Century of the Skyscraper", Council on Tall Buildings and Urban Habitat, Van Nostrand Reinhold, 1988.

[6] International Conference on Tall Buildings — a World View: Proceedings of 67th Regional Conference in Conjunction with ASCE Structures Congress XIV, Chicago, Illinois, USA, April 15–18, 1996. Bethlehem, Pa.: Council on Tall Buildings and Urban Habitat, Lehigh University, c1996.

[7] G. H. Douglas, *Skyscrapers: A Social History of the Very Tall Building in America*, McFarland & Co., c1995.

[8] S. B. Stafford, *Tall Building Structures: Analysis and Design*, Wiley, New York, 1991.

[9] M. Salvadori, *Why Buildings Stand Up*, WW Norton, New York, 1990.

[10] W. Schueller, *The Vertical Building Structure*, Van Nostrand Reinhold, 1990.

[11] R. D. Rush, *The Building System Integration Handbook*, Butterworth-Heinemann, 1986.

[12] V. Hartkopf, "The Office of the Future", Seminar Notes, CIDB Annual Seminar, October 1990.

[13] P. A. Stone, *Building Design Evaluation — Cost-in Use*, E & F.N. Spon, 1980.

[14] P. A. Stone, *Building Economy — Design, Production and Organisation*, Pergamon Press, 1976.

[15] A. J. Elder, *Guide to the Building Regulations 1985*, Anchor Press, 1989.

[16] Building Operations and Works of Engineering Construction Regulations, Singapore Government Printer, 1985.

[17] The Building Control Act, Singapore Government Printer, 1989.

[18] Local Government Integration Act, Singapore Government Printer, 1985.

[19] A Guide to the Factories Act, Ministry of Labour, 1988.

[20] Fire Safety Bureau, "Code of Practice for Fire Precaution in Buildings", Fire Safety Bureau, Singapore Civil Defence Force, Singapore, 1997.

[21] R. Venables, *Environmental Handbook for Building and Civil Engineering Projects*, CIRIA, London; Thomas Telford Services, 1994.

[22] Environmental Issues in Construction: A Review of Issues and Initiatives Relevant to the Building, Construction and Related Industries, CIRIA, London, 1993.

[23] M. N. Fabrick, *Environmental Planning for Design and Construction*, Wiley, New York, 1982.

[24] F. Moore, *Environmental Control Systems — Heating Cooling Lighting*, McGraw-Hill, 1993.

[25] H. J. Cowan, *Environmental Systems*, Van Nostrand Reinhold, 1983.

[26] B. B. P. Lim, "Environmental Design Criteria of Tall Buildings", Lehigh University, 1994.

[27] M. Y. L. Chew, L. H. Kang and C. W. Wong, *Building facades: A Guide to Common Defects in Tropical Climates*, World Scientific, Singapore, 1998.

[28] A. Warszawski, *Industrialisation and Robotics in Building*, Harper & Row, 1990.

[29] B. Russel, *Building Systems Industrialisation and Architecture*, Wiley, New York, 1981.

[30] B. J. Sullivan, *Industrialisation in the Building Industry*, Van Nostrand, New York, 1980.

[31] M. Ebihara, "The development of an apparatus for diagnosing the interior condition of walls", Proc. 5th Int. Symp. Robotics in Construction, Tokyo, 1988.

[32] S. Obayashi, "The surrounding of construction industry and problems of automatisation and robotisation in construction in Japan", Proc. 5th Int. Symp. Robotics in Construction, Tokyo, 1988.

CHAPTER 2

SAFETY AND HEALTH

It must be shocking for many to hear that it was common practice in the early years of tall building construction to assume in cost estimating that accidents would claim one life for say every two floors or each million dollars of general construction work performed [1]. It was a widespread myth then that construction accidents are inevitable.

Things have changed and more and more effort is put on minimising accidents not only because it is a moral obligation, but also because of the realisation that through effective safety management, tremendous cost saving is possible. Accidents may result in high direct and indirect costs. The direct costs include the medical costs, workers' compensation and other insurance benefits. The indirect costs include reduced productivity, job schedule delays, damage to equipment and facilities, low morale among workers, and possible additional liability claims [2–13].

2.1. Accident Prevention

Accidents are costly not only in terms of human suffering, it also results in loss of production, poor morale and lower productivity. Construction industry has always been one with among the highest accident rate. Table 2.1 lists the number and categories of accidents in the various industries in Singapore from 1985 to 1998.

In 1998, a total of 4247 accidents were reported to the Chief Inspector of Factories, of which 36% of the reported accidents occurred in the construction industry. Details of key industrial safety statistics can be found in (http://www.gov.sg/mom/dis/stats.html).

Table 2.1. Industrial accidents by degree of incapacity in selected industries, 1985–1998.

Degree of Incapacity in Selected Industries	1985	1986	1987	1988	1989	1990	1991	1992	1993	1994	1995	1996	1997	1998
ALL INDUSTRIES	4,357	3,856	4,155	4,328	4,835	4,889	5,154	4,698	4,257	4,003	3,947	4,306	4,422	4,247
Temporary Disablement Cases	4,162	3,711	3,971	4,127	4,637	4,714	4,954	4,512	4,076	3,833	3,801	4,126	4,195	3,995
Permanent Disablement Cases	134	102	140	165	136	119	134	121	104	117	82	107	124	161
Fatal Cases	61	43	44	36	62	56	66	65	77	53	64	73	103	91
Shipbuilding & Repairing	594	529	673	814	931	1,036	1,277	1,140	954	829	803	754	681	666
Temporary Disablement Cases	575	511	658	784	899	1,007	1,238	1,112	922	798	784	726	658	635
Permanent Disablement Cases	9	8	8	19	15	18	17	15	15	16	10	17	14	22
Fatal Cases	10	10	7	11	17	11	22	13	17	15	9	11	9	9
Construction	1,427	1,090	806	526	612	597	802	802	764	856	887	1,243	1,538	1,532
Temporary Disablement Cases	1,364	1,045	771	497	567	551	747	746	698	798	825	1,159	1,417	1,414
Permanent Disablement Cases	21	19	16	12	10	10	20	14	17	26	19	33	49	51
Fatal Cases	42	26	19	17	35	36	35	42	49	32	43	51	72	67
Other Industries	2,336	2,237	2,676	2,988	3,292	3,256	3,075	2,756	2,539	2,318	2,257	2,309	2,203	2,049
Temporary Disablement Cases	2,223	2,155	2,542	2,846	3,171	3,156	2,969	2,654	2,456	2,237	2,192	2,241	2,120	1,946
Permanent Disablement Cases	104	75	116	134	111	91	97	92	72	75	53	57	61	88
Fatal Cases	9	7	18	8	10	9	9	10	11	6	12	11	22	15

Note: Figures refer to number of cases.

Source: Industrial Safety Section, Ministry of Manpower, Republic of Singapore

It should be noted when analysing these figures that the accident rate is very sensitive to many factors other than the safety standard. Some of the major factors are:

— construction activity which can fluctuate frequently
— types of construction — public/private
— proportion of foreign unskilled workers
— height of buildings — system versus traditional construction technology

The temporary duration of work on sites and the rapidly changing character of the work and the workforce have been major problems in terms of safety, especially for small contractors. This has made long term investment on acquiring, installing and utilising proper safety equipment difficult. The rapidly changing character of construction work causes site hazards to change continuously. What is safe today may become hazardous the next day. This, coupled with the high usage of temporary workers, especially unskilled foreign workers, has made the job of maintaining high safety standard difficult.

A successful accident prevention scheme requires the following four basic activities:

a. A risk assessment analysis of all working areas to quantify and control physical or environmental hazards which can contribute to accidents.
b. A study of all operating methods and practices.
c. Provide education, instruction, training incentives and discipline (e.g. by imposing a fine) to minimize human factors which contribute to accidents.
d. Carry out thorough investigations of every accident, including accidents which do not result in personal injury (so called near-misses). Accident investigation is a defence against hazards that are overlooked in the first three activities, those that are less obvious, or hazards that are the result of combinations of circumstances that are difficult to foresee.

2.2. Safe Working Environment

Most accidents in a worksite are categorised under "falls of person" and "workers struck by falling objects" [14]. These two causes account for more than 50% of all cases and 70% of all fatal cases. In 1998, out of the 67 fatal cases in the construction industry, 34 cases were due to the fall of persons from height (51%) and 14 cases due to workers struck by falling objects (21%).

2.2.1. *Falls of Person*

To combat against fall of persons, two approaches can be adopted. (1) A proper working platform should be provided to workers whenever practicable. The working platform should be of adequate width, carrying capacity and with sufficient guardrails to afford a safe and steady foothold and handhold. The width should not be less than 635 mm and toe-boards must be provided (Figure 2.1). (2) In the case where a platform cannot be provided for reasons of space constraint and the like, safety belts and lifelines which are adequately anchored should be provided (Figure 2.2 and Figure 2.3).

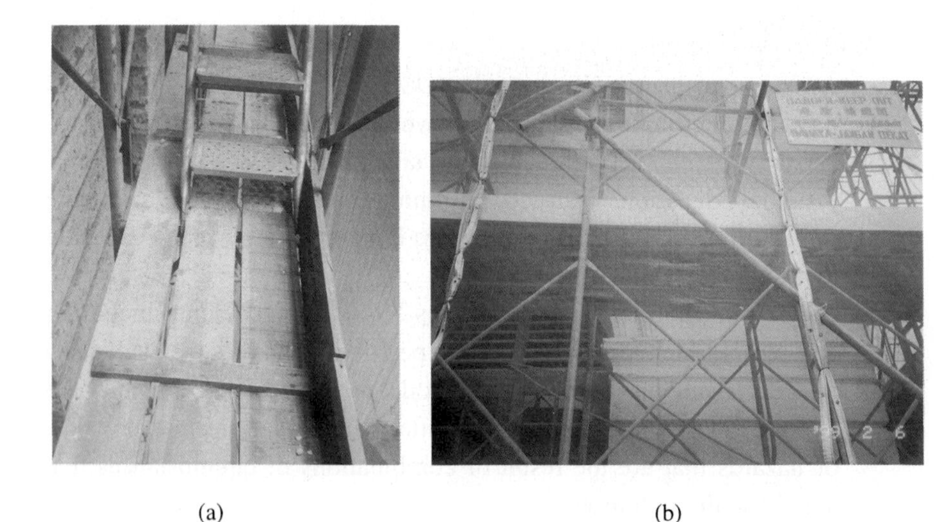

(a) (b)

Figure 2.1(a) and (b). Working platform of not less than 635 mm with toe-boards.

Figure 2.2. Reflective signage tide around the perimeter.

Figure 2.3. Provision of handrails incorporated in the system formwork.

Platforms would be difficult to provide if scaffolds and especially access scaffolds are not provided. Access scaffolds have transformed the method of construction, making building construction much safer. These are periphery scaffolds provided from the ground upwards to the floor under construction, enveloping the building. The envelopment facilitates the provision of platforms to formwork workers and has prevented many fatal falls of persons.

2.2.2. *Falling Objects*

Measures against person struck by falling objects include (a) good housekeeping and minimizing debris being generated, hence less falling material, (b) systematic and regular disposal of accumulated debris, provision of perimeter overhead shelters (Figure 2.4), (c) access and egress shelters to building, provision of safety nets (Figure 2.5), provision of pedestrian walkway or hoarding (Figure 2.6) and (d) the compulsory wearing of safety helmets. One can also segregate activities which are likely to generate falling objects away from potential victims working at the ground level [15–20].

Figure 2.4. Peripheral shelter (catch platform).

Figure 2.5. Provision of safety nets.

Figure 2.6. Safety pedestrian walkway (hoarding).

2.3. Measures Against Aggravated Accidents

"Aggravated accidents" are obvious hazards complicated or made more dangerous by conditions immediately surrounding it. As an example, the daily clearing of debris from the floors of the workspace and the edge of opensides can effectively contribute to the reduction of hazards. It is important to have a good housekeeping scheme for each project.

2.4. Safety Programme Organisation

The Ministry of Manpower (MOM) requires that all contractors, in their submission of the application for registration of a worksite to include the proper documents as shown in Table 2.2 and other related information pertaining to the safety organisation and the safety programme [21].

Safety programme organisation is defined as a method employed by the management to share and to assign responsibility for accident prevention and to ensure performance under that responsibility. The prevention of accidents and injuries is basically achieved through the control of the working

Table 2.2. Documents to accompany application for registration of a worksite.

- Photocopy of workers' SOC certificates/receipts for those have attended/intend to attend the Safety Orientation Course (SOC) conducted by the Occupational Safety and Health Training and Promotion Centre (OSHC) or the Construction Industry Training Centre.
- Work Safety Programme.
- Site Layout and Location Plans.
- Site Safety Supervisor Certificate.
- Site Responsibility Chart.
- Safety Committee Organisation Chart.
- Safety Provision Undertaking Form for Occupier.
- Safety Provision Undertaking Form for Site Safety Supervisor.
- Tower Crane Documents (only for worksites using tower crane).
- Public Utility Board (PUB) Electrical License for Temporary Electrical Installation Declaration of presence/absence of Asbestos-containing materials (for worksite involving demolition, repair or refurbishing works.

Source: Industrial Safety Section, Ministry of Manpower, Republic of Singapore

environment and of workers' behaviour. A company that has an effective safety programme will be able to conduct its operations economically, efficiently and safely [22].

2.5. Safety Policy

To achieve an acceptable safety performance, it is necessary to draw up a safety policy and have it publicised and promoted. The policy should be brief and clear and the management attitude well defined.

Safety policy should be top down and should involve everybody on the site. Top management's attitude and approach toward accident prevention is of paramount importance and should be reflected right from the planning stage. The duties and responsibilities of personnel involved in a typical project is shown in Table 2.3.

Table 2.3. Duties and responsibilities of personnel in a typical project.

Personnel	Duties and responsibilities
Senior Manager	(a) overall in charge of the safety management of the company. (b) to review and update the company's safety policy.
Project Manager	(a) overall in charge of the safety aspect related to his site. (b) Chairman for the site safety committee. (c) to ensure proper records such as personal protective equipment, machinery and equipment inspection records, etc. are kept at the site office. (d) to arrange for professional engineers to inspect machinery equipment at the site, as required by the Ministry of Manpower and other authorities.
Plant Manager	(a) overall in charge of the safety aspects involving plants and equipment. (b) to arrange for all equipment to be maintained and checked regularly to ensure that they are in good working order.
Safety Officer	(a) to advise the site staff on all matters related to safety. (b) to investigate accidents occurred at the worksites; to put up accident investigation reports and recommend corrective actions. (c) to compile relevant reports and records, e.g. accident statistics.

Table 2.3. (*Continued*).

Safety Officer	(d) to liase with authorities such as the Ministry of the Environment and the Ministry of Manpower. (e) centralised control of purchase of safety equipment.
Foreman	(a) to ensure that the provisions of the Factories Act and the Building Operations and Works of Engineering Construction Regulations and any other regulations made thereunder are complied with. (b) to promote the safe conduct of the work within the worksite. (c) to rectify any unsafe conduct of the work within the worksite. (d) to check the sub-contractor's work to ensure compliance with the Acts and Regulations.
Safety Supervisor	(a) to carry out daily safety inspection; a copy of the inspection report should be kept in the site office. (b) to act as secretary of the Site Safety Committee. (c) to investigate accidents occurred at worksite.
Mechanics	(a) to carry out regular maintenance and checking to all equipment. (b) to report to the plant engineer immediately if safety provisions are found to be tampered with or not provided to the machine. (c) to attend to the repair and maintenance of machines promptly.
All Staff	(a) to wear and make use of all personal protective equipment issued while carrying out their duties. (b) all equipment and hand tools must be kept in good condition. (c) must observe all safety precautions and report any unsafe conditions immediately. (d) must accept safety training. (e) to report all injuries, accidents or dangerous occurrence.
Sub-contractor	(a) to adhere to the Safety and Accident Prevention Programme. (b) to instruct workers under their charge to work safely. (c) to send their workers for the Safety Orientation Course conducted by Ministry of Manpower (MOM) or Building and Construction Authority (BCA).

Basic to a policy declaration are these statements:

a. Safety of employees, the public and company operations are paramount.
b. Safety will take precedence over expediency or short cuts.

c. That every attempt will be made to reduce the incidence of accidents.
d. That the company intends to comply with the Factories Act and all other safety regulations.

2.6. Implementing the Safety Policy

Once a safety policy statement is established, it should be publicised so that every employee becomes familiar with it. The effectiveness of any safety program hinges on the amount of support given to it by the management. Expressions of management support include enforcement of policy and programme (e.g. penalties), recognition of good safety records (e.g. rewards), review of safety reports, participation in meetings, safety campaigns and other events to show their support.

2.7. Assignment of Responsibilities — Safety Professionals

While top management has the ultimate responsibility for safety, it delegates authority for safe operation all the way down through all management levels. The supervisor is the key man in a safety programme because he is in constant contact with the employees. The safety professional acts in a staff capacity to help administer policy to provide technical support and information, to help train and to supply programme material. In Singapore, the employment of safety officers for some "factories" is mandatory while for construction worksites the employment of safety supervisors is compulsory by law. For more complex industries a more qualified safety professional may be in order.

In Singapore, the site safety supervisor is obliged by law to ensure that the provisions of the Factories Act and other safety regulations are complied with and is expected to promote the safe conduct of the work within the site.

He is also expected to correct any unsafe practice, rectify any unsafe place of work to ensure compliance with the Factories Act. He shares all these duties with the contractor's safety supervisor (safety supervisors employed by sub-contractors). A site safety supervisor is also required to conduct/attend meetings of the safety committee as its secretary.

2.8. Safety Committees

The main contractor of a worksite in which 50 or more persons are employed is required by law to establish a safety committee. This is one of the measures to counter the ill-effects of multi-tier contracting. A safety committee should comprise representatives of both the employee and the management. The number of representatives of the employees should be as large as possible in order to develop and maintain their interest in establishing safe and healthy working conditions. Only matters relating to safety and health in the worksites should be discussed at any meeting of a safety committee. Minutes of meetings should be kept and any decisions reached must be disseminated to all involved for implementation and follow-up. A safety committee is expected to inspect the worksite at least once a month and also immediately after any accident or dangerous occurrence had happened.

2.9. Acts and Regulations

The main Act of Law concerning industrial safety is the Factories Act [23]. Since 1921 when the first legislation was introduced to prevent accidents resulting from the use of machinery, the then Machinery Ordinance had since metamorphosised into what is now known as the Factories Act. As a matter of interest, the Machinery Ordinance originated from the concept that machinery could cause accidents, hence if machinery was safe, there would be no accidents [24, 25].

The Building Operations and Works of Engineering Construction Regulations (BOWEC) 1971 had been promulgated to spell out in greater detail the safety requirements for various operations carried out at worksites. In 1977, the Building Operations and Works of Engineering Construction Regulations 1977 was promulgated superseding the 1971 version. In the 1977 version, timber scaffolds were detailed to help contractors to erect safe timber (bintangor) scaffolds. Among other changes, the employment of safety supervisors at worksites and the formation of safety committees were made compulsory. In 1985, prompted by the construction of the Mass Rapid Transit (MRT) in Singapore as well as to answer to the need to upgrade and modernise the construction industry in Singapore, the Factories (Building Operations and Works of Engineering Construction) Regulations 1985 was promulgated, superseding the 1977 version.

Details on the relevant acts and regulations and a guide to the more important provisions of the Factory Act can be found in MOM's website under the Department of Industrial Safety (http://www.gov.sg/mom/dis/dis.html).

References

[1] R. E. Levitt and N. M. Samelson, *Construction Safety Management*, 2nd Edition, John Wiley and Sons, 1993.

[2] C. D. Reese, *Handbook of OSHA Construction Safety and Health*, Rewis, Boca Raton, 1999.

[3] National Safety Council of Australia, "Safety for the Building Industry", McGraw-Hill, Sydney, 1970.

[4] RIBA, "Model Safety Policy with Safety Codes: For Architects, Engineers and Surveyors", RIBA, London, 1996.

[5] R. X. Peyton, *Construction Safety Practices and Principles*, Van Nostrand Reinhold, New York, 1991.

[6] G. K. Cook, *Appraising Building Defects: Perspective on Stability and Hygro-thermal Performance*, Longman Scientific & Technical, Harlow, Essex, 1992.

[7] R. T. Ratay, "Construction Safety Affected by Codes and Standards", American Society of Civil Engineers, 1997.

[8] A. Civitello, *Construction Safety and Loss Control Program Manual*, M. E. Sharpe, New York, 1997.

[9] D. V. MacCollum, *Construction Safety Planning*, Van Nostrand Reinhold, New York, 1995.

[10] V. J. Davies, *Construction Safety Handbook*, 2nd Edition, Thomas Telford, London, 1996.

[11] H. K. Lee, "Construction Safety in Hong Kong", Concept Design & Project Management Ltd, Hong Kong, 1996.

[12] G. H. L. Seah, "Construction Safety Audit in Singapore", SBRE, National University of Singapore, 1997.

[13] D. Heberle, *Construction Safety Manual*, McGraw-Hill, New York, 1998.

[14] Code of Practice: Safety and Health at Construction Worksites, Parts 1 & 2, Labour Ministry, 1991.

[15] M. Grant, *Scaffold Falsework Design to BS5975*, Viewpoint Publication, 1982.

[16] J. R. Illingworth, *Temporary Works: Their Role in Construction*, T. Telford, London, 1987.

[17] R. T. Ratay, *Handbook of Temporary Structures in Construction: Engineering Standards, Designs, Practices & Procedures*, McGraw-Hill, New York, 1984.

[18] MOL, "Code of Practice for Examination and Test of Suspended Scaffolds for Approved Persons", Ministry of Labour, Singapore, 1989.

[19] S. Champion, *Access Scaffolding*, Chartered Institute of Building, Longman, Harlow, 1996.

[20] J. B. Fullman, *Construction Safety, Security and Loss Prevention*, Wiley, New York, 1984.

[21] Building Operations and Works of Engineering Construction Regulations, Singapore Government Printer, 1985.

[22] T. S. Ferry, *Safety Program Administration for Engineers and Managers: A Resource Guide for Establishing and Evaluating Safety Programs*, C. C. Thomas, Springfield, USA, 1984.

[23] The Building Control Act, Singapore Government Printer, 1989.

[24] Local Government Integration Act, Singapore Government Printer, 1985.

[25] A Guide to the Factories Act, Ministry of Labour, 1988.

CHAPTER 3

SITE INVESTIGATION

3.1. General

Many projects exceed their budgets and their completion dates due to unforeseen problems during the excavation and construction of their foundations. To ensure that these problems are kept to a minimum, a thorough site investigation is required, particularly for tall buildings. A site investigation is a study of the environment and the ground conditions required for any engineering or building structure. It is a process by which geological, geotechnical, topographical, social, environmental, economic and other relevant information are collated and analysed. The investigation may range in scope from the study of maps or aerial photographs, site reconnaissance, a simple examination of the surface soils with or without a few shallow trial pits, to a detailed study of the soil and ground water conditions to a considerable depth below the surface by means of boreholes and *in situ*/ laboratory tests on the materials encountered.

The extent of the work depends on the importance and foundation arrangement of the structure, the complexity of the soil conditions, and the information which may be available on the behaviour of existing foundations on similar soils. Thus it is not the normal practice to sink boreholes and carry out soil tests for single or two storey dwelling houses or similar structures since there is usually adequate knowledge of the required foundation depths and bearing pressures in any particular locality. Sufficient information to check the presumed soil conditions can usually be obtained by examining open sewer trenches or shallow excavations for roadworks, or from a few shallow trial pits or hand auger borings.

A detailed site investigation involving deep boreholes and laboratory testing of soils is always a necessity for heavy structures such as bridges, multi-storey buildings or industrial plants [1, 2].

3.2. Information to be Retrieved from a Site Investigation

Assuming a fairly detailed study is required, the following information should be obtained in the course of a site investigation:

(a) The general topography of the site as it affects foundation design and construction, e.g. surface configuration, adjacent property, the presence of watercourses, ponds, hedges, trees, rock outcrops, etc., and the available access for construction vehicles and plants.
(b) The location of buried services such as electric power and telephone cables, water mains and sewers.
(c) The general geology of the area with particular reference to the main geological formations underlying the site and the possibility of subsidence from mineral extraction or other causes.
(d) The previous history and use of the site including information on any defects or failures of existing or former buildings attributable to foundation conditions.
(e) Any special features such as the possibility of flooding, seasoning swelling and shrinkage, soil erosion, etc.
(f) A detailed record of the soil and rock strata and ground conditions within the zones affected by foundation bearing pressures and construction operations.
(g) Results of laboratory tests on soil and rock samples appropriate to the particular foundation design or constructional problems.

3.3. Stages of Site Investigation

The stages of a site investigation are similar for most projects although the content and number generally increase with the size of the project and the complexity of the ground conditions. Figure 3.1 shows the general

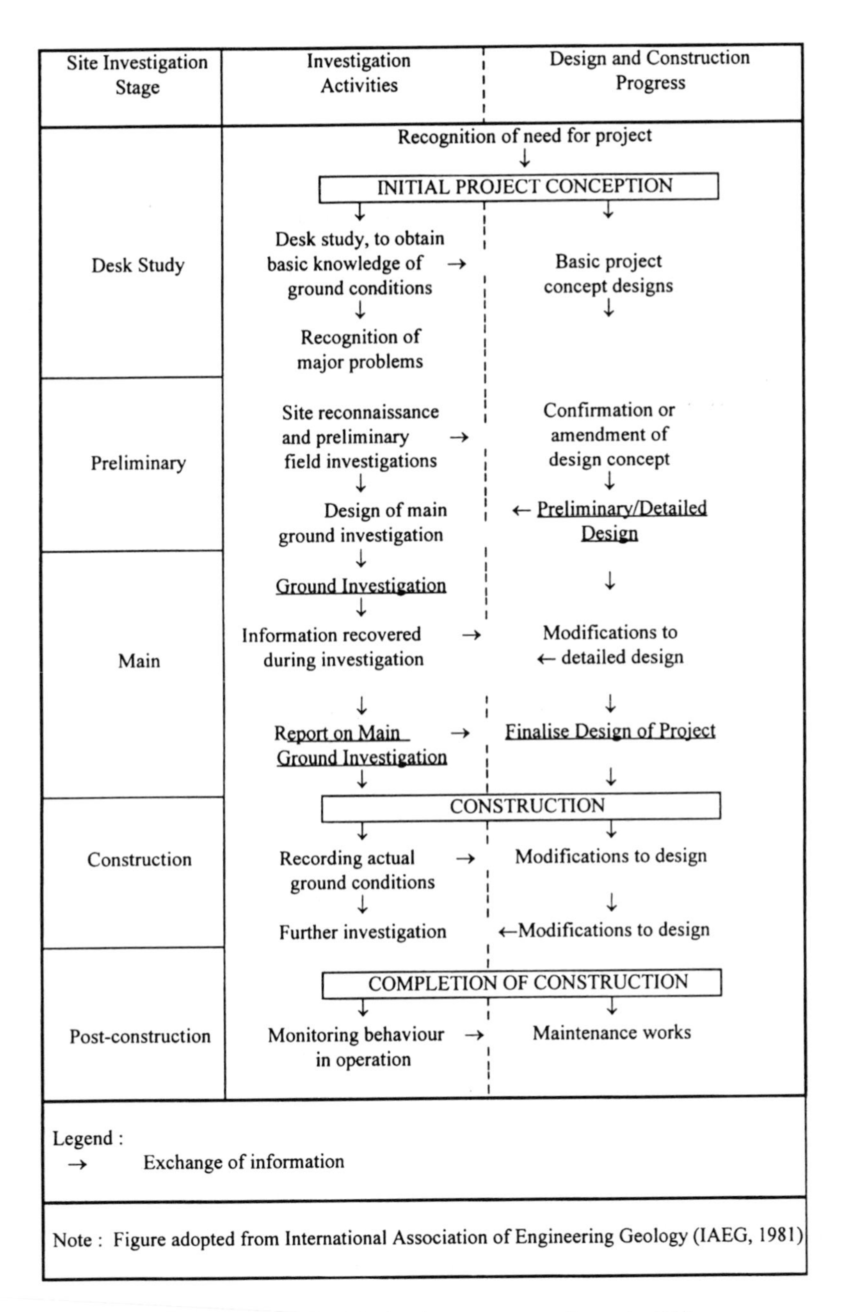

Figure 3.1. The investigation sequence of events [12].

investigation sequence of events from the inception to the completion of a project [3].

For a given site, some of the stages may be unnecessary, overlap or be taken out of the sequence. For example, site reconnaissance may take place before the completion of the desk study. There are generally four stages of site investigation, namely:

(a) Planning
(b) Desk Study
(c) Site Reconnaissance
(d) Ground Investigation

3.4. Planning

Planning is needed for all stages of a site investigation. The client's brief and the structural engineer's design thoughts should be coordinated before the desk study. Knowledge of the design proposals should include layout, alignments, function and probable loadings. Once these points have been established by discussion with the design engineer, the geotechnical or site investigation contractor can proceed to gather preliminary information.

3.5. Desk Study

The desk study involves the collection of available documentary materials relevant to the site, the immediate environment and the proposed structure in any of the following forms:

(a) Previous Ground Investigation
— Useful information may be extracted which will greatly assist the planning of the ground investigation, and may reduce the scope and extent of the investigation required. It should be noted, however, when interpreting these data, that the methodology, procedure and requirement at that time may be different.

(b) Topographical Maps
— This is useful for undeveloped areas such as forests and swampy or mountainous areas where information of the following general features can be obtained:

— Past uses of site.
— The location of roads, railways, tracks, rivers, canals, footpath, pipelines, overhead lines, contour lines, etc.

(c) Aerial Photos
— Aerial photos can provide a detailed and definitive picture of the topography, lines of communication (roads, railways and canals), surface drainage and urban development.

(d) Geological Maps and Memoirs
— These provide detailed information of the geology of the district, and are useful as a basis for evaluating the likely influence of the local geology on the proposed works and in the selection of the ground investigation methods. These are used frequently in Singapore and such maps are provided in a publication by the then Public Works Department (PWD) titled "Geology of the Republic of Singapore" [4].

(e) Historical Maps
— These provide information on the past uses of the site which could lead to identification of buried river courses or previous swampy ground. These would be important in delineating the possible problem areas which will need more detailed investigation during the main ground investigation. As a result of the extensive land reclamation and associated removal of high ground in Singapore over the last 160 years, the present topography is frequently unrepresentative of the actual geological formations. Some examples of the historical maps from the National Archives are listed in Table 3.1.

(f) Rainfall Records
— This is particularly important where basement construction and works requiring slope and major drainage system are required. Hydrological information is useful in drainage studies, including the assessment of flooding risk and the influence of proposed works on the local and downstream drainage regimes. These information can be obtained from the Meteorological Service Stations.

Table 3.1. List of some Historical Maps available in National Archives, Singapore.

Reference No.	Publication Date	Description
216	1836	Map of Town & Environment of Singapore by G. D. Coleman
99	1854	Old map of Singapore
2152	1845	Southern Part of Singapore
4	1865	Map of Island of Singapore & its dependencies drawn by J. Van Cuylenburg, Surveyor General Office, Singapore
217	1873	Map of Island of Singapore & its dependencies showing various administration districts
14	1873	Map of Island of Singapore & its dependencies drawn by Wajud Khan, Surveyor General Office, Singapore
106	1886	Map of Singapore & its dependencies drawn by Wajud Khan.
13	1858	Map showing Government House
12	1905	Map showing Government House
6	1506	Map of Singapore Town with Municipal Limits
15	1526	Singapore Road Map — FMS Surveys No. 34
11	1527	Singapore published under direction of Surveyor General FMS & SS
16	1532	Map of Singapore Town FMS Survey No. 15, 1532
415	1532	Singapore showing City area
416	1532	Singapore showing Kallang & Paya Lebar
405	1543	Map of Singapore Town — Tiong Bahru area
140	1951	Singapore by Survey Department Singapore
420	1957	Map of Singapore, Survey Department Federation of Malaya
17	1961	Singapore printed by the 84 Survey Squadron RE Far East

(g) Geographic Information System (GIS)

— The most important GIS at the national level is the Singapore Land Data Hub project undertaken by the Ministry of Law [5, 6]. The project was conceived in 1989 as a multi-ministry effort to establish a central repository of accurate and comprehensive land data. The digitised land data in the hub are contributed by more than 15 public sector agencies. These agencies have the relevant business processes within their operations to keep the digitised land data updated. The digitised land data include the survey maps (the legal land boundary); buildings and roads information; infrastructure data such as drainage and sewerage; utilities information such as electricity, gas and water network; electronic street directory data etc. [6]. Details can be found in the website of the Ministry of Law under Land Systems Support Unit (http://www.gov.sg/molaw/lssu/lssu.html).

3.6. Site Reconnaissance

This is to confirm, amplify and supplement the information collected earlier by (a) having a visual investigation on the site, and (b) collecting information from local inhabitants.

In the visual investigation, the following features are observed:

Topography: The general topography of the site is observed for evidence of the soils present, their distribution and their properties. Instability such as soil creep especially on sloping sites must be noted. Information on undulations, tilted tress, evidence of movement of buildings or deformation should be gathered.

Construction Materials and Labour: The cost of transporting bulk construction materials such as sand and aggregates can be excessive if there is no nearby source available. Similarly, the employment of labourers will be expensive as they will have to be brought in from other places. Accommodation, transport, food, amenities and facilities will have to be provided.

Groundwater Conditions: The presence of rivers, canals, springs and seepages, areas of wet ground, shallow wells, vegetational features can provide information on the water table conditions.

Site Access: This is to confirm the problems with access during the desk study and formulate solutions formulated.

Surrounding Structures: In cases where the neighbouring lands have been developed, the sensitivity of the proposed project affecting the existing structures must be evaluated. Authorities' restrictions on movement/vibration and noise allowable from a construction site in a way may determine the types of foundation works to be employed later. Minimum vibration is allowed, for instance, if there is an underground tunnel nearby. Minimum noise is allowed if there is a hospital, residential area nearby, etc. In Singapore, the Environmental Public Health Act (Chapter 95), the Environmental Public

Table 3.2. Maximum noise level permitted for construction sites (as an equivalent continuous noise level over a period of 12 hours) in decibels (A) [8].

Type of Building Affected	7.00am–7.00pm	7.00pm–7.00am
(a) Hospitals, schools, institutions of higher learning, homes for the aged sick, etc.	60	50
(b) Buildings other than the above	75	65

Health (Control of Noise from Construction Sites) Regulations 1990 stipulates that "the owner or occupier of any construction site shall ensure that the level of noise emitted from his construction site shall not exceed the maximum permissible noise levels' as set out in Table 3.2 [7, 8].

Useful information can also be collected from local authorities, local statutory boards and government bodies, local archives, local inhabitants, local contractors, local clubs and societies, schools, colleges and universities.

3.7. Ground Investigation

Geotechnical experts specialising in soil sampling and testing are engaged by project managers to establish the parameters that will be used in the design of the building foundation. The amount of testing depends on the size and complexity of the structure, the type of soil encountered, proximity of the proposed structure to existing buildings, and the level of the groundwater table.

Other considerations are the scope of the proposed work, the amount of existing information available, the probable nature and variability of ground conditions, the availability of plant and equipment, the cost of investigation, the manpower for operation and supervision and the access limitations and temporary works required.

The degree to which the proposed development affects the ground condition may provide a guide to the scope of investigation (Figure 3.2). In the case of a large structure where many piles are clustered together, the

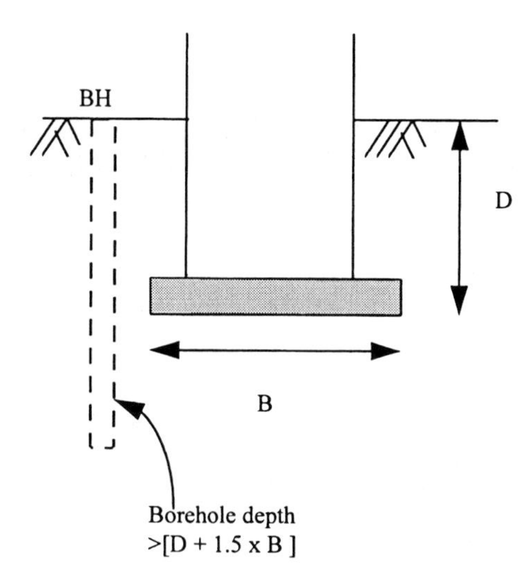

(a) Structure on isolated pad or raft

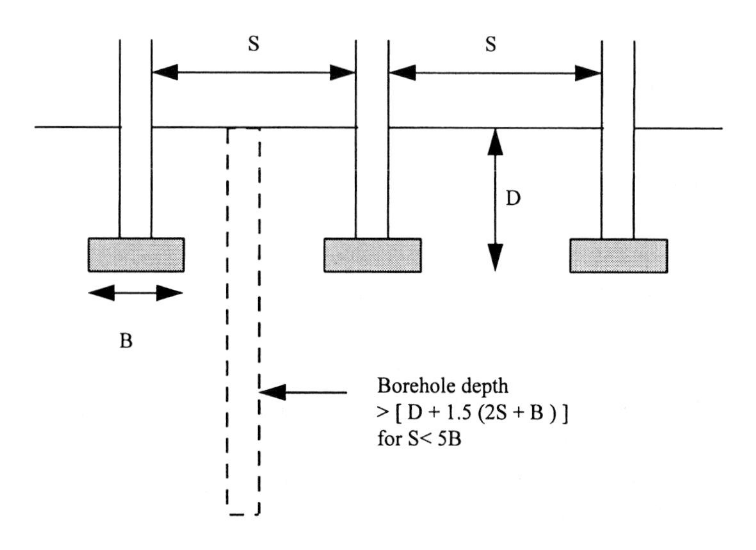

(b) Closely spaced strip on pad footings

Figure 3.2. Estimated borehold depths based on pressure bulb.

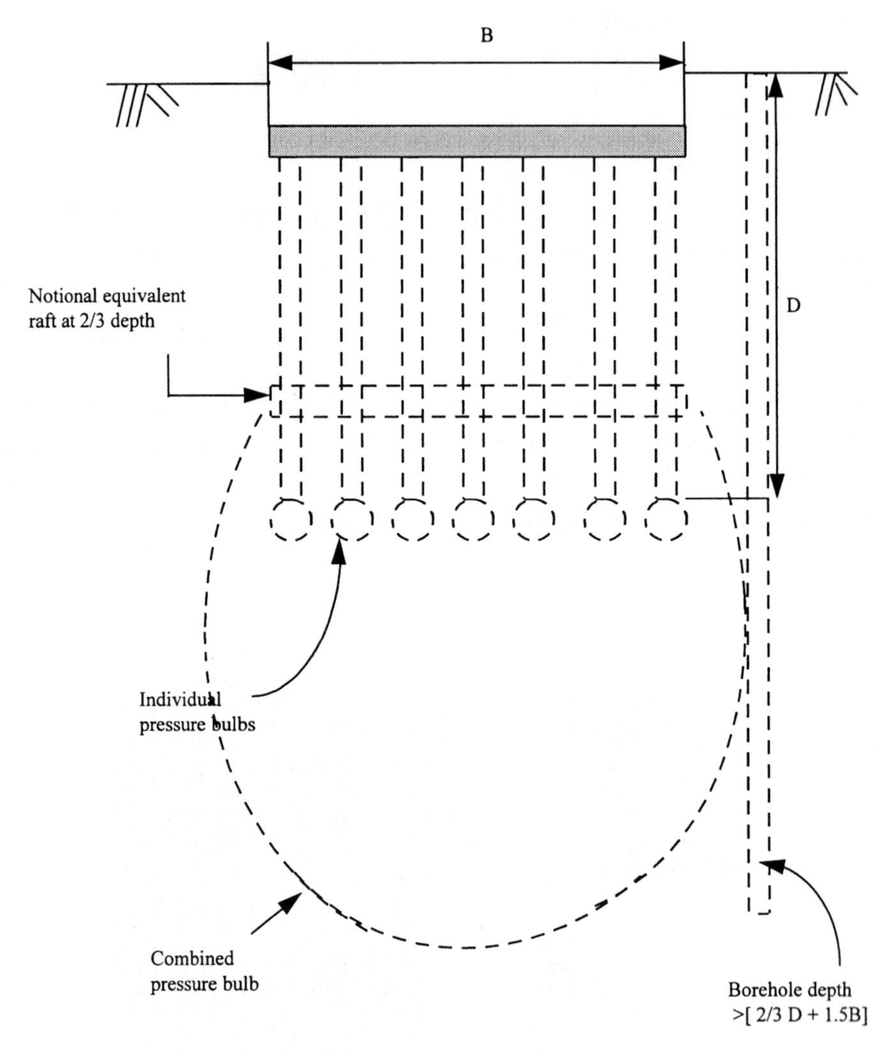

(c) Large structure on friction piles

Figure 3.2. (*Continued*).

much larger combined pressure bulb indicates that a much more detailed investigation is needed [9–12].

3.7.1. *On-Site Tests*

3.7.1.1. *Trial Pits*

Trial pits are used where only shallow depths are to be investigated. Trial pits up to 5 m can be quickly and cheaply dug using a hydraulic backactor. It allows a visual inspection of *in situ* soil conditions both laterally and vertically and allows a detailed examination of soil variability, structure and weathering profile.

Field testing can be carried out in a trial pit. Examples are (a) penetrometer (Figure 3.3), a spring loaded plunger which gives an indication of the indirect shear strength of the soil, and (b) Vane tester (Figure 3.4) which gives an indication of the torque property of the soil [13, 14].

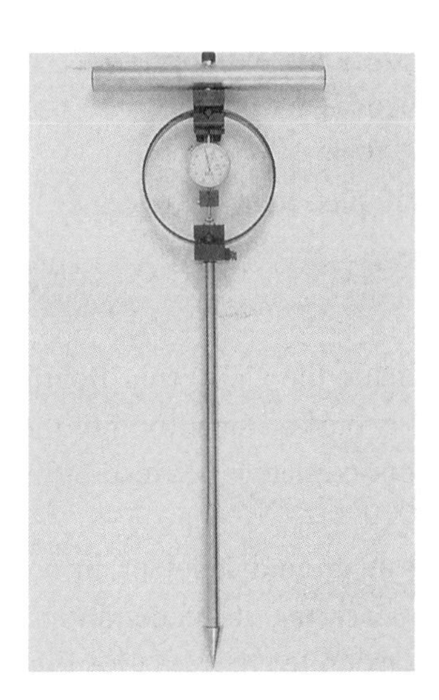

Figure 3.3. A penetrometer.

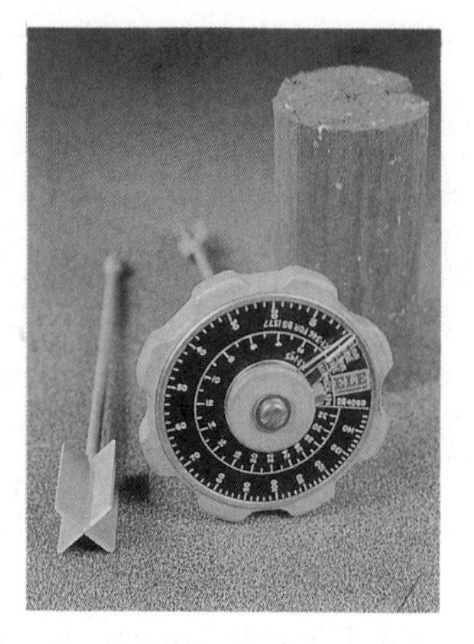

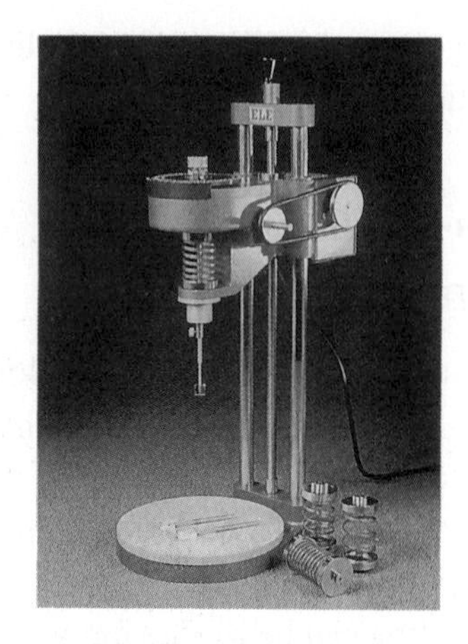

Figure 3.4. A Vane tester.

The field record of a trial pit should include a plan giving the location and orientation of the pit, and a dimensioned section showing the sides and the floor. Figure 3.5 shows an example of a trial pit log.

The limitations of trial pitting are obvious:

— Excavation below water table is difficult
— Not applicable for certain soil such as hard rock
— May cause some areas of ground disturbance
— Supports are needed for deep trial pits such as trenches

3.7.1.2. *Bearing Capacity*

A cone penetrometer can be used to estimate the California Bearing Ratio (CBR). A small cone on the end of a rod is pushed steadily into the ground at a constant rate of penetration and the approximate CBR read off a moving scale at the top (see Figure 3.6) [15].

Plate bearing test can be carried out at ground level or in a trial pit. The test involves loading a plate and measuring its penetration into the ground thus giving the stress–strain relationship for the soil (see Figure 3.7). A loaded lorry may be used as the fixed support for the hydraulic jack.

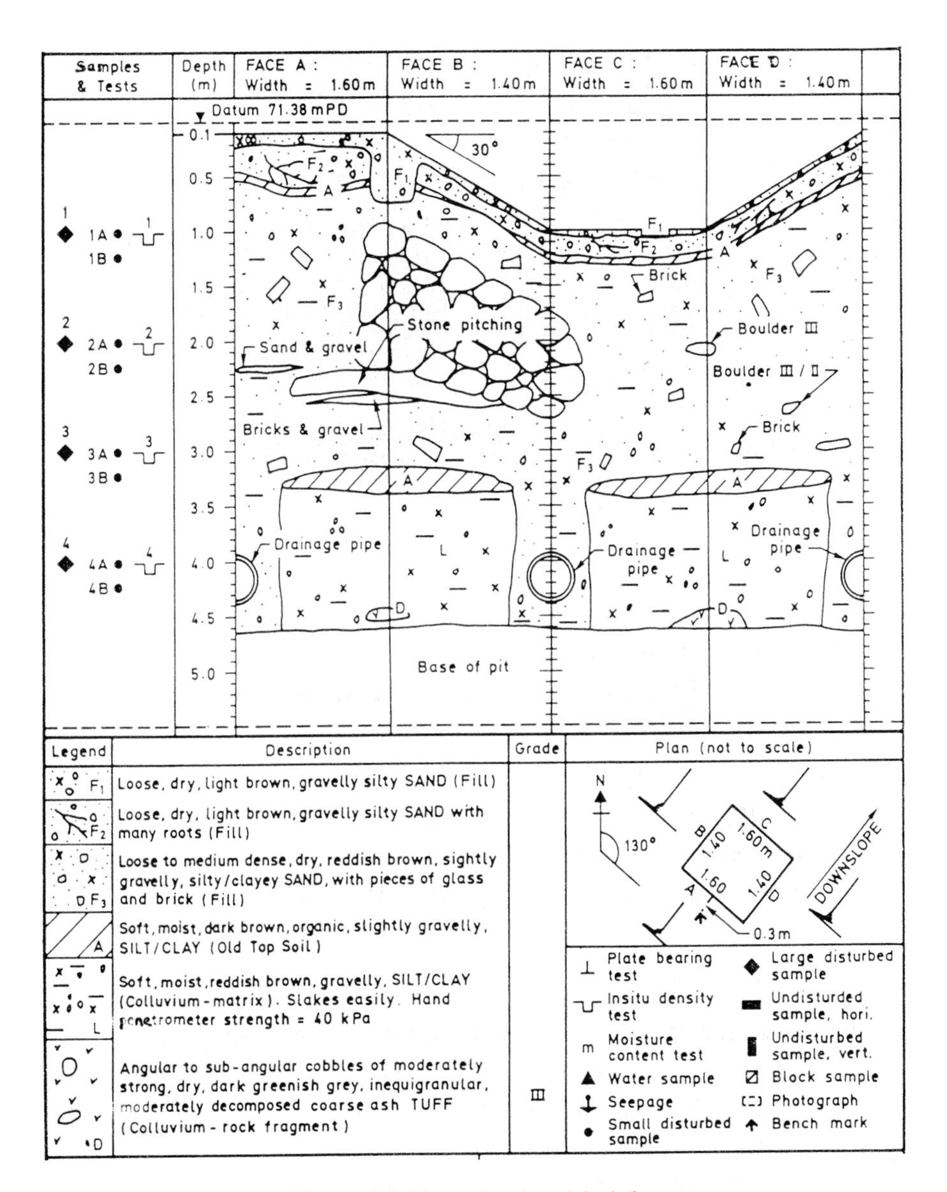

Figure 3.5. Example of a trial pit log.

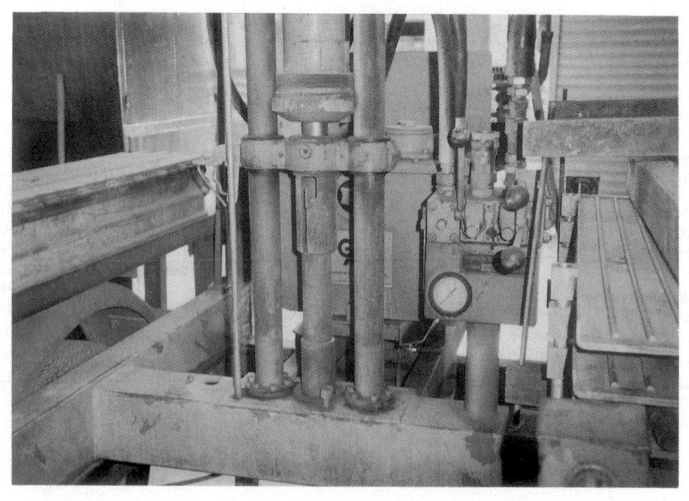

Figure 3.6. A cone penetrometer.

Figure 3.7. Plate bearing test.

3.7.1.3. *Probing*

This is a quick and cheap method to acquire an indication of the depth to a hard stratum. It consists of a steel rod with a bullet shaped tip. The rod is hammered into the ground using a constant weight and the number of blows required to drive the probe to a pre-determined depth is recorded. This number is commonly referred as the N value. A variety of small samplers, augers and probe heads can be attached to the probe for sampling purposes.

An example of a Mackintosh Probe is shown in Figure 3.8.

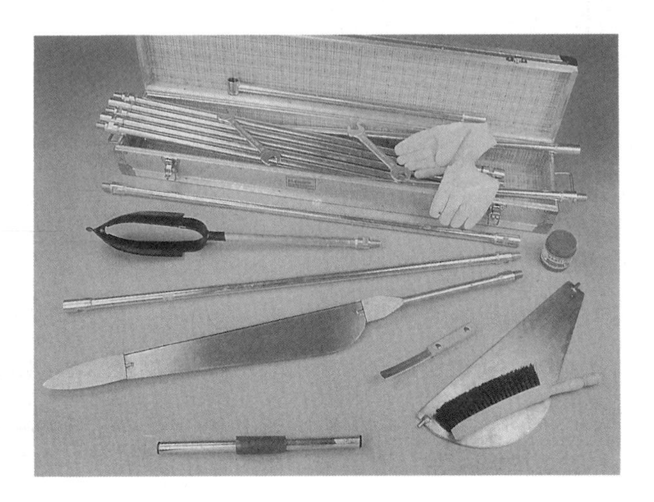

Figure 3.8. Mackintosh Probes.

3.7.1.4. *Shear Test*

A Vane tester (Figure 3.4) can be used to estimate the shear strength of clayish soils. The tester measures the torque required to shear a small cylinder of soil within the soil mass.

A pocket penetrometer is another easy method to give an estimation of the shear strength of clayish soils. The plunger is gently pushed in as far as the cut graduation line and the maximum reading read off a sliding indicator on a marked scale.

3.7.1.5. *Boreholes*

Boreholes are sunk to collect soil samples in the case where investigation requires deeper soil samples than can be excavated by trial pitting. Boreholes can be formed by hand augering (Figure 3.9), light cable percussion boring, mechanical augering (Figure 3.10), rotary open hole and rotary core drilling (Figure 3.11) [16–20].

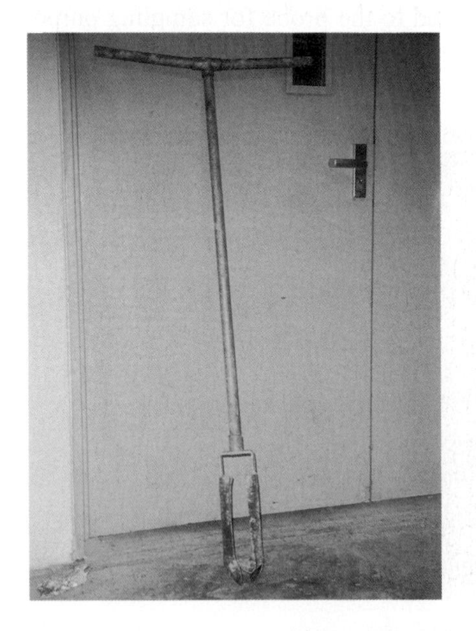
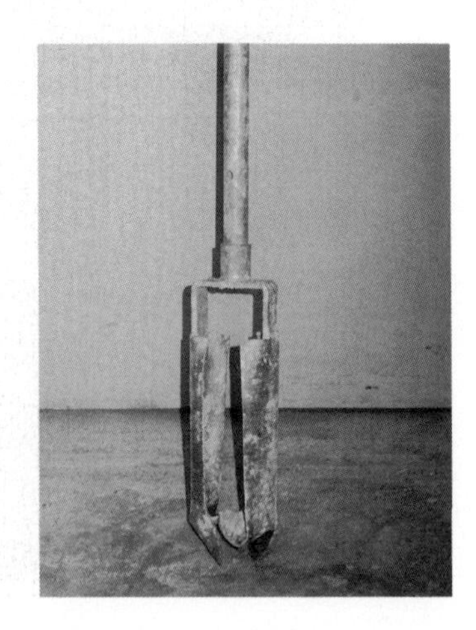

Figure 3.9. Hand augering.

Figure 3.10. Mechanical augering.

Figure 3.11. Rotary drilling.

3.7.2. *Laboratory Tests*

Laboratory tests for the collected samples are required to identify and classify the samples in terms of type, age, composition, weathering and moisture content, and to make an evaluation on the soil behaviour [21–27].

The various tests available are summarised in Table 3.3 to Table 3.5 [28–32].

3.8. Site Works and Setting Out

After a site is handed over, the task of (a) clearing the site, (b) setting out the building and (c) establishing a datum level can commence. A land surveyor is involved in the works for (b) and (c). The contractor is also required to provide the following:

Accommodation

— first aid
— washing facilities
— meal room
— sanitary facilities

Storage

— cover/no cover
— durability
— security

Fencing

— security

Hoarding

— safety

Electricity & Water Supply

— portable self-powered generator
— metered supply from local authority

Details of these are discussed in Chapter 6, "Material Handling and Mechanisation".

Table 3.3. Tests on soils and groundwater (group 1).

Category of Test	Name of Test	Remarks
Soil Classification Tests	Moisture content	Frequently used in the determination of soil properties, e.g. dry density, degree of saturation. Soils containing holloysitic clays, gypsum or calcite can lose water of crystallisation when heated, and should be dried at various temperatures to assess the effect on determination of moisture control.
	Liquid and plastic limits (Atterberg limits)	Used to classify fine-grained soil and as an aid in classifying the fine fraction of mixed soils. Soils containing holloysitic clays must be tested at natural moisture content.
	Linear shrinkage	Used to detect the presence of expansive clay minerals.
	Specific gravity	Frequently used in the determination of other properties, e.g. void ratio, particle size distribution by sedimentation.
	Particle size distribution: (a) Sieving	(a) Sieving gives the grading of soil coarser than silt. Core is required with soils derived from *in situ* rock weathering, to avoid crushing of soils grains during disaggregation. As a variation to the standard method of wet sieving (BSI, 1975b Test 7(A)), it will be appropriate to exclude the use of dispersants when determining particle size distribution for certain applications, e.g. for designing filters, and in selecting fill for reinforced fill structures.
	(b) Sedimentation	(b) The proportion of the soil passing the finest sieve (64 μm) represents the combined silt and clay fraction. The relative proportions of silt and clay can only be determined by sedimentation.
	Laboratory vane shear	A useful test for classifying silts and clays in term of consistency. See also Geoguide 3 (GCO, 1987b).

Table 3.4. Tests on soils and groundwater (group 2).

Category of Test	Name of Test	Remarks
Soil Permeability Tests	Permeability: (a) Constant head permeability test (b) Falling head permeability test (c) Triaxial permeability test (d) Rowe cell	The constant head test is suited only to soils of permeability roughly within the range 10^{-4} m/s to 10^{-2} m/s. For soils of lower permeability the falling head test is applicable. For various reasons, principally sample size and ground variability, laboratory permeability tests often yield results of limited value, and *in situ* tests should generally yield more representative data. The Rowe cell allows the direct measurement of permeability by a constant head, with a back pressure and confining pressures more closely consistent with the field state, and by both vertical and radial flow.
Soil Compaction Tests	Dry density/moisture contant relationship	Indicates the degree of compaction that can be achieved at different moisture contents and with different compactive effort. Test 12 is commonly carried out in conjunction with determinations of *in situ* dry density (ASTM, 1985b; ASTM 1985e; ASTM 1985h; BSI, 1975b Test 15).
Pavement Design Tests	California bearing ratio (CBR)	This is an empirical test used in the design of flexible pavements. The test can also be carried out *in situ*, but the results may be substantially different from the laboratory test due to the difference in the confining condition, especially for sands.
Soil Collapse Potential Tests	Double oedometer test	Assesses the potential for soils to collapse on wetting.
Soil Dispersion Tests	Double hydrometer test (dispersion test) Exchangeable sodium percentage test Emerson crumb test (furbidity test) Pinhole test	Used to identify dispersive soils, in order to assess the potential for dispersive piping and internal erosion to occur in slopes and earth structures. The different tests may not give consistent indications of dispersion, consequently it is advisable to use more than one test method.

Table 3.4. (*Continued*).

Category of Test	Name of Test	Remarks
Chemical and Corrosivity on Soils and Groundwater	Organic matter content	Detects the presence of organic matter, which can: (i) interfere with the hydration of Portland cement in soil–cement pastes. (ii) influence shear strength, bearing capacity and compressibility. (iii) influence the magnitude of the correction factor require when using nuclear methods to estimate the *in situ* moisture content of soils (ASTM, 1985h). (iv) promote microbiological corrosion of buried steel.
	Sulphate content: (a) Total sulphate content of soil (b) Sulphate ion content of groundwater and aqueous soil extracts	These tests assess the aggressiveness of soil and groundwater to buried concrete and steel.
	Total sulphide content of groundwater and soil extracts	Assesses the aggressiveness of soil and groundwater to buried steel.
	pH value	Assesses the aggressiveness of soil and groundwater to buried concrete and steel.
	Chloride ion contents	Assesses: (i) the aggressiveness of soil to buried concrete and steel. (ii) the suitability of fine aggregate for use in concrete.

Table 3.4. (*Continued*).

Category of Test	Name of Test	Remarks
Chemical and Corrosivity on Soils and Groundwater	Carbonate content	The reference describes the method using the Collins calcimeter.
	Resistivity	Assesses the potential for electrochemical corrosion of buried steel. The quoted reference gives a test method for compacted fill, as opposed to field measurement using the four electrode method. Corrosion of steel in soils is discussed in BS8004 (BSI, 1986) and King (1977).
	Redox potential	Assesses the likelihood of sulphate reducing bacteria being present, which promote microbiological corrosion of buried steel. The quoted reference gives a test method for compacted fill, as opposed to field measurement which is described in CP1021 (BSI, 1973).
	Bacteriological tests	Undisturbed specimens should be stored in air-sealed, sterilized containers [30].
Soil Strength Tests	Triaxial compression tests (a) Quick undrained (b) Consolidated drained (c) Consolidated undrained with measurement of pore water pressure	The quick undrained test gives undrained shear strength in terms of total stresses, and has application to short-term stability and bearing capacity analyses. For saturated clays with undrained shear strength less than about 75 kPa, the *in situ* penetration vane test, used in conjunction with the cone penetration test, will normally be the best method for measuring undrained shear strength.
	Direct Shear test	A useful and practical alternative to the consolidated drained triaxial test for shear strength measurements on fill, colluvium and soils derived from weathering of rock *in situ*. The test specimen can be oriented to measure shear strength on a pre-determined plane. The major limitation of the test is the specimen thickness, which governs the maximum particle size that can be tested. Common specimen sizes are 60 mm and 100 mm square by 20 mm thick.

Table 3.4. (*Continued*).

Category of Test	Name of Test	Remarks
Soil Deformation Tests	Consolidation (a) One-dimensional consolidation (Oedometer test) (b) Triaxial consolidation (c) Rowe cell	These tests yield soil parameters from which the amount and time scale of settlements can be calculated. The simple oedometer test is the one in general use. Although reasonable assessment of settlement can be made from the results of the test, estimates of the time scale have been found to be extremely inaccurate for some soils. This is particularly true for clays soils containing layers and partings of silt and sand, where the horizontal permeability is much greater than the vertical. In these cases, more reliable data may be obtained from tests in the Rowe cell, which is available in sizes up to 250 mm diameter and where a larger and potentially more representative sample of soil can be tested. Another alternative is to obtain values of permeability, k from *in situ* permeability tests, and combine them with coefficients of volume decrease, m_v, obtained from the simple oedometer test.
	Modulus of deformation	Values of the modulus of deformation of soil can be obtained from the stress-strain curves from triaxial compression tests, where the test specimens have been consolidated under effective stresses corresponding to those in the field. However, values obtained in this way frequently do not correlate well with *in situ* observations. It is now generally considered that the plate test, the pressuremeter and back analysis of existing structures yield more reliable results [31].

Table 3.5. Tests on rock [5].

Category of Test	Name of Test	Remarks
Rock Classification Tests	Water content, porosity, density, absorption, swelling, and slake durability	Used for classification and characterisation of rocks.
	Sonic wave velocity (sound velocity)	Used to measure velocities of compression and shear waves for the determination of elastic constants of isotropic and slightly anisotropic rocks. Tests are usually carried out on small specimens using ultrasonic frequencies.
	Thin section	Used for petrographic description of texture, fabric and state of alteration in rock material [32].
	Point load	Used to measure the point load strength index and strength anisotropy. The results are used as an index test for strength classification of rock material, and to predict its uniaxial compressive strength. The test can be carried out on pieces of drill core or irregular limps of rock. It can also be carried out in the field.

Table 3.5. (*Continued*).

Category of Test	Name of Test	Remarks
Rock Strength and Deformation Tests	Uniaxial compressive strength and deformability	Used for direct determination of uniaxial compression strength, and for determination of static Young's Modulus of Elasticity and Poisson's ratio. The results can be used in conjunction with information on the nature and spacing of discontinuities to assess allowable bearing stress and settlement in rock foundation design, stability of underground excavations and to design rock support measures.
	Triaxial compression	Used for determination of triaxial compressive strength, static Young's Modulus of Elasticity and Poisson's ratio. Test results are used to assess the stability of underground excavations and to design support measures.
	Direct and indirect tensite strength	Used in stability assessment of underground excavations. Specimens for direct tests are difficult to prepare, and indirect tests such as the 'Brazil Test' are more commonly performed.
Rock Discontinuity Strength Tests	Direct shear	Used to determine the shear strength characteristics of rock discontinuities. The Robertson Shear Box and the Golder Associates Shear Box are routinely used. Both are sufficiently portable for field use, but specimen preparation time is a disadvantage. The results are used in rock slope stability analysis, and for local stability calculations in tunnels.

Reference

[1] ASCE, "Subsurface Exploration for Underground Excavation and Heavy Construction", Proceedings Spec. Conference, 1974.

[2] ASCE, "Subsurface Investigation for Design and Construction of Foundations of Buildings", Manuals and Reports on Engineering Practice, No. 56, 1976.

[3] M. D. Joyce, *Site Investigation Practice*, E & F. N. Spon, 1982.

[4] PWD, "Geology of the Republic of Singapore", Geological Unit, Public Work Department, Singapore, 1976.

[5] K. Parthipan and E. H. Chia, "Geographical Information Systems: Implications for Regional Development", *The Naga Awakens: Growth and Change in South East Asia*, Times Academic Press, 1998, pp. 201–218.

[6] J. K. Chua, "Singapore Land Data Hub 21 — Vision for the Future", paper presented in LSSU Website, Ministry of Law, 1999.

[7] Environmental Public Health Act, "Environmental Public Health Act" (Chapter 95), Government Printer, 1990.

[8] The Code of Practice for Environmental Health, Ministry of Environment, 1992.

[9] R. C. Smith and C. K. Andres, *Principles and Practices of Heavy Construction*, 3rd Edition, Prentice-Hall, 1986.

[10] C. R. I. Clayton, M. C. Matthews and N. E. Simons, *Site Investigation*, Blackwell Science, Oxford, 1995.

[11] A. D. Robb, *Site Investigation*, T. Telford, London, 1982.

[12] A. Anagnostopoulos, "Geotechnical engineering of hard soils, soft rocks", Proceedings of an International Symposium under the Auspices of the International Society for Soil Mechanics and Foundation Engineering (ISSMFE), the International Association of Engineering Geology (IAEG) and the International Society for Rock Mechanics (ISRM), Athens, Greece, 20–23 September 1993.

[13] F. R. Adrian, *Vane Shear Strength Testing in Soils: Field and Laboratory Studies*, ASTM, Philadelphia, 1988.

[14] N. Ahmad, "Evaluation of In-Situ Testing Methods in Soils", University Microfilms International, Ann Arbor, Michigan, 1975.

[15] T. Lunne, *Cone Penetration Testing in Geotechnical Practice*, Blackie Academic & Professional, London, 1997.

[16] S. Hansbo, *Foundation Engineering*, Elsevier, New York, 1994.

[17] S. J. Greenfield, *Foundation in Problem Soils: A Guide to Lightly Loaded Foundation Construction for Challenging Soil and Site Conditions*, Prentice Hall, 1992.

[18] T. H. Hanna, *Field Instrumentation in Geotechnical Engineering*, Karl Distribution, 1985.

[19] R. J. Ebelhar, *Dynamic Geotechnical Testing II*, ASTM, Philadelphia, 1994.

[20] Centre for Civil Engineering Research and Codes, "Building on Soft Soils: Design and Construction of Earthstructures Both on and into Highly Compressible Subsoils of Low Bearing Capacity", A. A. Balkema, Netherland, 1996.

[21] E. W. Brand, *Sampling and Testing of Residual Soils: A Review of International Practice*, Scorpion Press, Hong Kong, 1985.

[22] R. T. Donaghe, *Advanced Triaxial Testing of Soil and Rock*, ASTM, Philadelphia, 1988.

[23] J. P. Bardet, *Experimental Soil Mechanics*, Prentice Hall, 1997.

[24] P. V. D. Berg, *Analysis of Soil Penetration*, Delft University Press, 1994.

[25] J. E. Bowles, *Engineering Properties of Soils and Their Measurement*, 4th Edition, McGraw-Hill, 1992.

[26] International Symposium on Cone Penetration Testing, Linkoping, Sweden, October 4–5, 1995, Swedish Geotechnical Society, Linkoping, Sweden, 1995.

[27] N. Wilson, *Soil Water and Ground Water Sampling*, Lewis Publishers, 1995.

[28] Engineering Development Department, "Geotechnical Manual for Slopes", 2nd Edition, Geotechnical Control Office, Engineering Development Department, Government Publication Centre, Hong Kong, 1984.

[29] T. W. Lambe, *Soil Testing for Engineers*, Wiley, New York, 1951.

[30] K. Alef and P. Nannipieri, *Methods in Applied Soil Microbiology and Biochemistry*, Academic Press, London, 1995.

[31] G. Ballivy, "The pressuremeter and its new avenues", Proceedings for the 4th International Symposium, Sherbrooke, Quebec, 17–19 May 1995, A. A. Balkema, Rotterdam, 1995.

[32] A. V. Carozzi, *Sedimentary Petrography*, Prentice Hall, 1993.

CHAPTER 4

FOUNDATION

4.1. General

The function of a foundation is to transfer the structural loads from a building safely into the ground. The structural loads include the dead, superimposed and wind loads. To perform the function, the foundation must be properly designed and constructed. Its stability depends upon the behaviour under load of the soil on which it rests and this is affected partly by the design of the foundation and partly by the characteristics of the soil. It is necessary in the design and construction of foundation to pay attention to the nature and strength of the materials to be used for the foundations as well as the likely behaviour under load of the soils on which the foundation rests.

4.2. Soil Characteristics

It is convenient to categorise soils and their properties according to their particle size as shown in Table 4.1.

The soil conditions in Singapore is basically made up of four different major geological formations: (a) Bukit Timah Granite, (b) Jurong Formation, (c) Old Alluvium and (d) Kallang Formation as shown in Figure 4.1.

Bukit Timah Granite

The Bukit Timah Granite is distributed in the central part of the island such as Bukit Timah, Thomson, Sembawang, Mandai, Bukit Panjang and

Upper Changi areas. A typical cross section of the formation is shown in Figure 4.2(a). The weathered granite residual soil is of sandy clay or silt nature and consists of between 25% to 65% silt and clay-sized particles with increasing stiffness and percentage of coarser fraction with depth. Large boulders are often encountered within this weathered profile. The residual granite soil covers one-third of the surface of central Singapore [1]. At some low-lying areas, a layer of alluvium or marine clay may overlie the granite formation.

Jurong Formation

The Jurong formation covers the western and southern coastal areas. This formation consists of sedimentary rocks formed in the Triassic period. Mudstone, sandstone and shales can be found interbedded in this formation at depth ranging from 5 m to 45 m. A typical cross section of the formation is shown in Figure 4.2(b).

Old Alluvium Formation

This formation lies on the north-eastern and north-western part of the island. It is a product of heavily weathered sedimentary rocks consisting primarily of clayey to silty coarse sand with layers of silty clay as shown in Figure 4.2(c). It is often thick, in excess of 30 m, with a total thickness of 195 m having been recorded [2].

Kallang Formation

This formation consists of soils with marine, alluvial, littiral and estuarine origins, which covers much of the coastal plain, immediate off-shore zone and deeply incised river valleys, which penetrates to the centre of the island. Basically, the formation comprises of two layers, the upper and lower marine clay, separated by a thin layer of stiff silty clay as shown in Figure 4.2(d).

Table 4.1. Soil characteristics and bearing capacities [3].

Subsoil types	*Condition of subsoil*	*Means of Field Identification*	*Particle size range*	*Bearing capacity kN/m²*	*Minimum width of strip foundations in mm for total load in kN/m of loadbearing wall of not more than*					
					20	*30*	*40*	*50*	*60*	*70*
Gravel	Compact	Require pick for excavation. 50 mm peg hard to drive more than about 150 mm	Larger than 2 mm	>600	250	300	400	500	600	650
Sand		Clean sands break down completely when dry. Particles are visible to naked eye and gritty to fingers. Some dry strength indicates presence of clay	0.06 to 2 mm	>300						
Clay	Stiff	Require a pick or pneumatic spade for removal	Smaller than 0.002 mm	150–300	250	300	400	500	600	650
Sandy clay		Cannot be moulded with the fingers Clays are smooth and greasy to the touch. Hold together when dry, are sticky when moist. Wet lumps immersed in water soften without disintegration	See Sand and Clay	150–300						
Clay	Firm	Can be excavated with graft or spade Can be moulded with strong finger pressure	See above	75–150	300	350	450	600	750	850
Sandy clay			See Sand and Clay	75–150						
Gravel	Loose	Can be excavated with a spade A 50 mm peg can be easily driven	See above	<200	400	600	For loading of more than 30.0 kN/m run on these types of soil, the necessary foundations do not fall within the provisions of Approved Document A, Section 1E from which these figures are taken Pad foundations generally and surface rafts are designed using the bearing capacities for soils given in this Table.			
Sand			See above	< 100						
Silty sand			See Silt and Sand	May need to be assessed by test						
Clayey sand			See Clay and sand	ditto						
Silt	Soft	Readily excavated Easily moulded in the fingers Silt particles are not normally visible to the naked eye. Slightly gritty. Moist lumps can be moulded with the fingers but not rolled into threads. Shaking a small moist pat brings water to surface which draws back on pressure between fingers. Dries rapidly. Fairly easily powdered	0.002 to 0.06 mm	<75	450	650				
Clay			See above	<75						
Sandy clay			See Sand and Clay	May need to be assessed by test						
Silty clay			See Silt and Clay	ditto						
Silt	Very soft	A natural sample of clay exudes between the fingers when squeezed in fist	See above	ditto	600	850				
Clay			See above	ditto						
Sandy clay			See Sand and Clay	May need to be assessed by test						
Silty clay			See Silt and Clay	ditto						
Chalk	Plastic	Shattered, damp and slightly compressible or crumbly	—	—	Assess as clay above					
Chalk	Solid	Requires a pick for removal	—	600	Equal to width of wall					

Sands and gravels: In these soils the permissible bearing capacity can be increased by 12.5 kN/m² for each 0.30 m of depth of the loaded area below ground level. If ground water-level is likely to be less than the foundation width below the foundation base the bearing capacities given should be halved. The bearing capacities given for these soils assume a width of foundation around 1.00 m but the bearing capacities decrease with a decrease in width of foundation. For narrower foundations a reduced value should be used: the bearing capacity given in the table multiplied by the width of the foundation in metres.

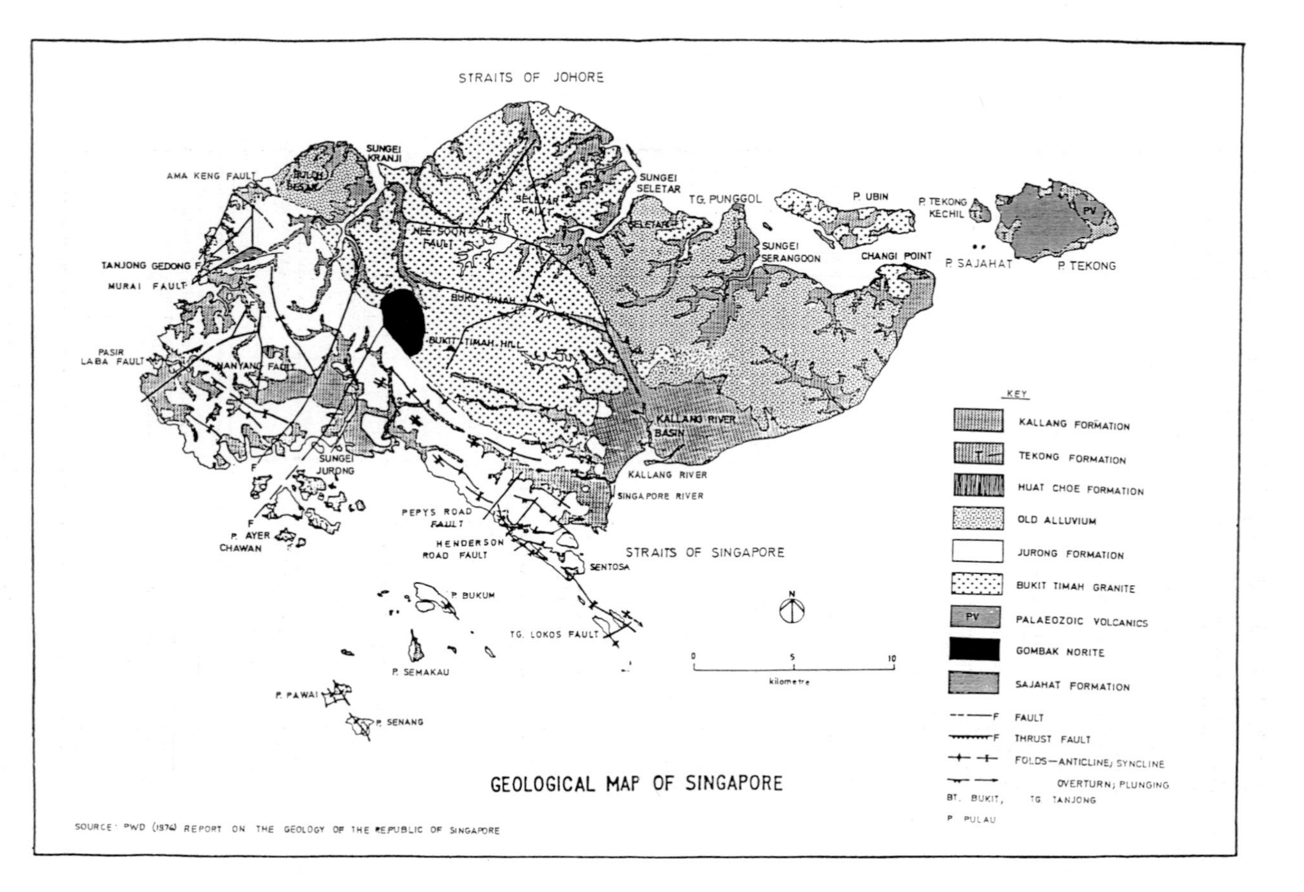

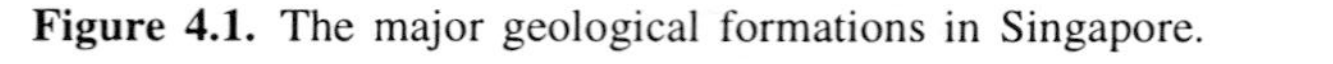
Figure 4.1. The major geological formations in Singapore.

	SOIL PROFILE	N-VALUE	MOISTURE (%)
	Clayey Silt Fill	20-30	20-40
	Stiff to very stiff sandy clayey silt	40-60	20-40
	Weathered granite	>80	

Figure 4.2(a). Typical soil profile of Bukit Timah Granite.

	SOIL PROFILE	N-VALUE	MOISTURE (%)
	Medium Stiff Sandy Silty Clay	10-20	20-40
	Soft Sandy Silty Clay	10-20	20-40
	Very Stiff Sandy Silty Clay	20-40	20-40
	Hard Silty Clay	60-80	20-40
	Very Dense Clayey Silt	>100	20-40
	Silty Clay with Shale	>100	20-40

Figure 4.2(b). Typical soil profile of Jurong Formation.

	SOIL PROFILE	N-VALUE	MOISTURE (%)
	Sandy silty clay	10-20	20-40
	Stiff silty clay	20-30	20-40
	Hard silty clay (with some sand)	40-50	20-40
	Very dense cemented silty sand	>100	10-20

Figure 4.2(c). Typical soil profile of Old Alluvium Formation.

SOIL PROFILE	N-VALUE	MOISTURE (%)
Clayey fill	10-20	20-30
Soft to firm marine clay (upper member)	<3	60-80
Stiff clay	20-30	
Firm marine clay (lower member)	<5	40-60
Medium to dense cemented clayey sand	>100	20-30

Figure 4.2(d). Typical soil profile of Kallang Formation.

4.3. Foundation Systems

Due to the greater compressibility, cohesive soils suffer greater settlement than cohesionless soils. In Singapore, with the rapidly increasing urbanisation and industrialisation, many structures are constructed on poor soils.

Figure 4.3 shows the various types of foundation system used for buildings in the Central Business District (CBD) in Singapore. The types of foundation system used by buildings are quite diversified, indicating that these systems are feasible and suitable to their respective buildings, at least at their time of construction.

The following ground materials are commonly encountered in a typical boring investigation: (a) fill layer, (b) Huat Choe Formation, (c) bouldery clay, (d) marine clay.

Fill Layer: The fill layer consists mainly of loose materials of sand, bouldery sand and boulders with clay. The colour of the fill is light brown to dark brown and is highly permeable to water.

Huat Choe Formation: Huat Choe Formation consists mainly of hard sandy silt and is light greyish white in colour. The formation is not porous and water cannot seep through the formation easily.

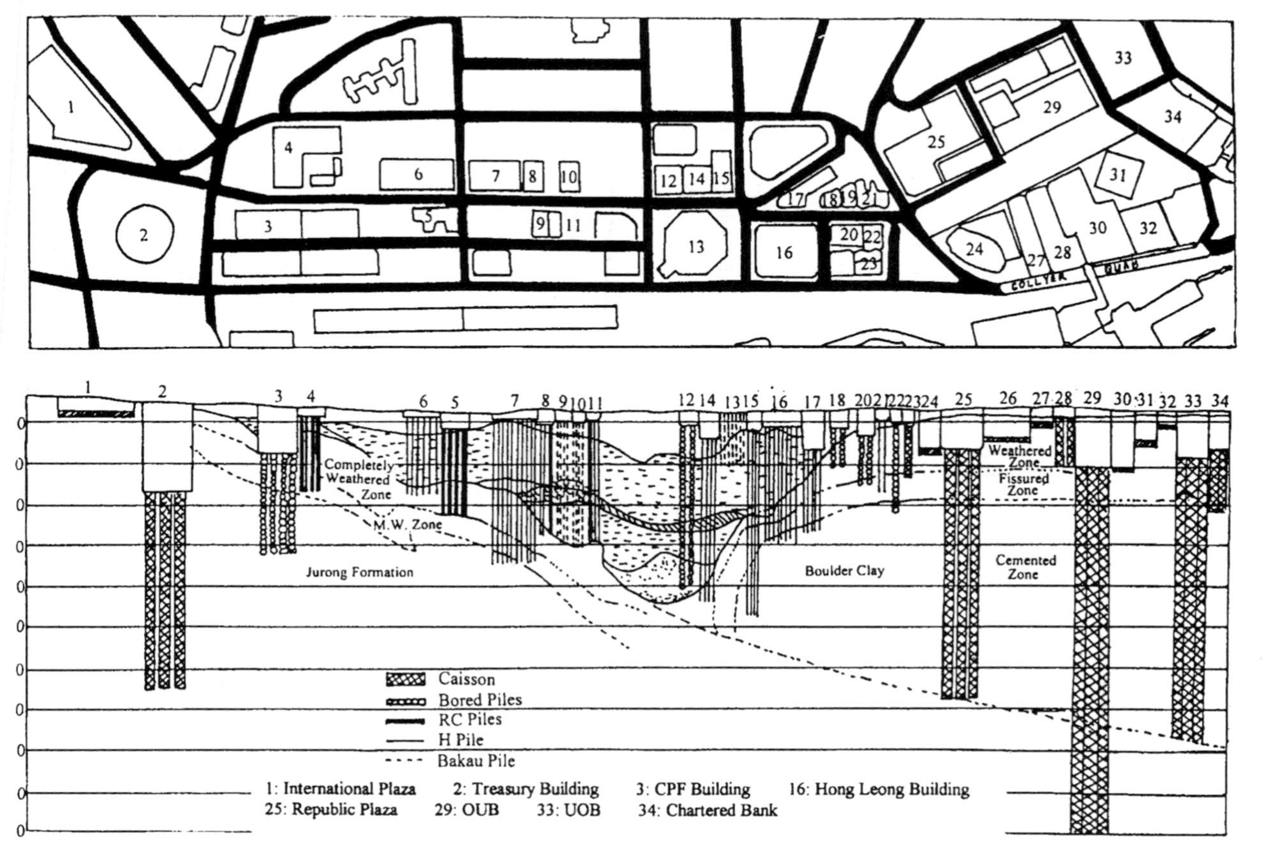

Figure 4.3. Foundation system of some major buildings in the Central Business District (CBD).

Bouldery Clay: This consists mainly of strong rounded boulders of varying size and stiff soils. The bouldery clay layer is usually composed of a few zones according to its weathering grade. An example is shown in Figure 4.4 with its properties shown in Table 4.2.

Marine Clay: Marine clay is kaolinite rich, pale-grey to dark blue-grey, soft silty clay, with sandy, silty, peaty and shelly fragments. The fragmented shells indicate its marine origin. The *in situ* moisture content is close to the liquid limit and the cohesive strength is low. The marine clay does not become siltier with depth, hence retaining its essential characteristic of 65% to 70% clay throughout [3]. It is usually consolidated with average shear strength of 10 kN/mm^2 and 40 kN/mm^2 for the upper and lower marine member respectively.

Geologists believe that marine clay is formed from materials brought down by the river and with the retreat of the ice in the Pleistocene period, the land rose because of the consequent relief in load. Weathering and denudation then took place and the materials formed were deposited by the rivers in swamps and estuaries with the heavier particles i.e. sand and gravels at the bottom and the light ones such as clays above. These sediments were subsequently covered by sea water as a result of the submergence of the

Table 4.2. Contents of bouldery clay.

Weathered Zone	Most of the boulders are weathered and decomposed into weakly cemented sand or friable sandstone. Matrix is stiff clay and weakly cemented. There are no fissures or open discontinuities in this zone.
Fissured Zone	Some boulders are weathered and turn into weak sandstone but some of the boulder reserves are still very strong in compressive strength. Most of the boulders show traces of weathering in colour and strength. Matrix is very hard, massive clay and have many fissures and open discontinuities in this zone. Near the open discontinuities, the clay appears weakened by swelling. Ground water seeps out of the fissures and discontinuities are discovered.
Cemented Zone	Boulders are white; strong sandstone with thin weathering zone at the surface area. Matrix is hard and dense massive clay with some fine gravel. There are some shear zones and slicken sides in the matrix but all fissures are closed and permeability is considered low.

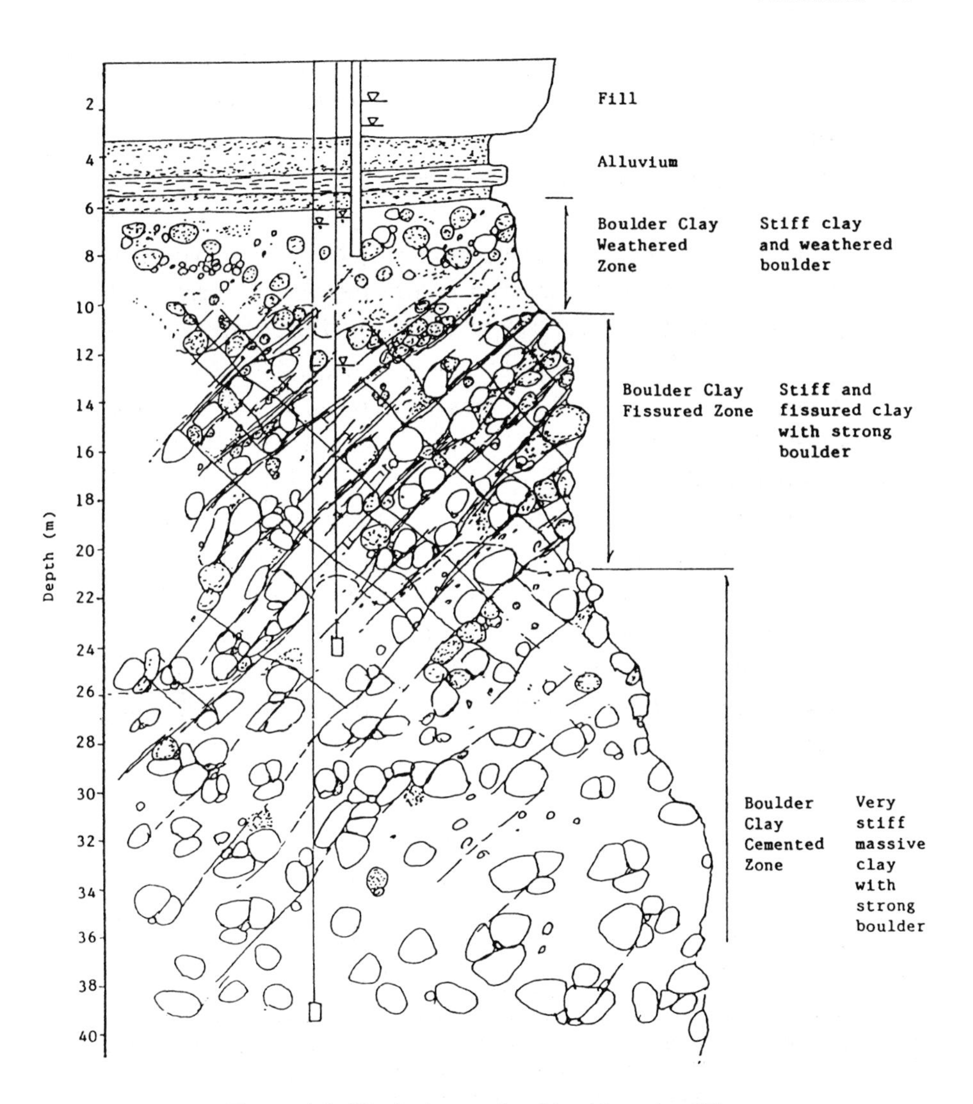

Figure 4.4. Weathering grade of bouldery clay [2].

Figure 4.5. Foundation works on marine clay.

landmass under the seas. With further emergence of the land and regression of the seas, more erosion and deposition took place. The soft marine clay is generally blue or grey in colour. Its natural moisture content is high at about 80%–100% depending on the depth [4]. The soil is very compressible and has a very low shear strength. Figure 4.5 shows an example of foundation works on marine clay.

In Singapore, soft marine clay covers a large area mainly along the estuaries of the present river system (Figure 4.6). The clay varies in thickness from place to place, up to about 30 m thick along Nicoll Highway (Figure 4.7).

The low shear strength coupled with high compressibility of the thick marine clay suggests that deep foundation is necessary. Problems anticipated when dealing with these kinds of soil include ground heave, adjacent ground surface settlement, long term settlement and low bearing capacity. Negative skin friction may develop [4, 5].

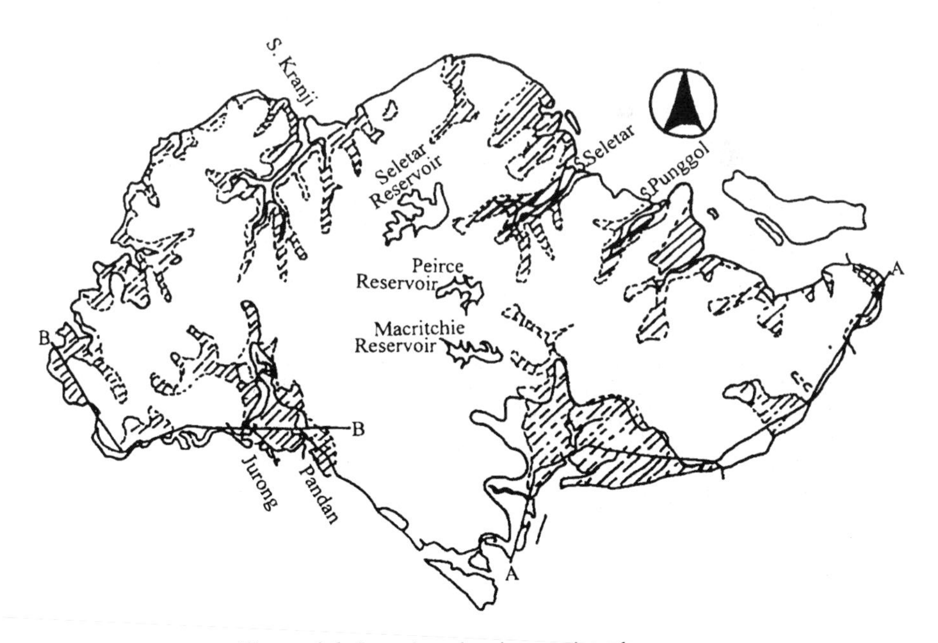

Figure 4.6. Location showing marine clay areas.

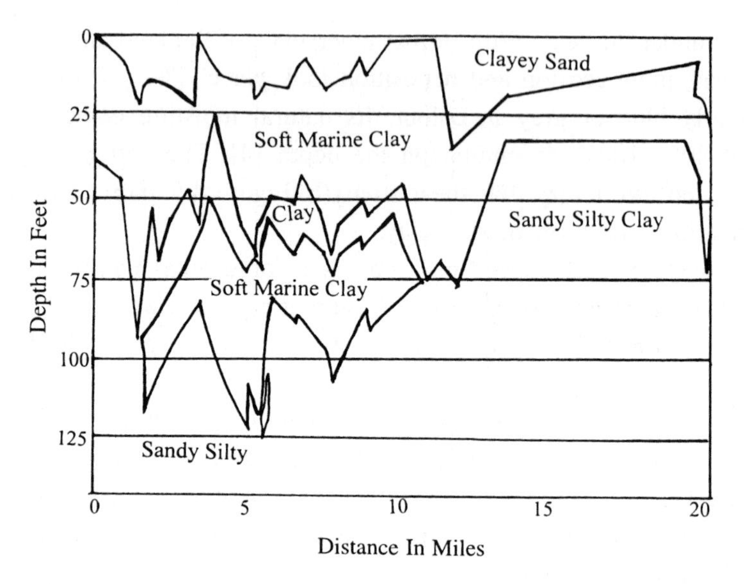

Figure 4.7. Cross-section from East Lagoon through Shenton Way, Nicoll Highway, Kallang River Basin, Geylang to Changi.

4.4. Types of Foundation

On the selection of a suitable foundation system for a building, various factors must be taken into consideration. Among them are soil conditions, load transfer pattern, shape and size of the building, site constraints, underground tunnels and/or services, environmental issues, etc. [6]. There are two basic types of foundations:

— Shallow foundations: those that transfer the load to the earth at the base of the column or wall of the substructure.

— Deep foundations: those that transfer the load at a point far below the substructure.

A general classification of foundation systems is shown in Figure 4.8.

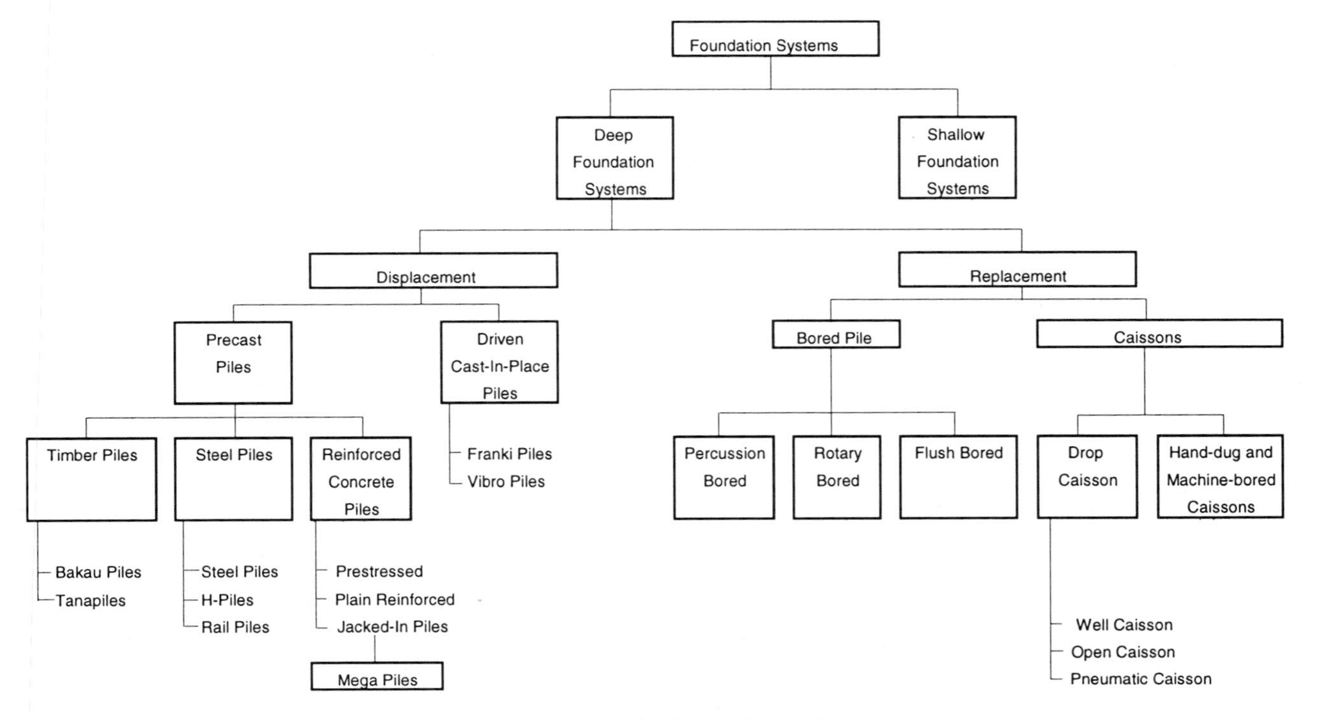

Figure 4.8. General classification of foundation systems.

4.5. Shallow Foundation

Shallow foundations transfer loads in bearing close to the surface. They either form individual spread footings or mat foundations, which combine the individual footings to support an entire building or part of it. The two systems may also act in combination with each other, for example, where a service core is seated on a large mat while the columns are founded on pad footings [7].

Spread footings are divided into isolated footings (e.g. column footings), strip footings (e.g. wall footings), and combined footings. Figure 4.9 shows some common shallow foundations. A column or isolated footing is a square pad of concrete transferring the concentrated load from above, across an area of soil large enough that the allowable stress of the soil is not exceeded. A strip footing or wall footing is a continuous strip of concrete that serves the same function for a loadbearing wall.

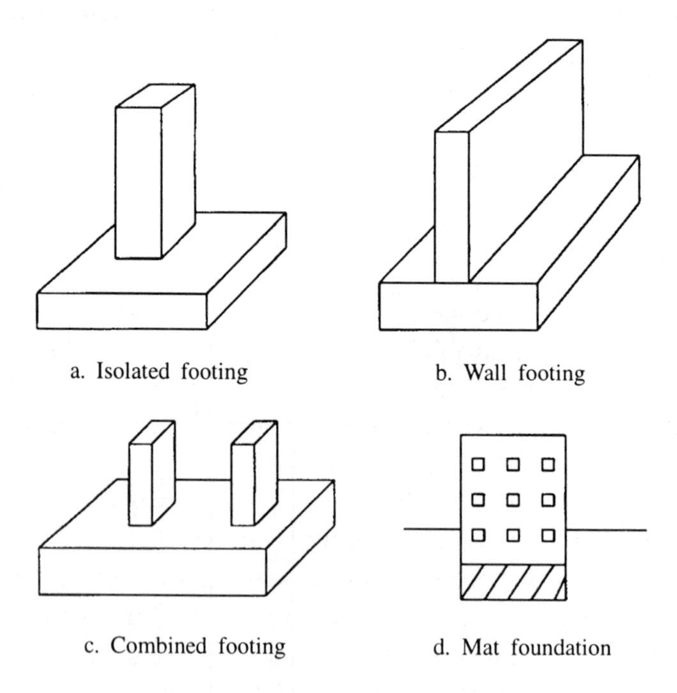

Figure 4.9. Types of shallow foundation.

In situations where the allowable bearing capacity of the soil is low in relation to the weight of the building, column footings may become large enough so that it is more economical to merge them into a single mat or raft foundation that supports the entire building. A mat foundation is basically one large continuous footing upon which the building rests. In this case, the total gross bearing pressure at the mat-soil interface cannot exceed the allowable bearing strength of the soil. Mat or raft foundation, however, is not suitable for tall buildings where the soil condition is not adequate. The highly concentrated and eccentric column loads means the thickness would need to be very large (often > 2 m) [8–10].

4.6. Deep Foundation

Deep foundations are used when adequate soil capacity is not available close to the surface and loads must be transferred to firm layers substantially below the ground surface. The common deep foundation systems for buildings are piles and caissons [11–19]. Figure 4.10 illustrates the various deep foundation types used according to the bearing layers. The common types of deep foundation used in Singapore are caisson, bored pile, driving reinforced concrete and steel piles.

4.7. Caisson

There have been a noticeable gain in the use of caissons for tall buildings recently, e.g. UOB building (12 caissons), OUB building (7 caissons), AIA Tower (65 caissons), DBS Building (4 caissons), Republic Plaza (14 caisson), Capital Tower (6 caissons) etc. There have also been cases where the bored pile method was initially specified but was later switched to the use of caisson. Among the reasons are difficulties in boring, increase in superimposed loads, requirement for speed of construction, etc.

The word "caisson" comes from French "caissee" meaning a chest or case. It was originally used to refer to a water-tight chamber within which foundation work could be carried out underwater. In common usage, it has also come to mean a type of deep foundation unit.

Condition	Type of Pile	RC Pile			Spun Concrete Pile			H-Steel Pile			Steel Pile			Bored Pile				Caisson	
		250	350	450	Ø250	Ø350	Ø450	H300x 300	H388x 351	H388x 402	Ø304	Ø406.4	Ø508	Ø800	Ø1000	Ø1200	Ø1500	Ø3000	Ø5000
Soil Layer for Piling	Soft Clay	♥	♥	♥	♥	♥	♥	♥	♥	♥	♥	♥	♥	♥	♥	♥	♥	♥	♥
	Loose Sand	♥	♥	♥	♥	♥	♥	♥	♥	♥	♥	♥	♥	ρ	ρ	ρ	ρ	ρ	ρ
	Gravel	X	ρ	♥	♥	♥	♥	♥	♥	♥	♥	♥	♥	♥	♥	♥	♥	♥	♥
	Boulder	X	X	X	X	X	X	X	X	X	X	X	X	ρ	ρ	ρ	ρ	♥	♥
	High Permeable Layer	♥	♥	♥	♥	♥	♥	♥	♥	♥	♥	♥	♥	♥	♥	♥	♥	ρ	ρ
	Inclined Hard Layer	X	X	X	X	X	X	X	X	X	X	X	X	ρ	ρ	X	X	♥	♥
Capability of Penetration by Pile	N > 30	♥	♥	♥	♥	♥	♥	♥	♥	♥	♥	♥	♥	♥	♥	♥	♥	♥	♥
	N > 40	ρ	♥	♥	ρ	♥	♥	♥	♥	♥	♥	♥	♥	♥	♥	♥	♥	♥	♥
	N > 50	X	X	ρ	X	X	ρ	ρ	♥	♥	♥	♥	♥	♥	♥	♥	♥	♥	♥
	N > 100	X	X	X	X	X	X	X	ρ	ρ	X	ρ	ρ	♥	♥	♥	♥	♥	♥
	N > 200	X	X	X	X	X	X	X	X	X	X	X	X	ρ	ρ	ρ	ρ	♥	♥
	Strong Rock	X	X	X	X	X	X	X	X	X	X	X	X	X	X	X	X	ρ	ρ
Length of Pile	Less 20m	♥	♥	♥	♥	♥	♥	♥	♥	♥	♥	♥	♥	♥	♥	♥	♥	♥	♥
	20-30m	ρ	♥	♥	♥	♥	♥	♥	♥	♥	♥	♥	♥	♥	♥	♥	♥	♥	♥
	30-40m	X	ρ	♥	X	♥	♥	ρ	♥	♥	♥	♥	♥	♥	♥	♥	♥	♥	♥
	40-50m	X	X	ρ	X	X	♥	X	ρ	♥	X	♥	♥	♥	♥	♥	♥	♥	♥
	50-60m	X	X	X	X	X	X	X	X	ρ	X	X	ρ	♥	♥	♥	ρ	♥	♥
	More 60m	X	X	X	X	X	X	X	X	X	X	X	X	ρ	ρ	ρ	X	♥	♥

♥ Suitable
X Not Suitable
ρ Possible

Figure 4.10. Types of deep foundation according to the bearing layer.

A caisson is a shell or box or casing which, when filled with concrete, will form a structure similar to a cast-in-place pile but larger in diameter. A caisson is similar to a column footing in that it spreads the load from a column over a large enough area of soil that the allowable stress in the soil is not exceeded. It differs from a column footing in that it reaches through strata of unsatisfactory soil beneath the substructure of a building until it reaches a satisfactory bearing stratum such as rock, dense sands and gravels or firm clay (see Figure 4.17). Over the years, however, the term has come to mean the complete bearing unit [20–25].

Factors affecting the choice of using caisson piles:

(a) On site where no firm bearing strata exists at a reasonable depth and the applied loading is uneven, making the use of a raft inadvisable.
(b) When a firm bearing strata does exist but at a depth such as to make a strip, slab or pier foundation uneconomical, i.e. at depths of over 3 to 4.5 m, but not so deep as to make the use of a raft essential.
(c) When pumping of ground water would be costly or shoring to excavation becomes too difficult to permit the construction of spread foundation.
(d) When very heavy loads must be carried through water-logged or unstable soil down to bed rock or to a firm strata, and having large number of piles with large pile caps are not economical.
(e) When the plan area of the required construction is small and the water is deep.

Bored Caisson: A bored caisson is one in which a hole of the proper size is bored to depth and a cylindrical casing or caisson is set into the hole.

This is the common type of caisson used for building construction in Singapore (see Figure 4.3). It is basically a concrete-filled pier hole for the support of columns. In Singapore, bored caissons of diameters ranging from 600 mm to 6 m are common. The hole can be bored by using a bucket drill, mini excavator or hand-dug (Figure 4.11 and Figure 4.12).

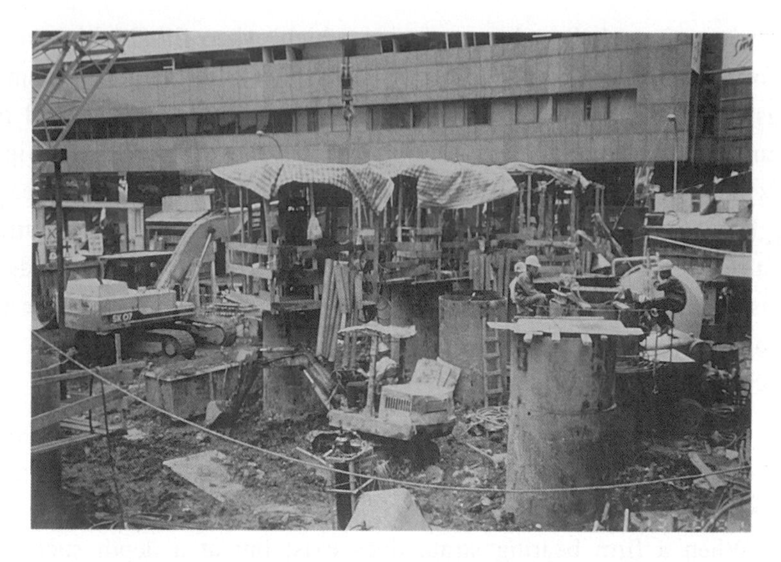

Figure 4.11. Closely spaced hand-dug caissons ($\phi > 600$ mm).

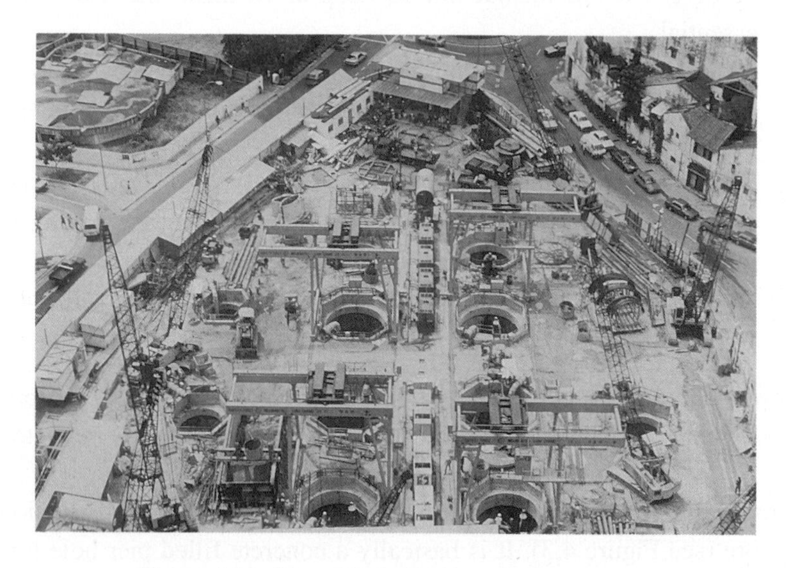

Figure 4.12. 12 mechanically-dug caissons ($\phi > 6$ m).

Figure 4.13(a). Construction of 6 bored caissons ($\phi = 6$ m).

Figure 4.13(b). Two mobile gantry cranes each serving 3 caissons slide along tracks cast on the concrete floor. Note safety nets on the handrails, artificial ventilators and lighting in each caisson.

A bored caisson depending on the size may be hand-dug or excavated with the use of mini excavators. In both cases, excavation takes place inside the caisson shafts instead of occupying additional space as compared to that of the conventional bored pile methods. It requires less space and hence enables a tidier site. As excavation is carried out bit by bit within a shaft, vibration, noise and dust can be reduced. Each stage of excavation is followed immediately by the casting of the concrete lining to resists the surrounding soil and water penetration.

The work sequence of 6 bored caissons of a recent project (Figure 4.13) is as follows:

- Establish level and position of caissons.
- Excavate to a depth of 1.5 m by means of mini excavators. Workers are lowered to excavate manually in hard to reach places (Figure 4.14).
- Lay circular reinforcement mesh for the caisson lining (Figure 4.15).
- Erect steel ring wall formwork (Figure 4.16). There are 3 segments of circular steel formwork connected with bolt and nut and locking pin. For the first caisson lining, assemble the segments on ground and hoist down for fixing.
- Cast concrete with hoisting buckets sequentially round the ring wall and progressively upward.
- Excavate further 1.5 m downward, lay reinforcement mesh. Detach formwork from the cast concrete by removing the locking pin and the use of chain block.
- Lower down the formwork to the next lower position by the use of hoisting crane and chain block for positioning and alignment.
- Fix formwork slightly inclined outward to enhance concrete placement (Figure 4.17).
- Place concrete leaving a gap to be grouted later (Figure 4.18).
- Material handling using gantry crane, mobile crane etc. (Figure 4.19).
- Provide adequate ventilation and lighting (Figure 4.20).
- Complete caisson lining to founding level.
- Carry out jack-in-test when required to assess the skin frictional resistance by inserting jacks in gap between 2 linings. Apply jacking

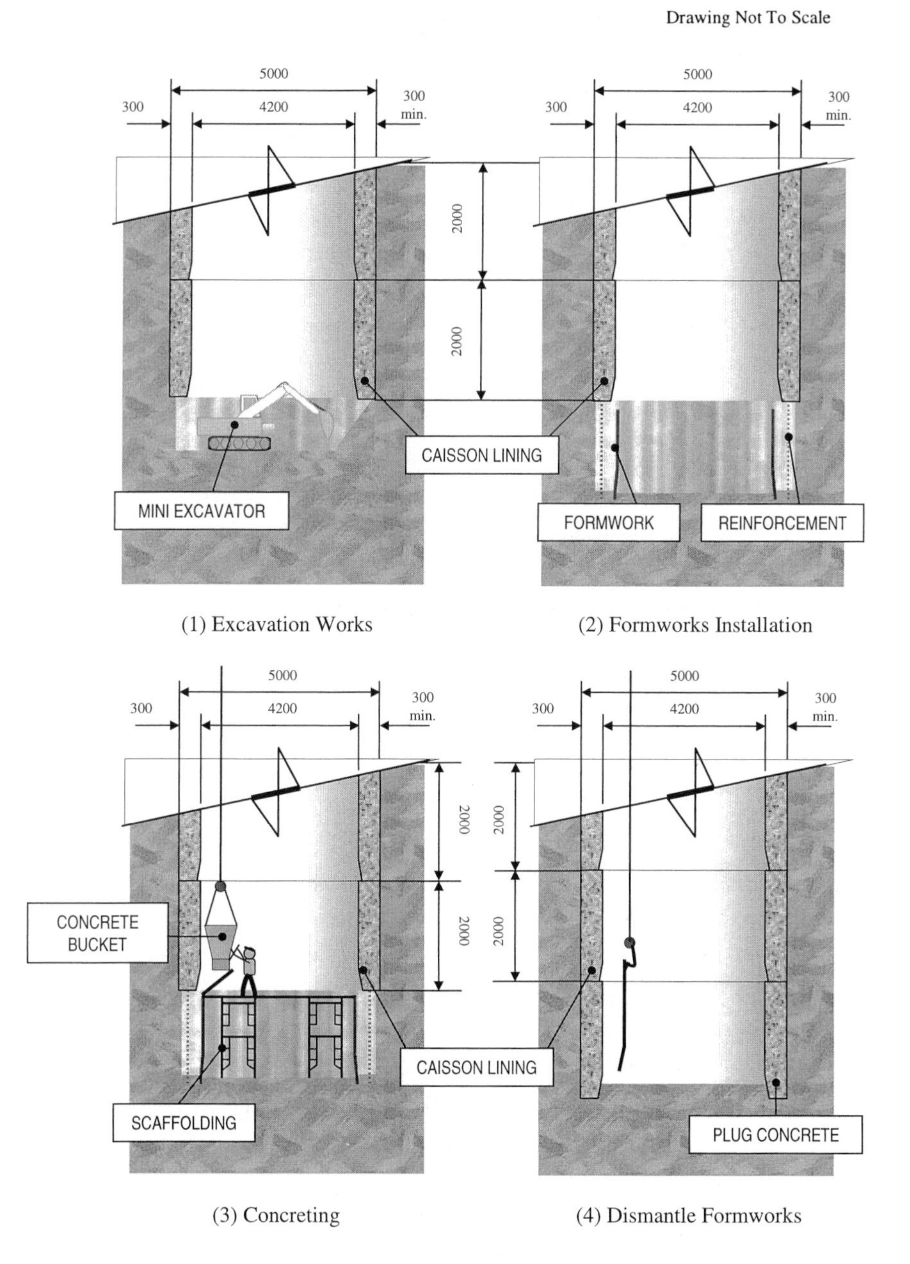

Figure 4.14. Construction sequence of a typical caisson lining.

Circular Reinforcement are put into the excavated ground, allowing sufficient concrete cover.

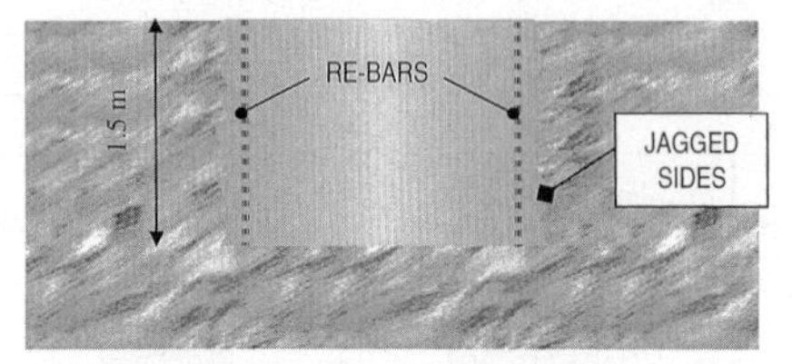

Cross-Sectional View of Caisson

Figure 4.15. A section of the circular reinforcement mesh for the caisson lining.

until the target load capacity is reached. Monitor pressure and movement using pressure gauges, level meters, strain and dial gauges etc. (Figure 4.21). Align eight jacks as shown in Figure 4.22.

- Monitor the dewatering system. Control ground water pressure around and beneath the caisson excavation as required. Provide instrumentation to monitor pore pressure changes and ground settlement when required. Provide recharge wells when necessary.
- Erect caisson reinforcement. Arrange concreting platform (Figure 4.23). Cast caisson through tremie pipes in one operation. Provide adequate vibration. Monitor temperature to avoid micro cracking due to the possible high heat of hydration. Cast concrete to the required height (Figure 4.24).

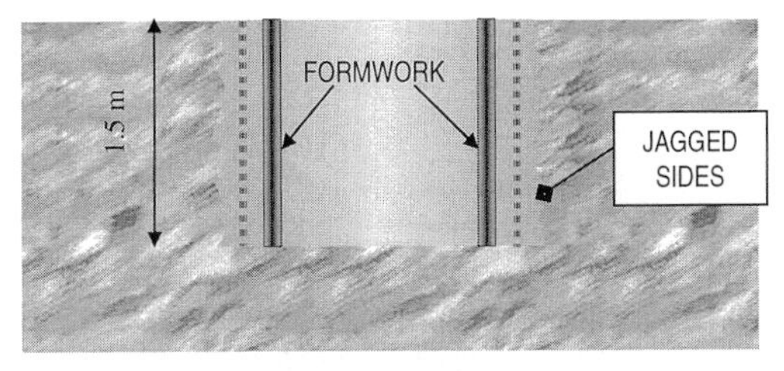

Cross-Sectional View of Caisson

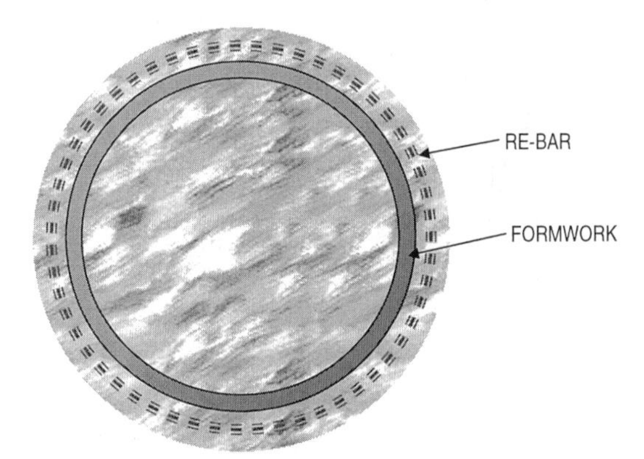

Plan View of the caisson

Figure 4.16. Erection of steel ring wall formwork.

Drawing Not To Scale

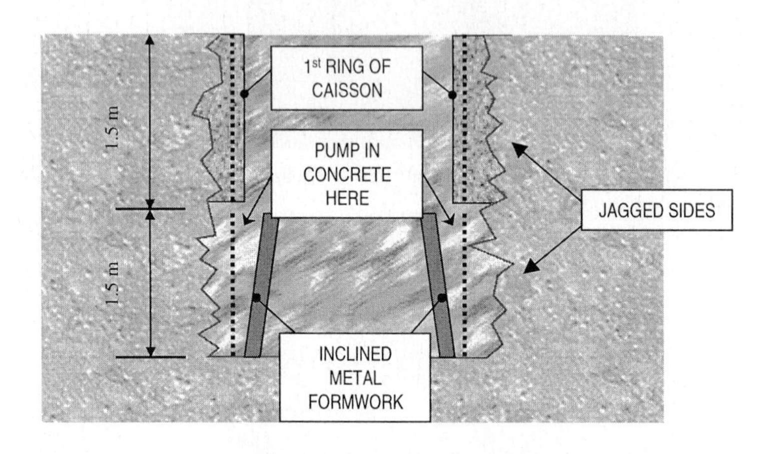

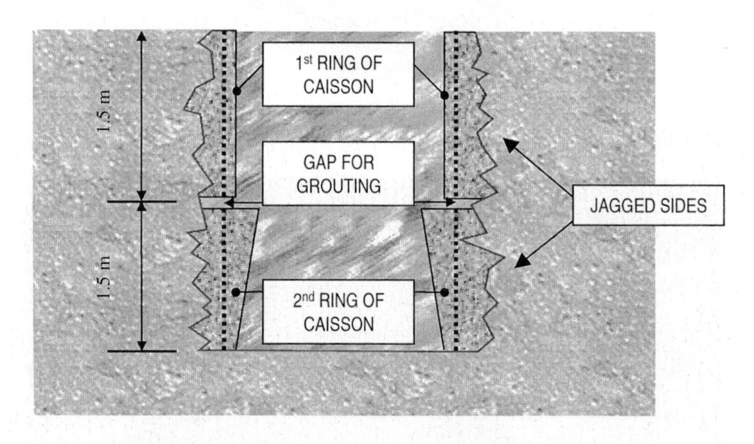

Figure 4.17. Caisson lining for subsequent layers.

Figure 4.18(a). Completed caisson lining with a gap between each layer.

Figure 4.18(b). A closer view of the gap. Note the pins and starter bars.

Figure 4.19(a). Lowering a mini excavator using the gantry crane.

Figure 4.19(b). Material handling using a gantry crane, mobile crane, etc.

Figure 4.20(a). Provision of adequate artificial ventilation through a duct. Note the staircase on the right.

Figure 4.20(b). Provision of a blower.

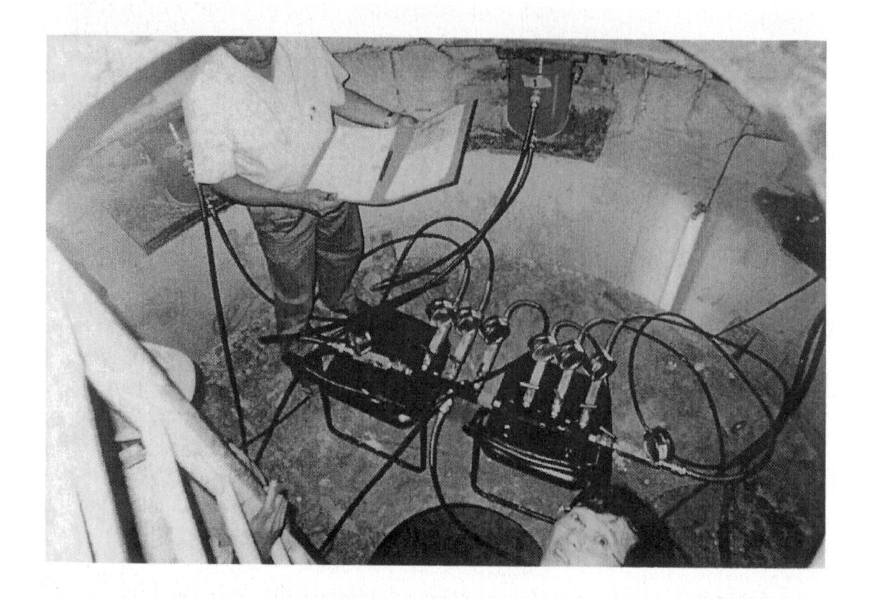

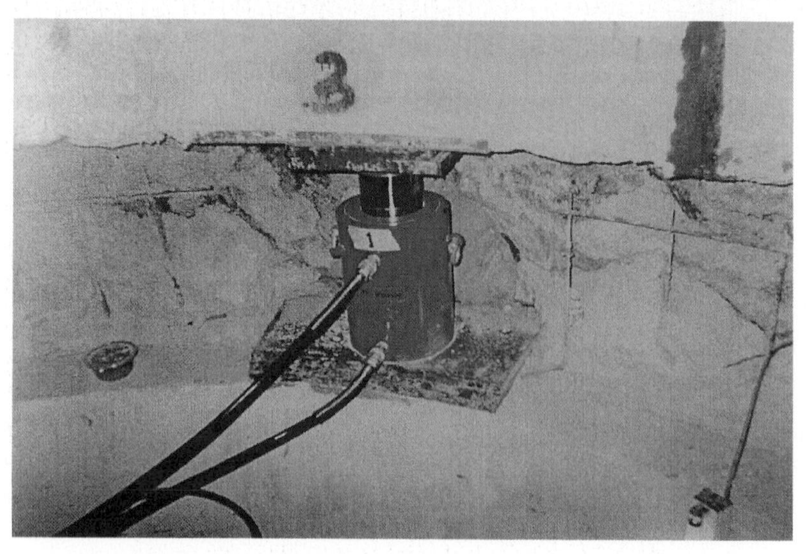

Figure 4.21(a). Hydraulic jacks placed in between 2 caisson rings.

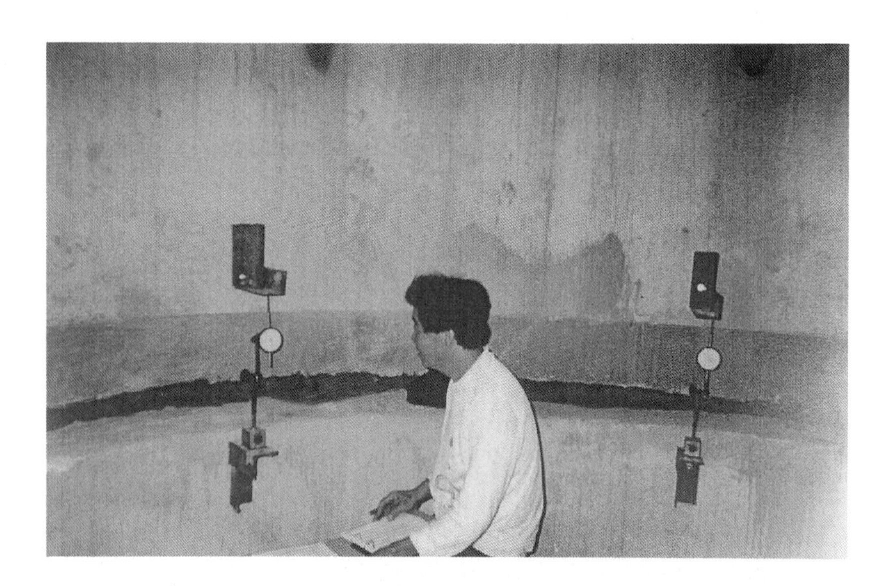

Figure 4.21(b). Monitor displacement using dial gauges.

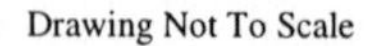

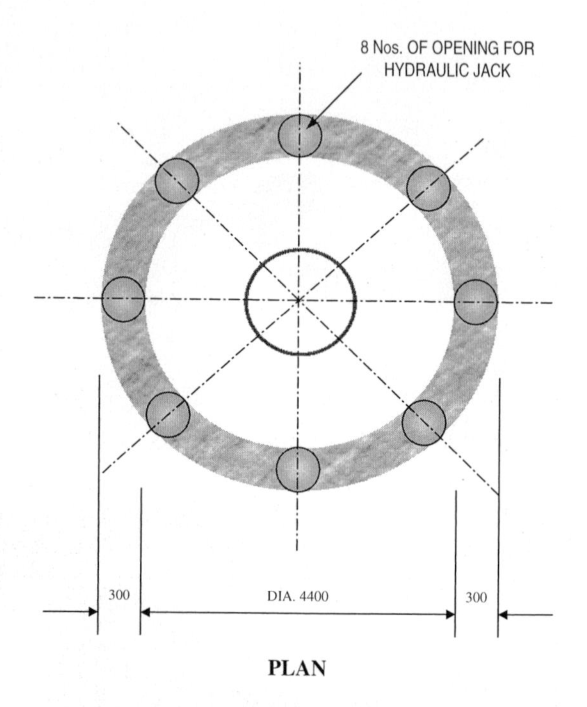

Figure 4.22(a). Layout of jacking test.

The same principles apply to small caissons, except that the whole operation has to be carried out manually (Figure 4.25 and Figure 4.26).

Gow Caisson: Figure 4.27 shows the construction sequence of a gow caisson. The largest cylinder is first sunk by excavating below the cutting edge and driving the cylinder down. After the first cylinder is positioned, the second with a smaller diameter is sunk in the same manner and others in succession to the desired depth.

A bell is excavated if the soil conditions permit. After the excavation is completed, the bell is filled with concrete and the cylinders are withdrawn as concreting proceeds until the pier is completed.

Drawing Not To Scale

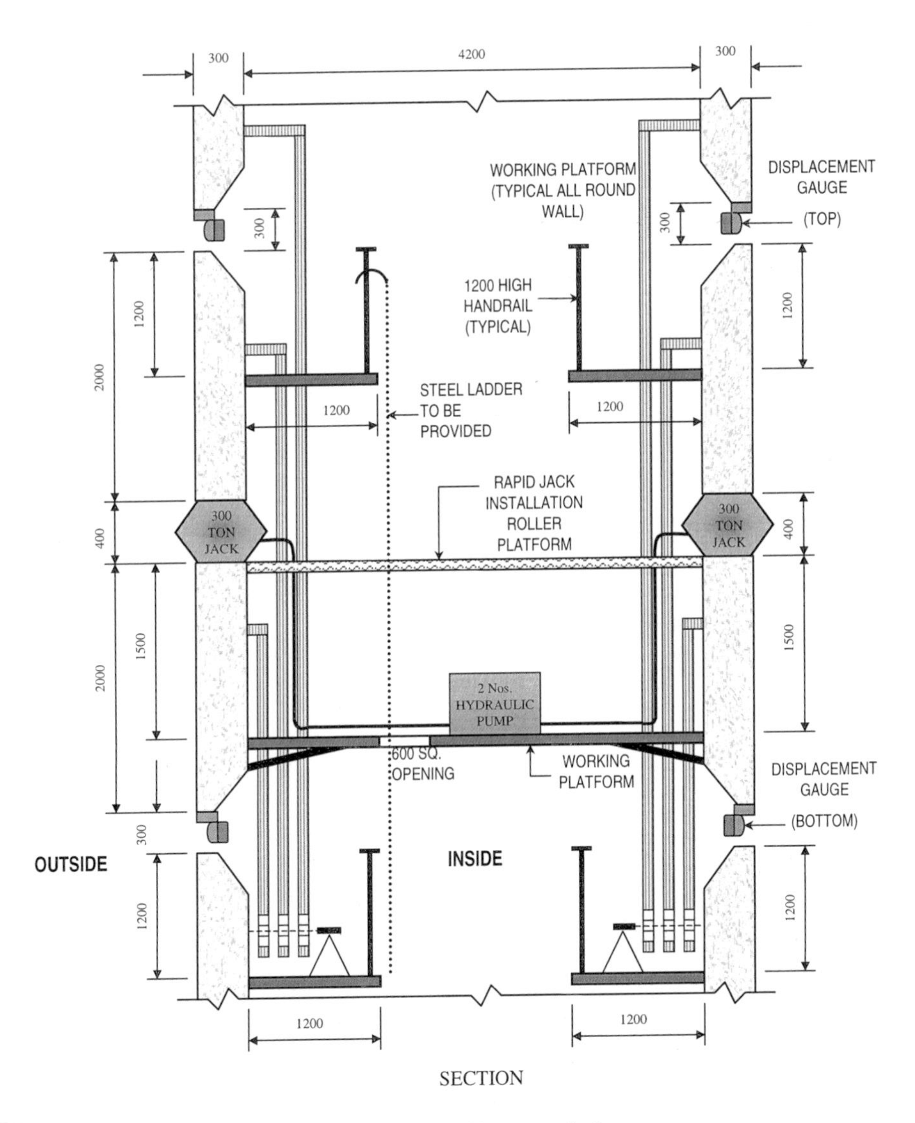

Figure 4.22(b). Jacking test platform.

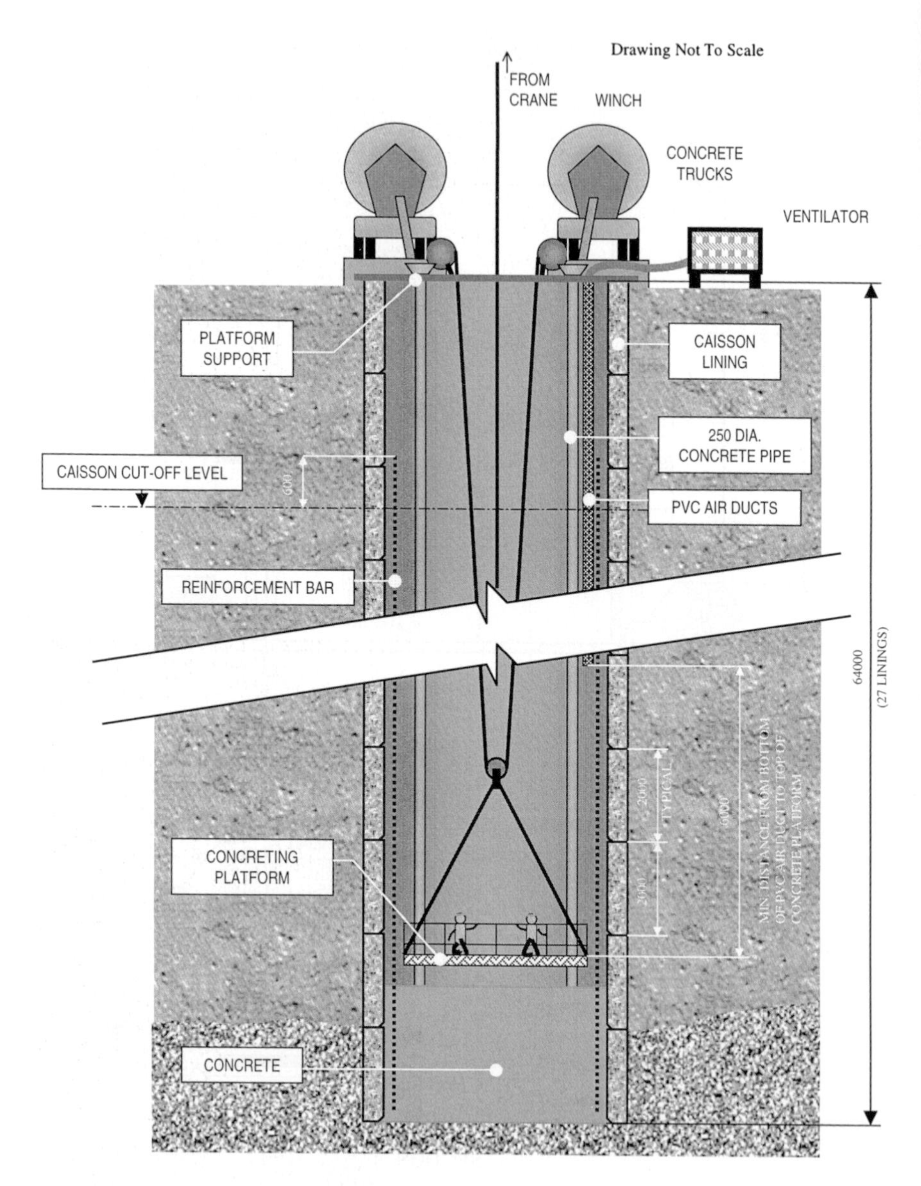

Figure 4.23. Mass concreting of a caisson.

Figure 4.24. Cast caisson to the required height covered with polystyrene for curing.

Figure 4.25. A worker being lowered into the casing of a hand-dug caisson.

Figure 4.26. Concreting through a tremie pipe into a hand-dug caisson.

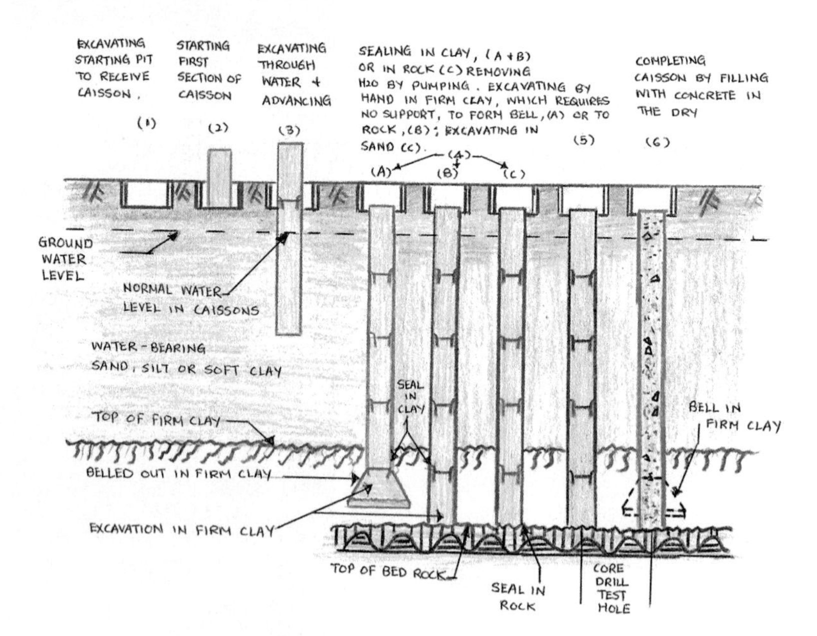

Figure 4.27. Construction sequence of gow caissons.

Socketed Caisson: A socketed caisson is one that is drilled into rock at the bottom, rather than belled. Its bearing capacity comes not only from its end bearing, but from the frictional forces between the sides of the caisson and the rock as well (Figure 4.28). A steel pipe is driven in concurrently with augering. When the bedrock is reached, a socket is churn-drilled into the rock that is slightly smaller in diameter than the caisson shell. The unit is then reinforced and concreted.

Box Caisson: Box caissons are structures with a closed bottom designed to be sunk into prepared foundations below water level. Box caissons are unsuitable for sites where erosion can undermine the foundations, but they are well suited for founding on a compact inerodible gravel or rock which can be trimmed by dredging (Figure 4.29a). They can be founded on an irregular rock surface if all mud or loose material is dredged away and replaced by a blanket of sound crushed rock (Figure 4.29b). Where the depth of soft material is too deep for dredging they can be founded on a piled raft (Figure 4.29c).

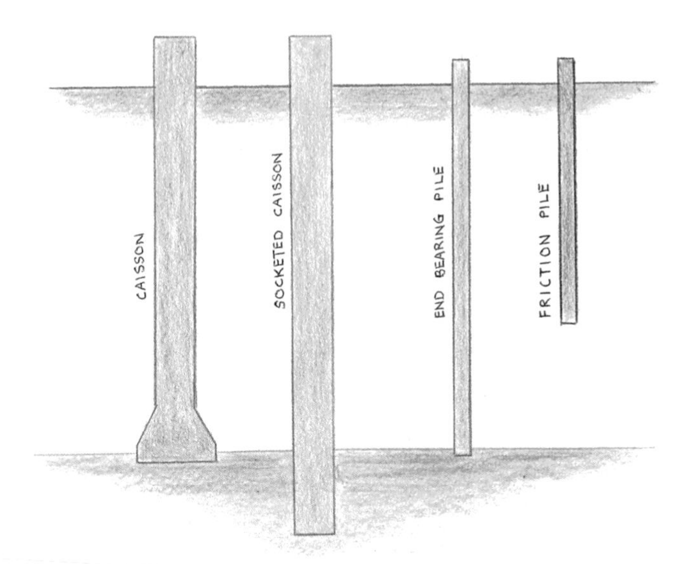

Figure 4.28. Socketed caissons.

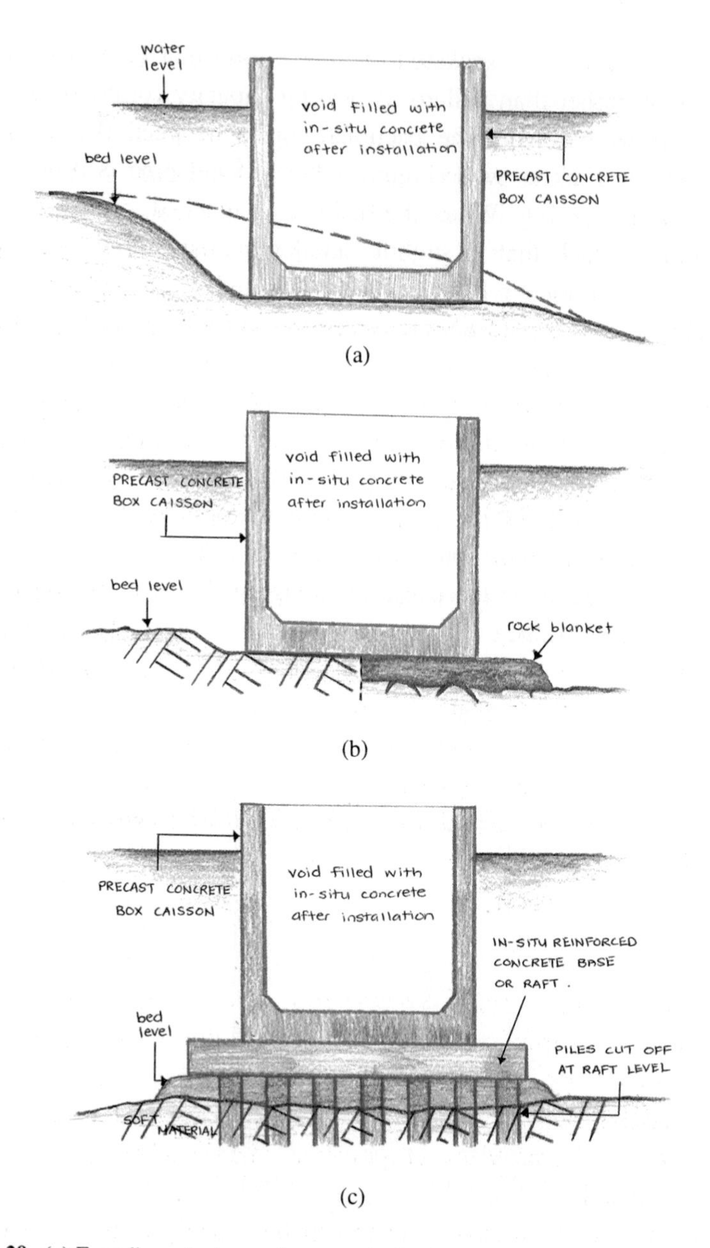

Figure 4.29. (a) Founding of a box caisson on dredged gravel or rock, (b) founding of a box caisson on crushed rock blanket over rock surface, (c) founding of a box caisson on piled raft.

After the box caisson is sunk to the required location, crushed rocks or sand is backfilled into the box to act as an additional weight on the bearing stratum. This is to ensure the rigidity of the foundation for the structure. An example of the use of box caisson is the Kepple-Brani Road Link as shown in Figure 4.30. Figure 4.31(a) shows the construction of a box caisson in a floating dock. Figure 4.31(b) shows the tug boat in operation. Figure 4.31(c) shows the filling of sand into a caisson. Other examples of the use of box caissons include the Jurong Island Linkway and the Causeway to Johore Bahru.

Pneumatic Caisson: A concrete box built with an airtight chamber at the bottom is constructed on ground. The air is compressed balancing with the ground water pressure to prevent the ground water from getting into the box. As soil is excavated and removed, the box is gradually sunk into the ground. Steel shafts are connected to the pressurised working chamber as access for workers and excavation machinery. The shafts are equipped with locks to regulate the difference between the atmospheric pressure on the ground and the pressure in the chamber. Excavate 4 m downwards, construct the caisson lining and sink further. Repeat the process to the desire depth. At the designed depth, test for soil bearing capacity, remove the equipment from the working chamber and fill with concrete. Details of the method can be found in Shiraishi homepage (http://www.shiraishi.com/index.html).

4.8. Piles

A pile can be loosely defined as a column inserted in the ground to transmit the structural loads to a lower level of subsoil. Piles have been used in this context for hundreds of years and until the twentieth century were at large of driven timber. Today, a wide variety of materials and methods are available to solve most of the problems encountered when confronted with the need for deep foundation.

The construction process of piles can be broadly characterised by the installation and testing. However, there are many proprietary types of piles and the installing process for each type differed.

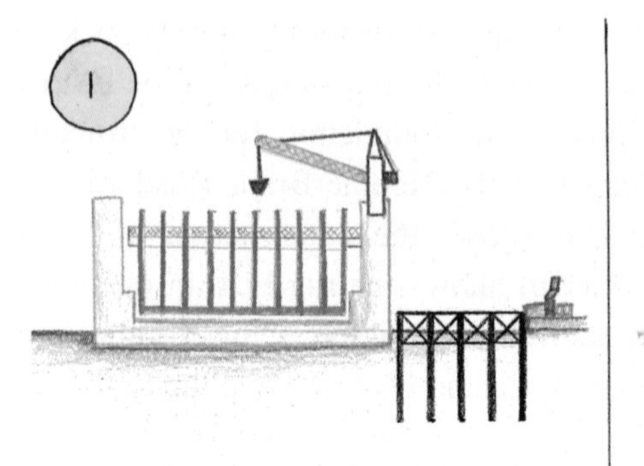

CAISSON IS CAST ON U-SHAPED FLOATING DOCK. EACH CAISSON WEIGHS A MINIMUM OF 2,280 TONNES.

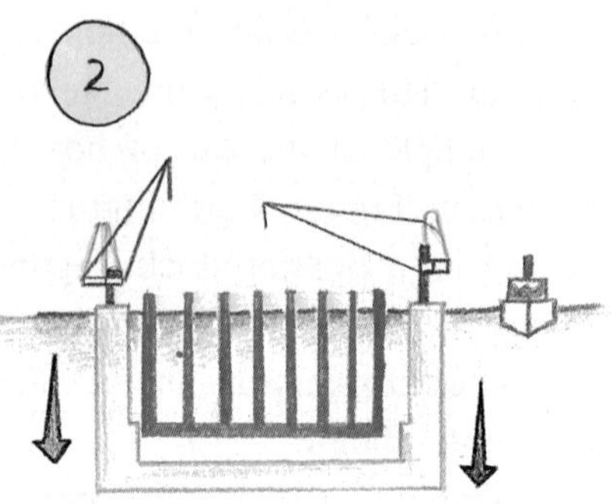

FLOATING DOCK IS TOWED BY TUGBOAT TO THE LAUNCHING AREA. THE DOCK IS THEN SUNK, LEAVING THE CAISSON TO FLOAT

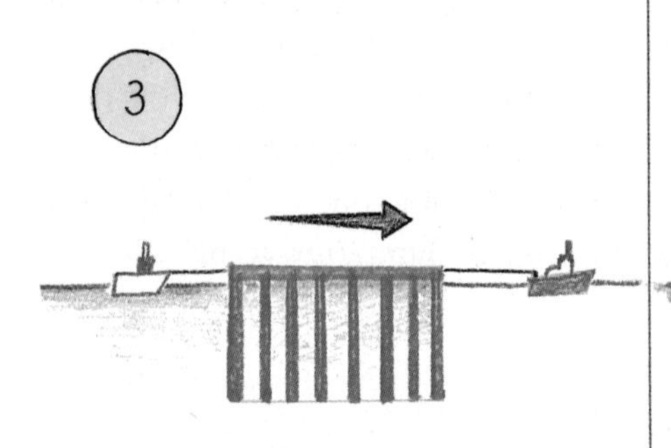

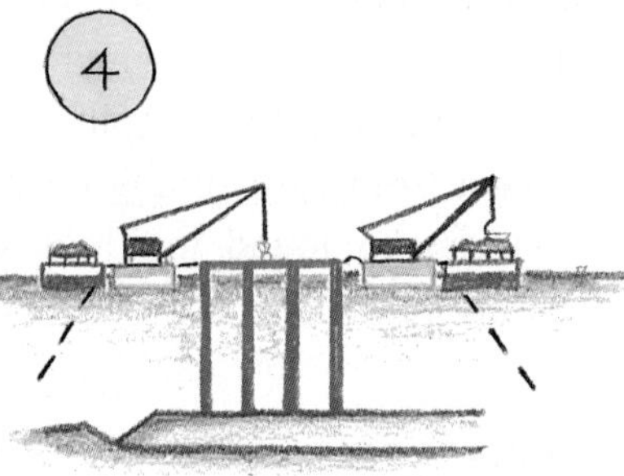

FLOATING CAISSON IS TOWED TO THE LOCATION BY TUGBOATS USING WIRE ROPE.

4
ONCE IN POSITION, THE CAISSON IS FILLED WITH WATER FROM AN UNDERWATER PUMP. THE WEIGHT OF THE WATER CAUSES CAISSON TO SINK TO ITS FINAL, ALINGED POSITION. IT IS THEN BACKFILLED WITH SAND. THEN, WORK STARTS ON THE NEXT CAISSON.

Figure 4.30. Typical construction sequence for a causeway using caissons.

Figure 4.31(a). Construction of a box caisson in a floating dock by the yard. Each caisson is 6 m × 6 m × 12 m.

Figure 4.31(b). A tug boat in operation and a submerged caisson filled with water.

Figure 4.31(c). Filling of sand into a caisson.

Piles may be classified by the way there are formed i.e. *displacement piles* and *non-displacement piles*.

The classification of displacement and non-displacement piles is shown in Figure 4.32. The displacement in the soil is the pressure that the pile exerts on the soil as a result of being driven into the soil. In deciding upon the type of piles to use for a particular construction, the following should be considered:

- Superstructure design and the site area.
- Soil conditions and surrounding buildings and structures (e.g. underground tunnels).
- Availability of equipment and site constraints.
- Knowledge of the pros and cons of various piling systems.

Piles may be classified as either end-bearing or friction piles, according to the manner in which the pile loads are resisted.

- End bearing: The shafts of the piles act as columns carrying the loads through the overlaying weak subsoils to firm strata into which the pile toe has penetrated. This can be a rock strata or a layer of firm sand

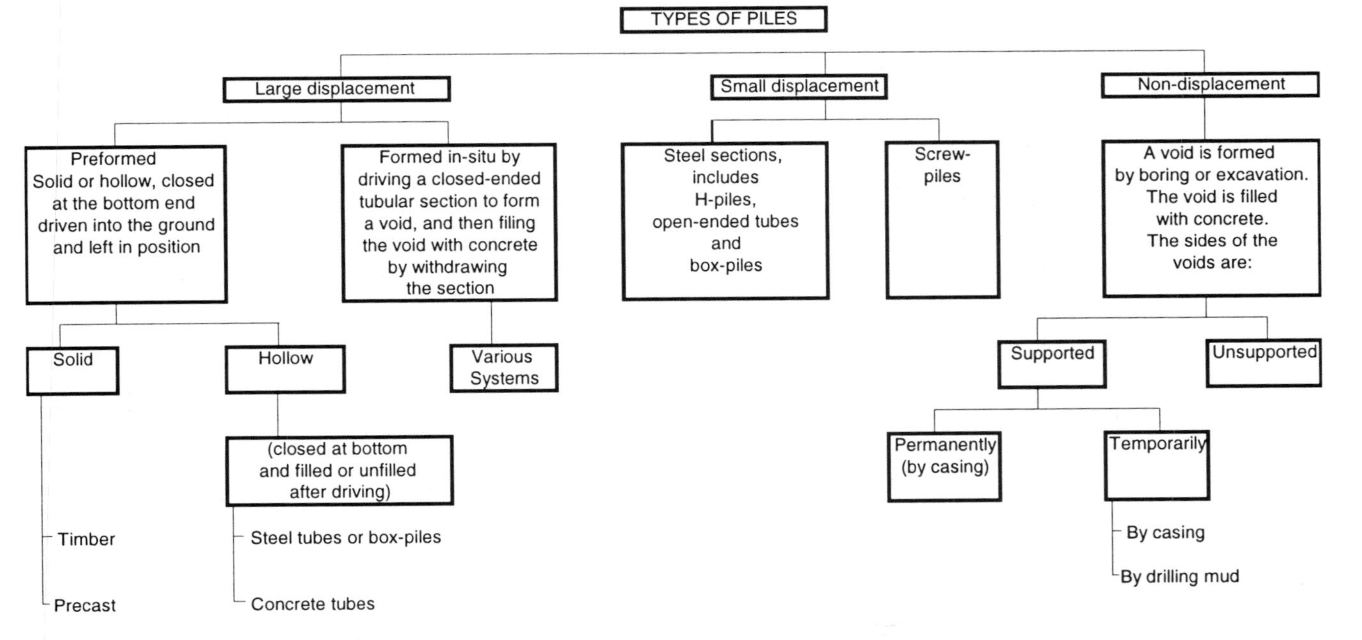

Figure 4.32. The classification of displacement and non-displacement piles.

or gravel which has been compacted by the displacement and vibration encountered during the driving.

- Friction: Any foundation imposes on the ground a pressure which spreads out to form a pressure bulb. If a suitable load bearing strata cannot be found at an acceptable level, particularly in stiff clay soils, it is possible to use a pile to carry this pressure bulb to a lower level where a higher bearing capacity is found. The friction or floating pile is mainly supported by the adhesion or friction action of the soil around the perimeter of the pile shaft.

However, in actual practice, virtually all piles are supported by a combination of skin friction and end bearing [26–34].

4.8.1. *Non-Displacement Pile*

Sometimes referred to as replacement piles but more commonly as bored piles. They are formed by boring/removing a column of soil and replaced with steel reinforcement and wet concrete cast through a funnel or tremie pipe. For soft grounds and where the water table is high, bentonite may be used during boring to resist the excavation and water inflow before casting. Bored piles are considered for sites where piling is being carried out in close proximity to existing buildings where vibration, dust and noise need to be minimised. They are also used instead of displacement piles in soils where negative friction is a problem. Bored piles of diameter ranges from 100 mm (micro-piles) to 2.6 m are common.

The construction sequence of a typical bored pile is shown in Figure 4.33:

- Set out and peg the exact location of the piles.
- With the boring rig fitted with an augering bit, bore the initial hole for the insertion of the temporary casing.
- Place the casing using a vibratory hammer with the top slightly higher than the ground level (Figure 4.34). The casing serves to align the drilling process as well as to prevent the collapse of the soil from the ground surface.

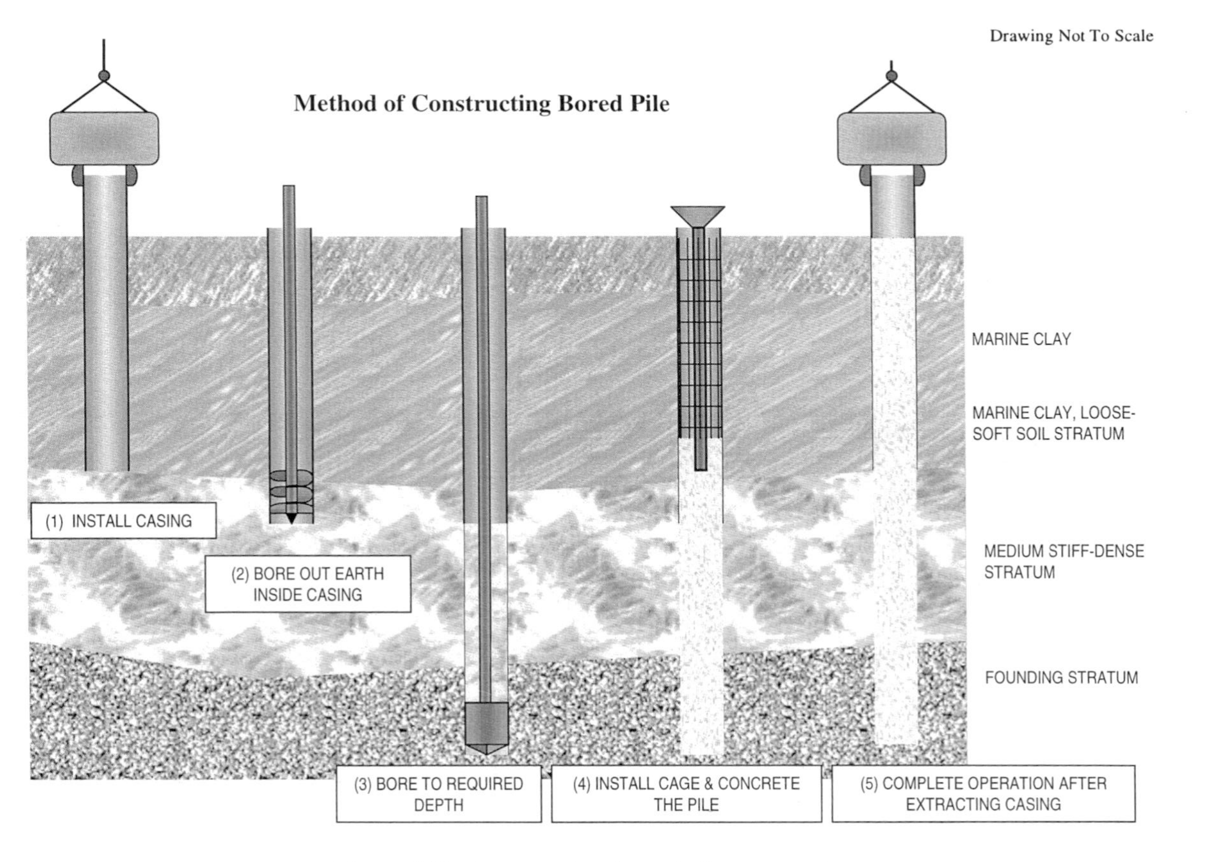

Figure 4.33. An example of bored pile construction.

Figure 4.34(a). Insertion of steel casing using vibro hammer operated by the crane operator.

Figure 4.34(b). Protruded steel casing for safety reasons.

Figure 4.35(a). Augering bits of different diameters.

Figure 4.35(b). An augering bit with industrial diamond "teeth" for cutting rock.

- Proceed with boring using an augering bit (Figure 4.35 and Figure 4.36). The auger can be of Cheshire or helix auger which has 1½ to two helix turns at the cutting end (Figure 4.37). The soil is cut by the auger, raised to the surface and spun off the helix to the side of the borehole. Alternatively, a continuous or flight auger can be used where the spiral motion brings the spoil to the surface for removal.

Figure 4.36. Removal of soils using an auger.

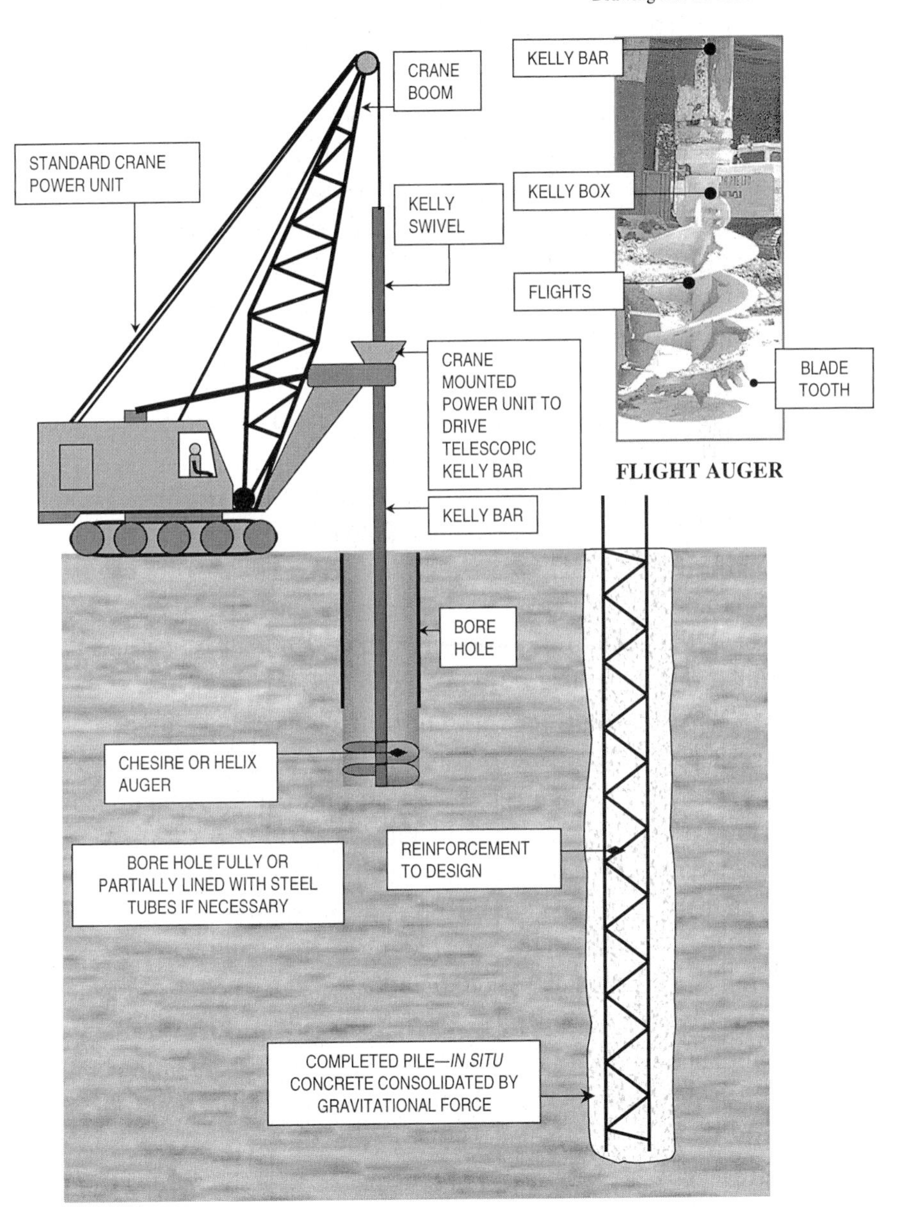

Figure 4.37. Typical rotary bored pile details.

- The base or toe of the pile can be enlarged or underreamed up to three times the shaft diameter to increase the bearing capacity of the pile (Figure 4.38).
- For deeper boreholes (> 20 m), a boring bucket may be used (Figure 4.39a). The boring bucket is designed with a shuttle to drill into the soil and carry the soil to the surface without it being washed away in the process. On the surface, the shuttle is opened to drop the soil into a soil pit (Figure 4.39b).
- Bentonite is commonly used to resist the excavation and water seepage. Bentonite mix ratio of 1:20 is common (See Chapter 5).
- If boulders or intermediate rock layers are encountered, a core bit (Figure 4.40), a chisel drop hammer (Figure 4.41), or cold explosive may be used to break up the rocks.
- Lighter debris is displaced by the bentonite. A cleaning bucket is used to clear the crush rocks and flatten the surface at the bottom (Figure 4.42).
- At the rock strata, a reverse circulation rig (RCD) may be used to bore into the rock. Piles socketed into hard rock with penetration ranging from 800 mm to 1.6 m is common.
- Reverse circulation drilling is common for mineral exploration works. It offers a cheaper option to a good quality sampling that nearly equals that of diamond coring. The reverse circulation rig is equipped with tungsten carbide percussion drill bits/teeth (Figure 4.43). Air tubes are inserted near the drilling to circulate the drilled rocks in the bentonite slurry which are then pumped out for filtering/processing.
- Before the insertion of the reinforcement cage and concrete casting, the high-density contaminated bentonite needs to be changed or it would mix and weaken the concrete. The high-density contaminated bentonite being heavier is pumped out from the bottom of the borehole and fresh bentonite pumped in from the top. An air pipe is inserted near the bottom to stir and circulate the slurry.
- The fresh, low-density bentonite is then tested for purity. Common tests include the sand content test, viscosity test, alkalinity (pH) test.
- Insert reinforcement cage with proper spacers (Figure 4.44).

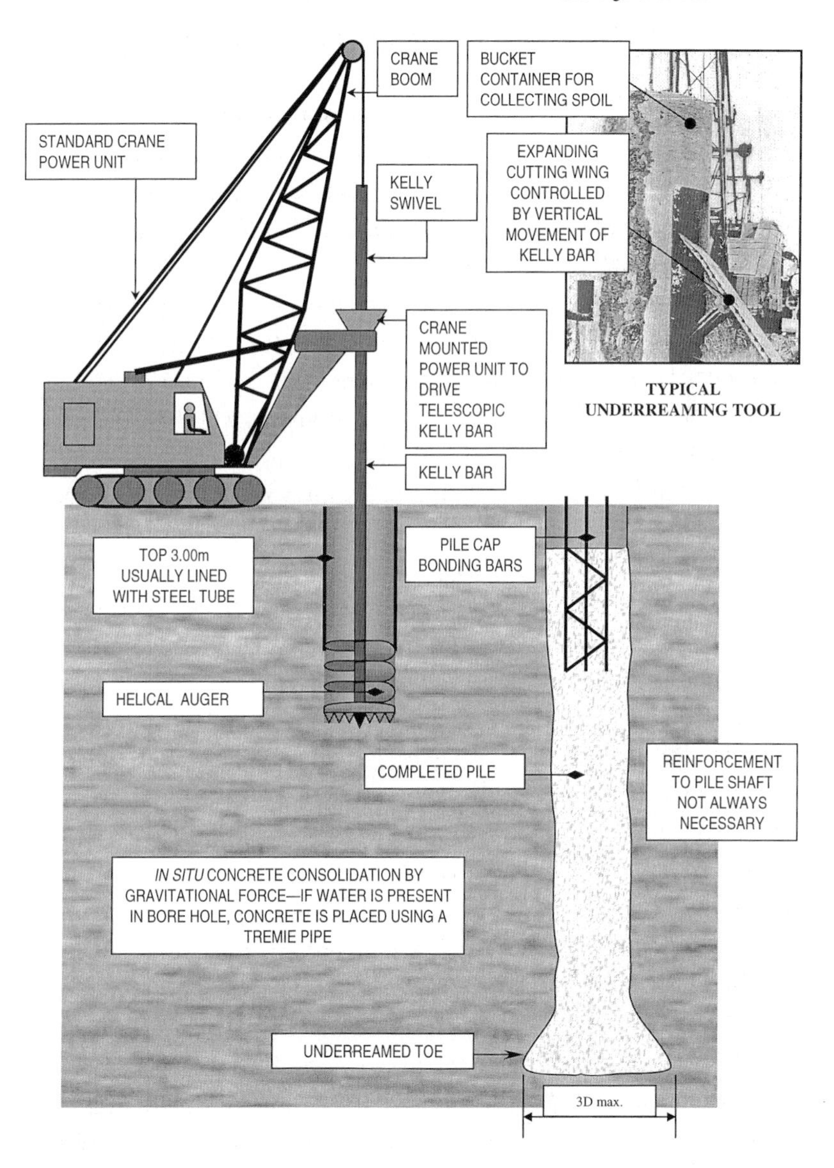

Figure 4.38. Typical large diameter bored pile details.

Figure 4.39(a). A boring bucket.

Figure 4.39(b). Shuttle opened to dump waste onto a soil pit.

Figure 4.40. A boring core bit for the boring process.

- Pour concrete with the use of a hopper and tremie pipes (Figure 4.45) to about 1 m higher than the required depth. The excess concrete contains contaminants displaced by the denser concrete and will be hacked off.
- Remove steel casing (Figure 4.46) after concreting.
- Excavate to cut-off level. Hack off excess contaminated concrete on the top exposing the reinforcement. Pour lean concrete around the bored pile. Install formwork for the pile cap and cast concrete (Figure 4.47).

Advantages:

(a) Length can readily be varied to suit the level of bearing stratum.
(b) Soil or rock removed during boring can be analysed for comparison with site investigation data.
(c) *In situ* loading tests can be made in large diameter pile boreholes, or penetration tests made in small boreholes.

Figure 4.41. Chisels used to break up boulders

Figure 4.42. A cleaning bucket to clear the crush rocks and flatten the surface at the bottom of a bore hole.

Figure 4.43. Drilling bits of the reverse circulation rig (RCD).

Figure 4.44(a). Insertion of reinforcement cage. Note the 1.2 m lapping between one cage section to another.

Figure 4.44(b). High tensile steel rods are tied to circular rings attached with concrete spacer.

Figure 4.45(a). A hopper to be connected to the top section of the tremie pipe. The hopper acts as a funnel to facilitate the concrete pouring.

Figure 4.45(b). Sections of tremie pipes to be connected to the length required.

Figure 4.45(c). Tremie pipe hoisted up and down to provide some vibration.

Figure 4.46. The removal of casing after concreting.

Figure 4.47(a). Excavate to the cut-off level.

Figure 4.47(b). Hack off the top contaminated excess concrete and expose the reinforcement.

Figure 4.47(c). Pour lean concrete around the bored pile.

Figure 4.47(d). Construct formwork and cast the pile cap.

(d) Very large (up to 6 m diameter) bases can be formed in favourable ground.
(e) Drilling tools can break up boulders or other obstructions which cannot be penetrated by any form of displacement pile.
(f) Material forming pile is not governed by handling or driving stresses.
(g) Can be installed in very long lengths.
(h) Can be installed without appreciable noise or vibration.
(i) No ground heave.
(j) Can be installed in conditions of low headroom.

Disadvantages:

(a) Concrete in shaft susceptible to squeezing or necking in soft soils where conventional types are used.
(b) Special techniques needed for concreting in water-bearing soils.
(c) Concrete cannot be readily inspected after installation.
(d) Drilling a number of piles in group may cause loss of ground and settlement of adjacent structures.

4.8.2. *Displacement Piles*

Displacement piles refer to piles that are driven, thus displacing the soil, and include those piles that are preformed, partially preformed or cast in place. This is the most cost efficient piling method but is losing its popularity in areas sensitive to noise, vibration and dust. The presence of boulders can also hinder the use of driving piles.

Precast Reinforced Concrete Piles — Come in different sizes and lengths, they are driven by drop hammers or vibrators using a piling rig as shown in Figure 4.48. They provide high strength and resistance to decay. They are however heavy, and because of its brittleness and low tensile strength, cares in handling and driving is required. Cutting requires the use of pneumatic hammers, cutting torches, etc. The construction sequence of a typical precast reinforced concrete pile is as follows:

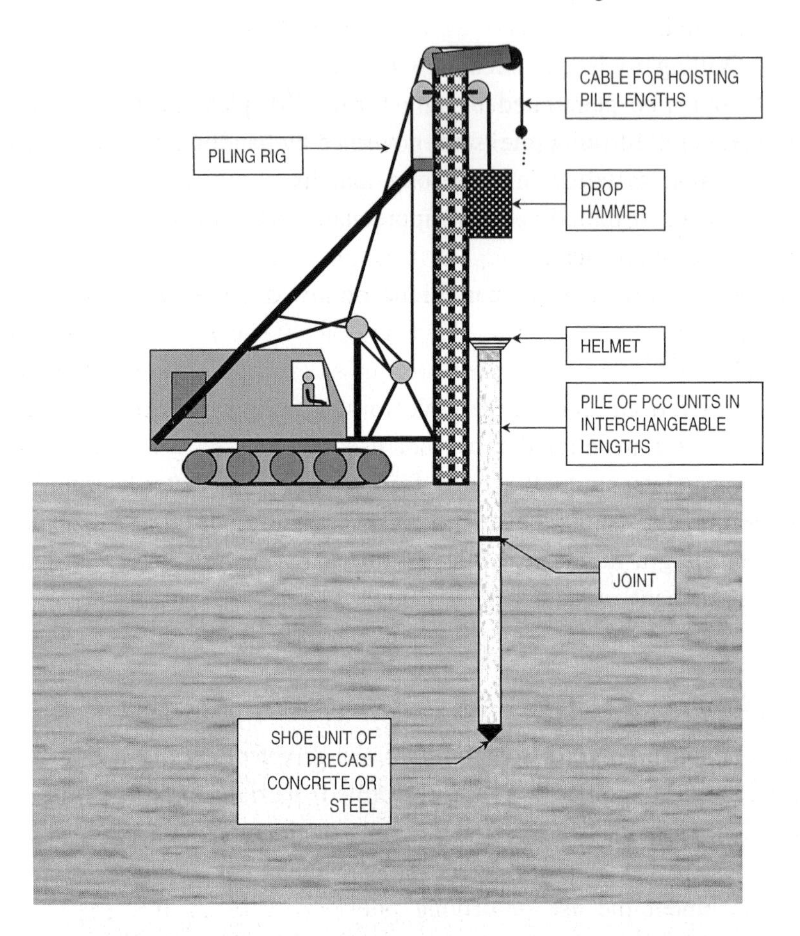

Figure 4.48. Precast reinforced concrete piles driven by a drop hammer.

- Set out the position of each pile and to establish the temporary benchmarks (TMB) on site for the determination of the cut-off levels of piles.
- Check the verticality of the leader of the piling rig using a plumb or spirit level.

- Provide markings along the pile section to enhance recording of penetration and to serve as a rough guide to estimate the set during driving.
- Install mild steel helmet (Figure 4.49). Protect pile head/joint plate (Figure 4.50) with packing or cushioning within e.g. a 25 mm thick plywood between the pile head and the helmet.
- Hoist up and place the pile in position (Figure 4.51).
- Check on verticality regularly (Figure 4.52).
- Proceed with the hammering. Monitor pile penetration according to the markings on the pile. When the rate of penetration is low, monitor pile penetration over 10 blows (Figure 4.53). Hole one end of a pencil supported firmly on a timber board not touching the pile. The other end of the pencil marks the pile displacements on a graph paper adhered on the pile over 10 blows.
- Stop piling if the displacement is less than the designed displacement over 10 blows. Otherwise, continue with the piling process.

Figure 4.49. A mild steel helmet with a 25 mm thick plywood between the helmet and the pile head.

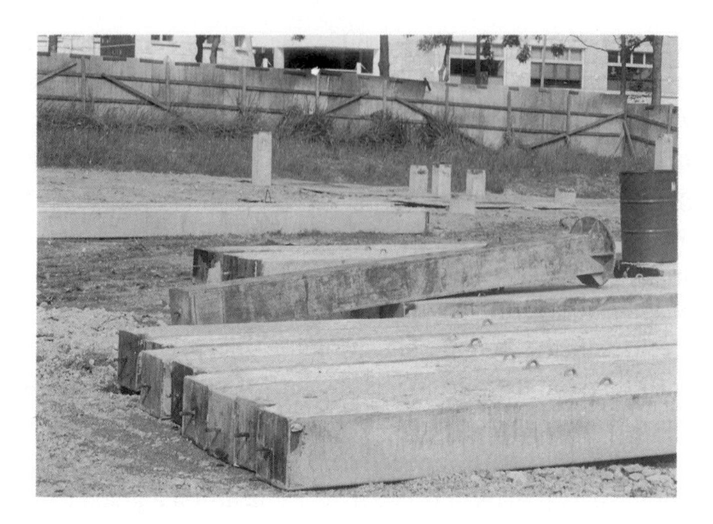

Figure 4.50(a). Pile heads/jointing plates inserted.

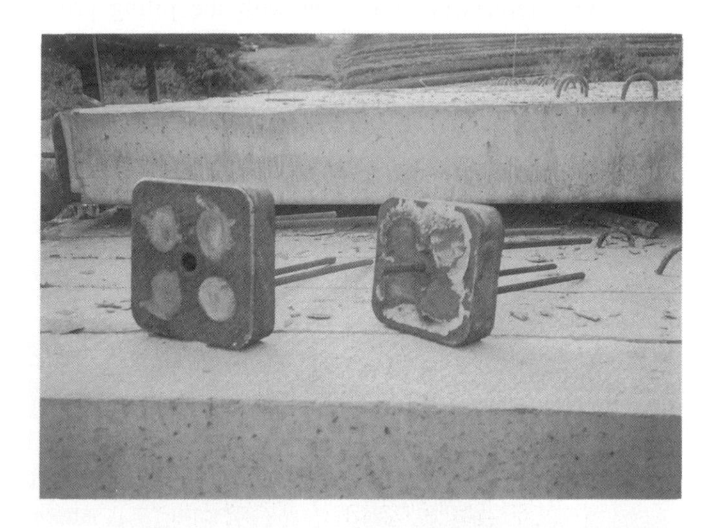

Figure 4.50(b). Pile heads with dowels (male and female).

Figure 4.51(a). Hoist up the pile and insert the helmet.

Figure 4.51(b). Place pile in position into the piling rig.

Figure 4.52. Guide the verticality using a spirit level and the leader of the rig.

- Lengthening of pile can be done by means of a mild steel splicing sleeve and a dowel inserted in and drilled through the centre of the pile. The connection is sealed with grout or epoxy resin. It can also be done by welding the pile head/joint plate which were pre-attached to both ends of a pile in the manufacturing process (Figure 4.54 and Figure 4.55).

Steel Preformed Piles — H-piles or universal steel beam in the form of wide-flange is commonly used (Figure 4.56). They do not cause large displacement and is useful where upheaval of the surrounding ground is a problem. They are capable of supporting heavy loads, can be easily cut and can be driven to great depth. The driving method for steel piles is

Figure 4.53(a). Marking pile displacements over the first 10 blows.

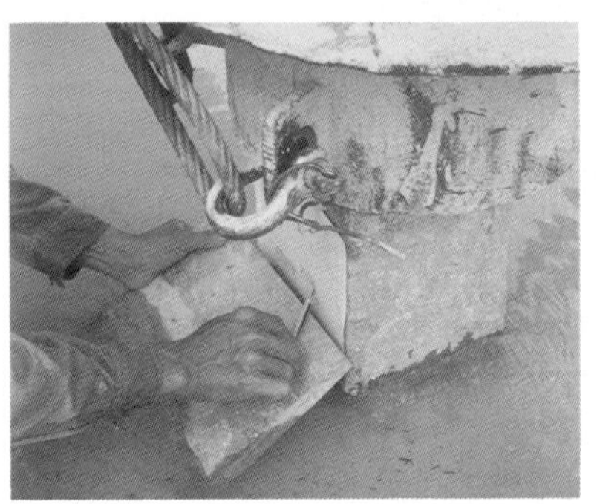

Figure 4.53(b). Marking pile displacements over the second 10 blows.

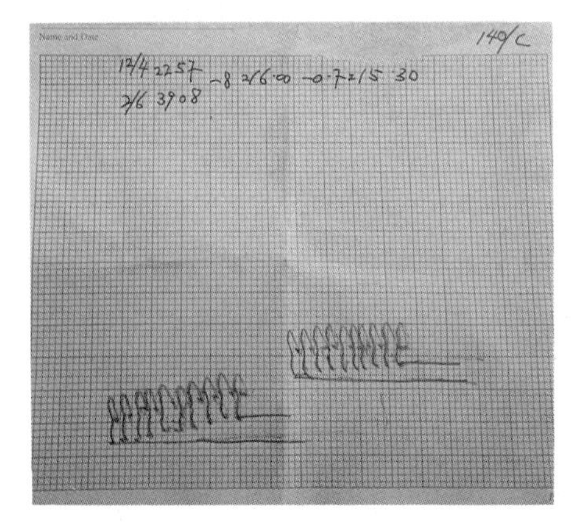

Figure 4.53(c). Markings on the graph paper.

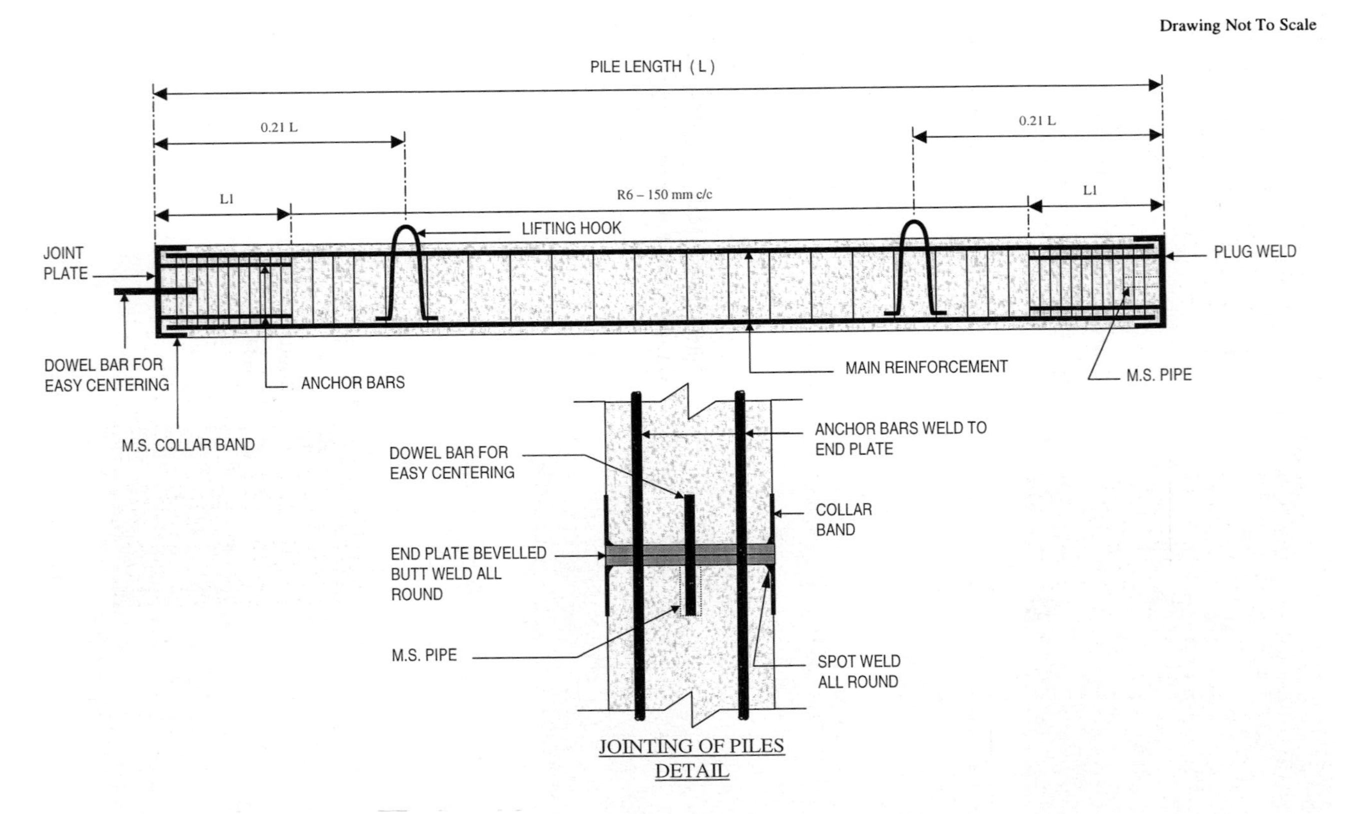

Figure 4.54. Precast reinforced concrete pile and splicing details.

Figure 4.55(a). Lifting of a new pile.

Figure 4.55(b). Insert the new pile to the old.

Figure 4.55(c). Align the new pile using a steel fork.

Figure 4.55(d). Weld the joint plate together.

Figure 4.55(e). A welded joint.

Figure 4.56. H-piles.

Figure 4.57. Joint welding of steel piles.

similar to that of precast reinforced concrete piles. The handling and lifting of steel pile is less critical due to the high tensile strength. Lengthening of steel piles is through welding (Figure 4.57). Care must be taken in the welding of joints to ensure that they are capable of withstanding driving stresses without failure. A protective steel guard should be welded at the joints when necessary.

Composite Piles — also referred to as partially preformed piles and are formed by a method which combines the use of precast and in-situ concrete or steel and in-situ concrete. They are useful for works where the presence of running water or very loose soils would render the use of bored or preformed piles as unsuitable. There are many commercial proprietary systems available. The common generic types are the shell piles and cased piles (Figure 4.58 and Figure 4.59).

Driven In-situ/Cast-in-place Piles — the pile shaft is formed by using a steel tube which is either top driven (Figure 4.60) or driven by means of an internal drop hammer working on a plug of dry concrete/gravel as in the case of Franki piles (Figure 4.61). A problem which can be encountered with this form of pile is necking due to ground water movement washing away some of the concrete thus reducing the effective diameter of the pile shaft and consequently the concrete cover.

Advantages:

(a) Material forming piles can be inspected for quality and soundness before driving.
(b) Not susceptible to squeezing or necking.
(c) The pile's carrying capacity can be monitored or 'felt' during the piling process.
(d) The construction operation is not affected by ground water.
(e) Projection above ground level advantageous to marine structures.
(f) Can be driven to long lengths.
(g) Can be designed to withstand high bending and tensile stresses.

Disadvantages:

(a) May break during driving, necessitating replacement piles.
(b) May suffer unseen damage which may reduce carrying capacity.

Drawing Not To Scale
CABLE FOR HOISTING PILE LENGTHS
PILING RIG
DROP HAMMER
HELMET OR DRIVING HEAD CONNECTED TO STEEL MANDREL
ACCESS PLATFORM ON FAR LEADER TUBE
PILE LOCATED BETWEEN PAIR OF STEEL TUBE LEADERS
STEEL LADDER
380 to 500 diameter
STANDARD LENGTH SHELL
SHELL PILE BEING DRIVEN
HELICAL BINDING
STEEL REINFORCEMENT TO DESIGN
900
POLYPROPYLENE REINFORCED CONCRETE SHELLS
STEEL MANDREL INSIDE SHELL CORE
900
IN SITU CONCRETE FILLING TO CORE
STEEL JOINTING BAND
SOLID CONCRETE DRIVING SHOE
TYPICAL PILE DETAILS

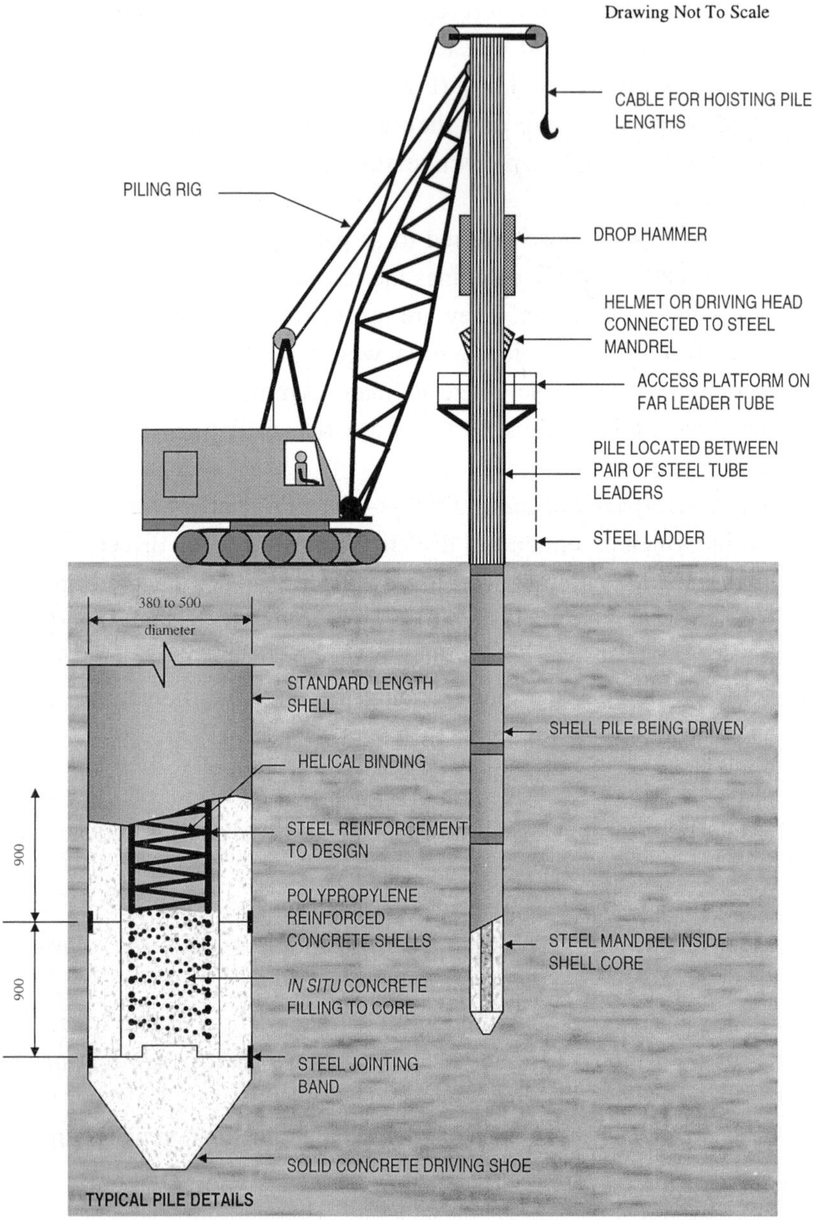

Figure 4.58. A shell pile using precast and *in-situ* concrete.

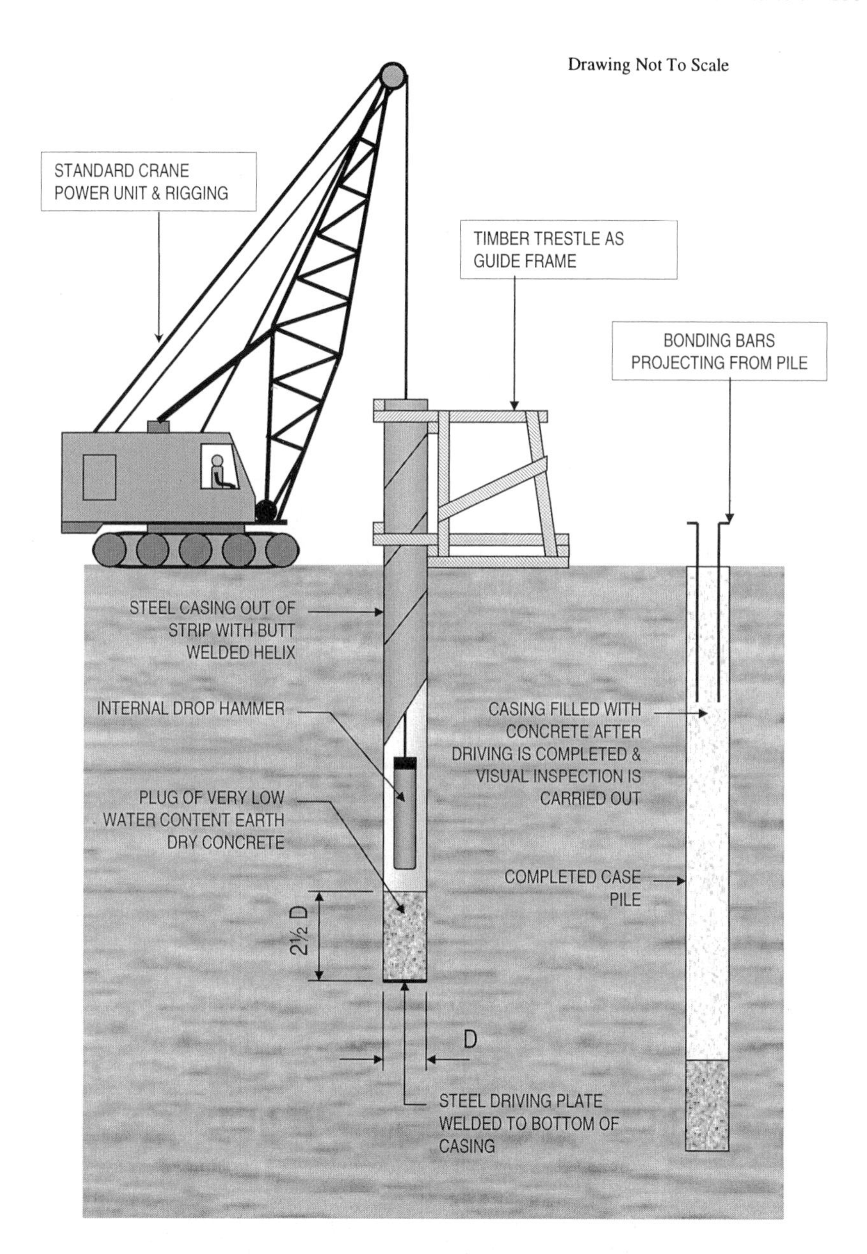

Figure 4.59. A cased pile using steel and *in-situ* concrete.

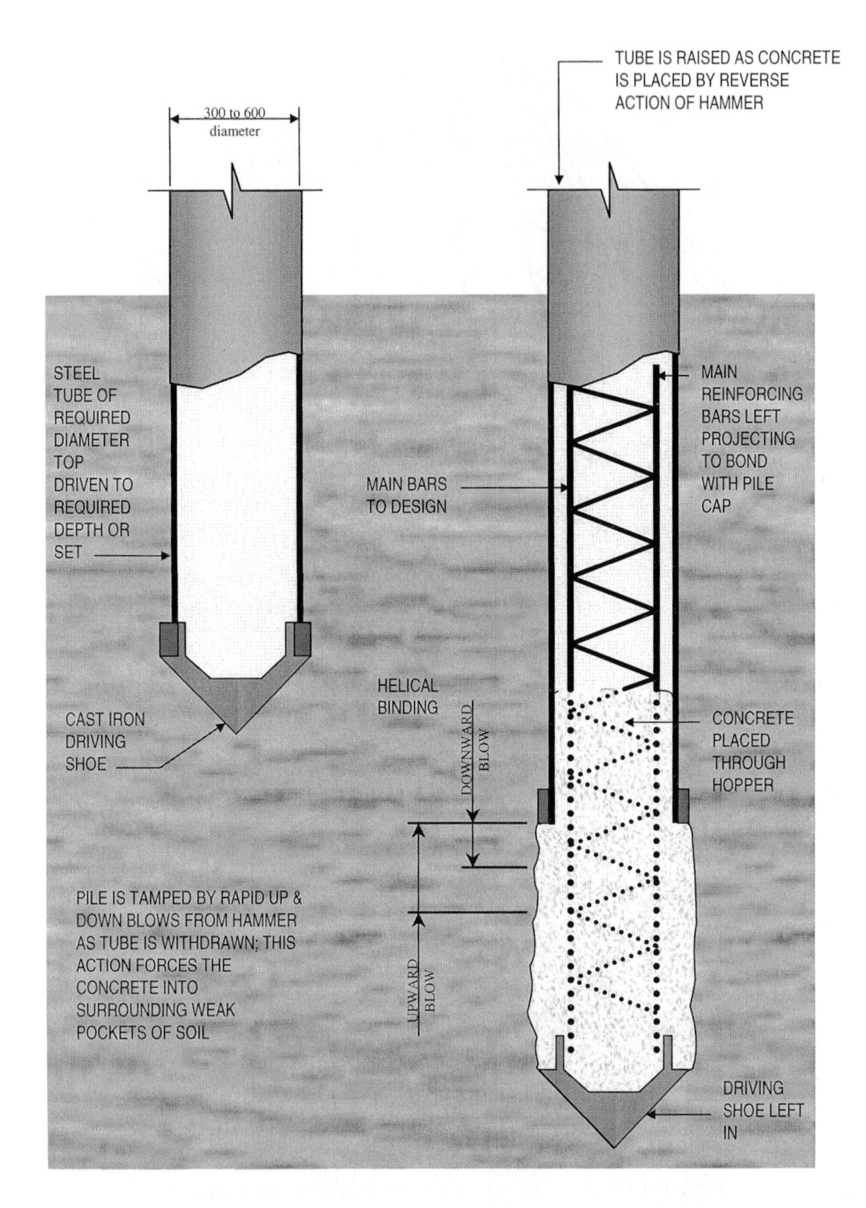
Drawing Not To Scale
TUBE IS RAISED AS CONCRETE IS PLACED BY REVERSE ACTION OF HAMMER
300 to 600 diameter
STEEL TUBE OF REQUIRED DIAMETER TOP DRIVEN TO REQUIRED DEPTH OR SET
MAIN BARS TO DESIGN
MAIN REINFORCING BARS LEFT PROJECTING TO BOND WITH PILE CAP
HELICAL BINDING
CAST IRON DRIVING SHOE
DOWNWARD BLOW
CONCRETE PLACED THROUGH HOPPER
PILE IS TAMPED BY RAPID UP & DOWN BLOWS FROM HAMMER AS TUBE IS WITHDRAWN; THIS ACTION FORCES THE CONCRETE INTO SURROUNDING WEAK POCKETS OF SOIL
UPWARD BLOW
DRIVING SHOE LEFT IN

Figure 4.60. Top-driven cast *in-situ* piles.

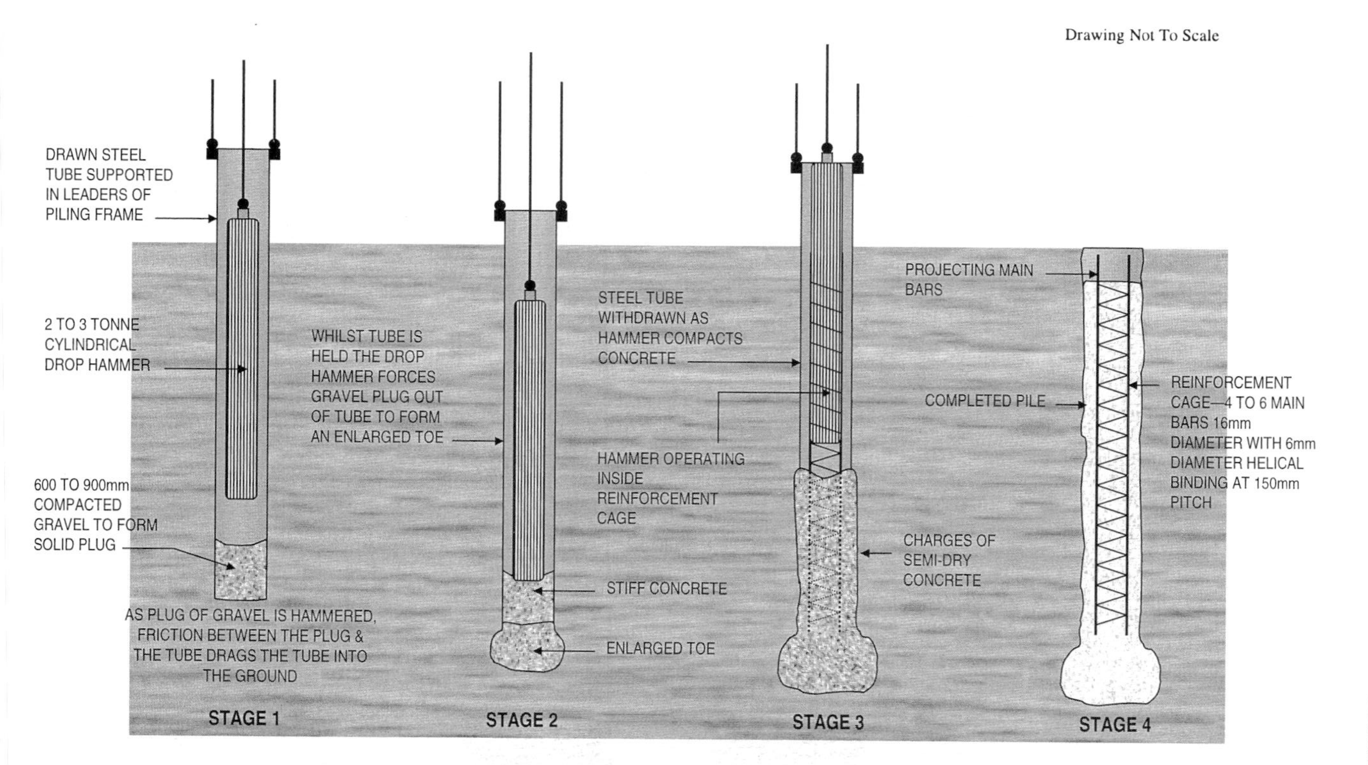

Figure 4.61. Franki driven in-situ piles.

(c) Uneconomical if cross-section is governed by stresses due to handling and driving rather than compressive, tensile or bending stresses caused by working conditions.
(d) Noise, vibration and dust levels due to driving may be unacceptable.
(e) Displacement of soil during driving may lift adjacent piles or damage adjacent structures.
(f) Not suitable for situations with low headroom.

4.9. Pile Testing

The main objective of forming test piles is to confirm that the design and formation of the chosen pile type is adequate. Pile load tests give information on the performance of the pile, installation problems, lengths, working loads and settlements.

4.9.1. *Types of Loading*

Maintained Load Test — The load is increased at fixed increments up to 1.5 to 2.5 times its working load. Settlement is recorded with respect to time of each increment. When the rate of settlement reaches the specified rate, the next load increment is added. Once the working load is reached, maintain the load for 12 hours. Thereafter, reverse the load in the same increment and note the recovery.

Ultimate Load Test — the pile is steadily jacked into the ground at a constant rate until failure. The ultimate bearing capacity of the pile is the load at which settlement continues to increase without any further increase of load or the load causing a gross settlement of 10% of the pile diameter. This is only applied to test piles which must not be used as part of the finished foundations but should be formed and tested in such a position that will not interfere with the actual contract but is nevertheless truly representative of site conditions.

(a)

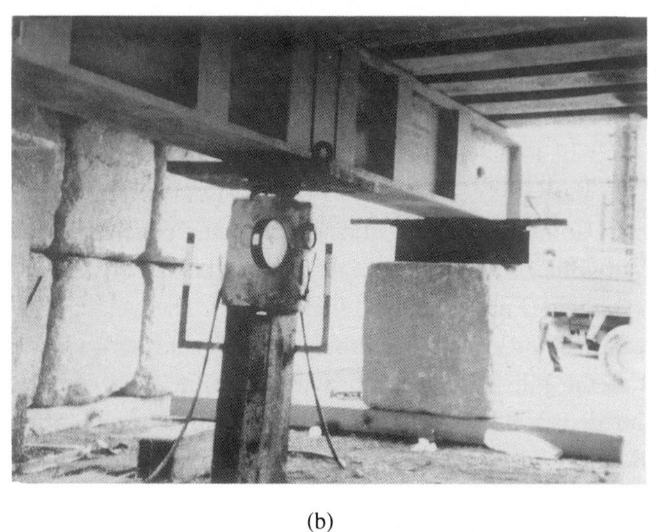

(b)

Figure 4.62. Kentledge pile test, with (a) stone blocks stacking up on steel channels to act as the weight, (b) a hydraulic jack is installed in between the tested pile and the loaded channel. During testing, the hydraulic jack is extended progressively. The applied force versus the displacement of the tested pile is monitored.

Figure 4.63. A kentledge set up using H-channels.

4.9.2. *Reaction Systems*

Kentledge — Comprises either stone blocks, cast concrete blocks, pig iron blocks, or any other suitable materials that can be safely stacked up to form the reaction weight (Figure 4.62 and Figure 4.63). The weight of the kentledge is borne on steel or concrete cribbings. Main and secondary girders are connected to the pile head in such a way that the load can be distributed evenly. The distance between the test pile and the supporting cribbings should be kept as far as possible (Figure 4.64). The system should be firmly wedged, cleated or bolted together to prevent slipping between members. The centre of gravity of the kentledge should be aligned with that of the test pile to prevent preferential lifting on either side which may lead to toppling.

Tension piles — Three piles are formed and the outer two piles are tied across their heads with a steel or concrete beam. The object is to jack down the centre or test pile against the uplift of the outer piles. It is preferable

Drawing Not To Scale

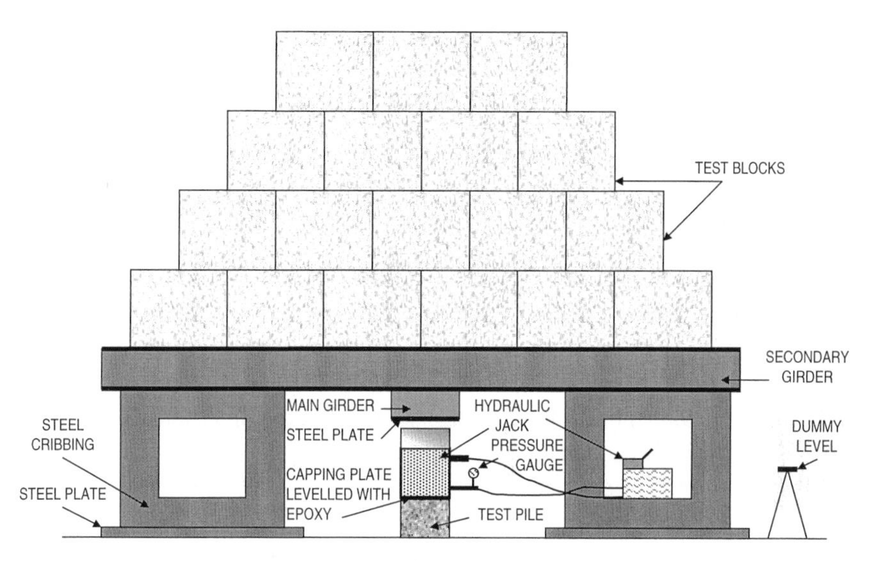

Figure 4.64(a). Front elevation of a typical kentledge set up.

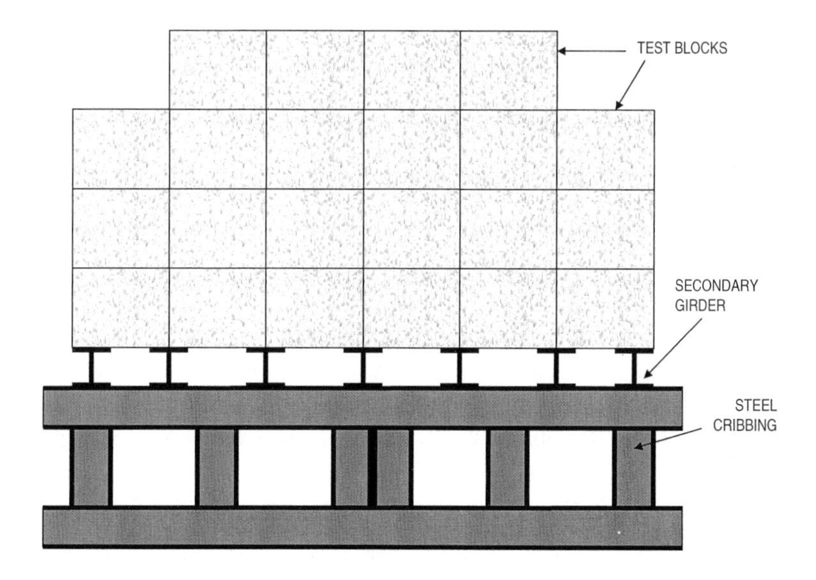

Figure 4.64(b). Side elevation of a typical kentledge set up.

when possible to utilise more than two outer piles to avoid lateral instability and to increase the pull-out resistance.

Ground anchors — Also known as rock anchors and are useful when testing piles which are en-bearing on rock, or if rock exists at shallow depths below pile toe level.

References

[1] K. B. Poh, H. L. Chuah and S. B. Tan, "Residual granite soils of Singapore", Proceedings of the 8th Southeast Asian Geotechnical conference, Kuala Lumpur, 1985, pp. 3.1–3.9.

[2] S. B. Tan, W. C. Loy and K. W. Lee, "Engineering geology of the old alluvium in Singapore", Proceedings of 6th Southeast Asian Conference on Soil Engineering, Taipei, 1980, pp. 673–684.

[3] J. Pitt, "A survey of engineering geology in Singapore", Geotechnical Engineering, Vol. 15, 1984, pp. 1–20.

[4] S. B. Tan, "Foundation Problems in Singapore Marine Clay", Asian Building & Construction, November 1972, pp. 30–33.

[5] F. E. S. Alexander, "Shear strength characteristics of the recent marine clays in Southeast Asia", *Journal S.E.A.S.S.E.*, Vol. 1, No. 1, p. 1.

[6] J. S. Foster, *Structure and Fabric*, Longman, 1994.

[7] W. Schueller, *The Vertical Building Structure*, Van Nostrand Reinhold, 1990.

[8] J. E. Bowles, *Foundation Analysis and Design*, 5th Edition, McGraw-Hill, 1996.

[9] D. P. Coduto, *Foundation Design: Principles and Practices*, Prentice Hall, 1994.

[10] B. M. Das, *Principles of Foundation Engineering*, 3rd Edition, PWS Pub. Co., Boston, 1995.

[11] W. F. Kane and J. M Tehaney, "Foundation upgrading and repair for infrastructure improvement", Proceedings of the Symposium Sponsored by the Deep Foundations Committee of the Geotechnical Engineering Division of the American Society of Civil Engineers in conjunction with the ASCE Convention in San Diego, California, October 23–26, 1995.

[12] M. J. Tomlinson, *Foundation Design and Construction*, 6th Edition, Longman Scientific & Technical, Harlow, 1995.

[13] 3rd International Conference on Deep Foundation Practice incorporating Piletalk International '94, Singapore, 19–20 May 1994.

[14] A. Anagnostopoulos, "Geotechnical engineering of hard soils, soft rocks", Proceedings of an International Symposium under the Auspices of the International Society for Soil Mechanics and Foundation Engineering (ISSMFE), the International Association of Engineering Geology (IAEG) and the International Society for Rock Mechanics (ISRM), Athens, Greece, 20–23 September 1993.
[15] R. Holmes, "Introduction to Civil Engineering Construction", College of Estate Management, 1983.
[16] Public Works Department, Singapore, "Geology of the Republic of Singapore", Singapore, 1976.
[17] R. W. Brown, *Foundation Behavior and Repair: Residential and Light Construction*, 3rd Edition, McGraw-Hill, 1997.
[18] M. I. Esrig, "Deep Foundation Improvements: Design, Construction & Testing", American Society for Testing & Materials (ASTM), Special Technical Publication Ser.; No. 1089, 1991.
[19] J. E. Bowles, *Foundation Analysis and Design*, 5th Edition, McGraw-Hill, New York, 1996.
[20] C. N. Baker (ed.), "Drilled Piers and Caissons II: Construction under Slurry/ Non-destructive Integrity Evaluation/Load Testing/Geotechnical Behaviour Under Load", Denver, Colorado, May 1, 1985, ASCE, New York, 1985.
[21] P. P. Xanthakos, *Bridge Substructure and Foundation Design*, Prentice Hall PTR, New Jersey, 1995.
[22] D. H. Lee, *An Introduction to Deep Foundations and Sheet-Piling*, Concrete Publications, London, 1961.
[23] J. N. Cernica, *Geotechnical Engineering — Foundation Design*, Wiley, New York, 1995.
[24] B. B. Bengt, editor, "Conference on Deep Foundation Practice", 30–31 October, 1990, CI-Premier, Singapore, 1990.
[25] J. M. Roesset, "Analysis, Design, Construction, and Testing of Deep Foundations: Proceedings of the OTRC'99 Conference", April 29–30, 1999, The Offshore Technology Research Center, Geo Institute of the American Society of Civil Engineers, 1999.
[26] M. J. Tomlinson, *Pile Design and Construction Practice*, 4th Edition, E & FN Spon, London, 1994.
[27] C. H. Dowling, *Construction Vibrations*, Prentice Hall, New Jersey, 1996.

[28] W. F. Imoe, editor, "BAP II: Proceedings of the 2nd International Geotechnical Seminar on Deep Foundations on Bored and Auger Piles", Ghent, Belgium 1–4 June 1993, A. A. Balkema, Rotterdam, 1993.

[29] ASCE, "Standard Guidelines for the Design and Installation of Pile Foundations", American Society of Civil Engineers, ASCE, New York, 1997.

[30] British Steel Corporation, "Piling Handbook", 6th Edition, The Corporation, Scunthorpe Humberside, 1988.

[31] American Society of Civil Engineers, "Practical Guidelines for the Selection, Design, and Installation of Piles", Committee on Deep Foundations, Geotechnical Engineering Division, ASCE, New York, 1984.

[32] Deep Foundation Institute, "4th International Conference on Piling and Deep Foundations", Stresa, Italy, 7–12 April 1991, A. A. Balkema, Rotterdam, 1991.

[33] R. A. Bullivant, *Underpinning: A Practical Guide*, Blackwell Science, Cambridge, MA, 1996.

[34] F. Harris, *Modern Construction and Ground Engineering Equipment and Methods*, 2nd Edition, Longman Scientific & Technical, England; Wiley, New York, 1994.

CHAPTER 5

BASEMENT CONSTRUCTION

5.1. General

Basements are common in tall buildings as carparks, storage of services and underground shopping centres. The term "basement" has been regarded as synonymous to the term "deep pit", which applies to excavations over 4.5 m deep [1].

The main purpose of constructing basements are:

(a) to provide additional space,
(b) as a form of buoyancy raft,
(c) in some cases, basements may be needed for reducing net bearing pressure by the removal of the soil.

In most cases, the main function of the basement in a building is to provide additional space for the owner, and the fact that it reduces the net bearing pressure by the weight of the displaced soil may be quite incidental. In cases where basements are actually needed for their function in reducing net bearing pressure, the additional floor space in the sub-structure is an added bonus [2–8].

5.2. Construction Methods

There are essentially three general methods of constructing a basement:

1. Open cut method.
2. Cut and cover method.
3. Top down method.

Whichever method is chosen, it is essential that the excavation is adequately supported, and the ground water properly controlled. Shoring should be provided for any excavation that is more than 1.8 m deep. The three common methods of supporting an excavation either in isolation or combination are:

- Excavations supported by sheet piling.
- Excavation supported by reinforced concrete diaphragm wall constructed in advance of the main excavation.
- Excavations supported by contiguous bored piles or secant piles constructed in advance of the main excavation.

5.2.1. *Construction in Excavations Supported by Sheet Piling*

This is a suitable method for sites where the space around the excavation is insufficient for sloping back the sides. If the soil conditions permit withdrawal of sheet piling for re-use elsewhere, this method of ground support is very economical compared with the alternative of a diaphragm wall.

Sheet piling comprises a row of piles which interlock with one another to form a continuous wall which may be temporary or permanent. It is manufactured by different firms and in general consists of rolled steel sections with interlocking edge joints. The interlocking edges allow each sheet pile to slide into the next with relative ease, and together they form a steel sheet wall that serves the purpose of retaining the soil and to some extent, exclusion of ground water. The standard length of sheet pile is 12 m. Longer piles are achieved by joining sections together by welding.

Figure 5.1 shows sheet piling in a guard rail using a hydraulic vibro hammer. The sheet piles are spliced into one another in alternate order. A power operated steel chain is used to adjust the positioning of the adjacent piles to ensure proper connection at joints. Figure 5.2 shows the interlocking at joints of sheet piles. Figure 5.3 shows H-guide beams acting as rakers to sheet piles, retaining the soil from the perimeter with horizontal strutting. The strutting also provides support to the working platform.

(a)

(b)

Figure 5.1. (a) The driving of a sheet pile using a vibratory hammer. (b) Piles in a guard rail are spliced into one another in alternate order. A power operated chain is used to ensure proper connection between piles.

Figure 5.2(a). Interlocking of sheet piles.

Figure 5.2(b). A closer view of the interlocking joints.

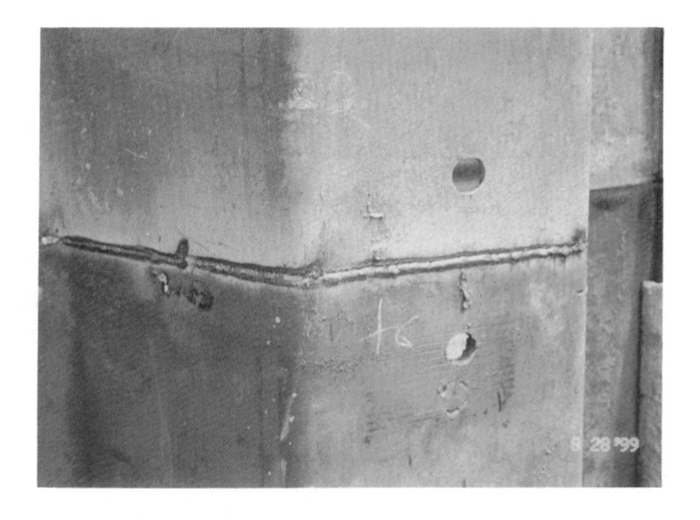

Figure 5.2(c). Joining sheet piles by welding.

Figure 5.2(d). Joining sheet piles by splicing.

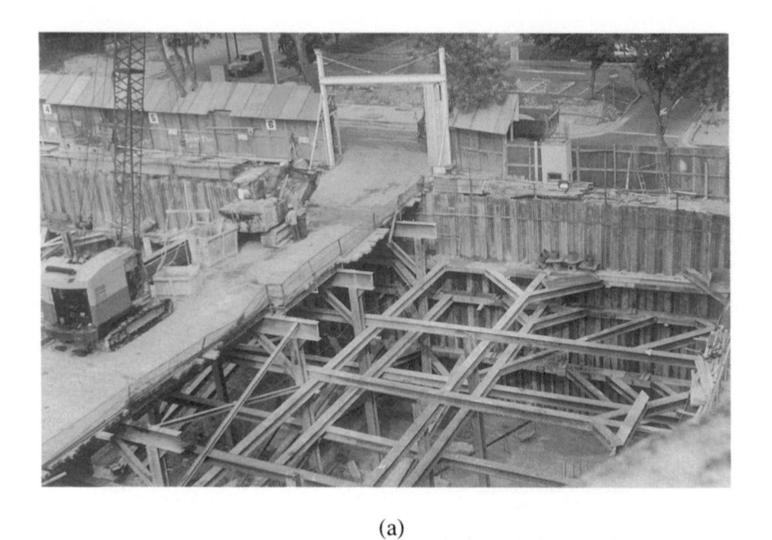

(a)

(b)

Figure 5.3. (a) An overall view of a congested site with sheet piling retaining the excavation. (b) A closer view of the horizontal strutting supporting excavation and working platform.

The main disadvantages of using sheet piles are similar to that of driving piles as mentioned in Chapter 4, i.e. (a) noise and (b) direct effect of the driving on the subsoil immediately surrounding the site [9–14].

5.2.2. *Construction in Excavations Supported by Reinforced Concrete Diaphragm Wall*

A diaphragm wall is constructed by excavation in a trench which is temporarily supported by a bentonite slurry. On reaching founding level steel reinforcement is lowered into the trench, followed by concrete to displace the bentonite.

This method is suitable for sites where obstructions in the ground prevent sheet piles from being driven and where the occurrence of ground water is unfavourable for other methods of support.

The method is also suitable for sites where considerations of noise and vibration preclude driving sheet piles and where ground heave and disturbance of the soil beneath existing foundations close to the margins of the excavation are to be avoided [15–17].

The bentonite slurry has the following properties:

- Supports the excavation by exerting hydrostatic pressure on the wall.
- Has the ability to form almost instantaneously a membrane with low permeability.
- Suspend sludgy layers building up at the excavated base.
- Allows clean displacement by concrete, with no subsequent interference with the bond between reinforcement and set concrete.

Figure 5.4 shows the set-up of bentonite slurry for diaphragm wall construction. Guide walls are normally constructed to improve trench stability and to serve as guides for excavation. Figure 5.5 shows the use of a cable operated clamshell grab hang to a mobile crane and a trench cutter for trench excavation. Excavation is carried out within the bentonite slurry immersion which supports the excavation by exerting hydrostatic pressure on the trench walls. Where rock is encountered, a drop chisel may be used. The typical excavation sequence of diaphragm walls is shown in Figure 5.6(a).

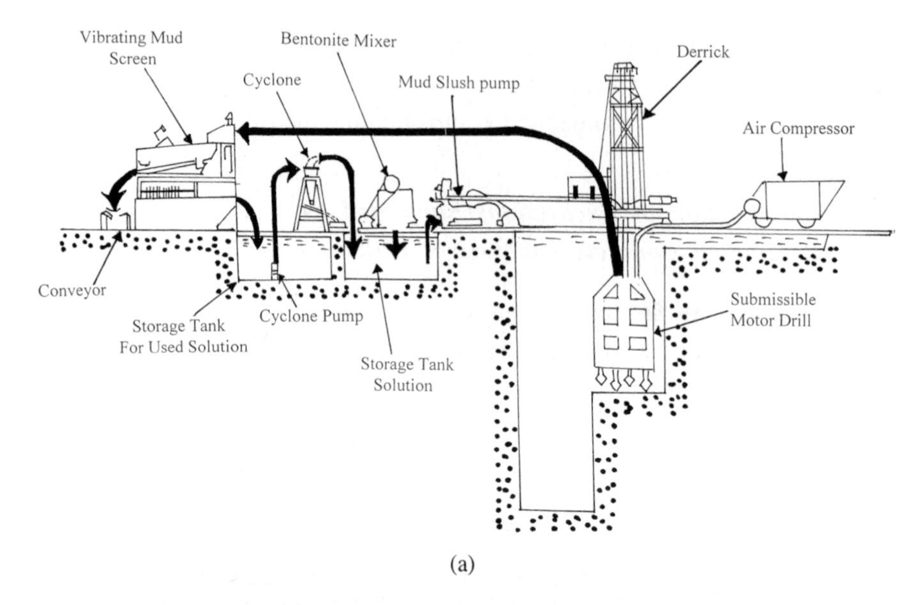

(a)

(b)

Figure 5.4. (a) A schematic diagram showing the typical set-up of bentonite slurry for diaphragm wall construction. (b) A typical set-up of bentonite slurry for diaphragm wall construction in an actual site.

(a)

(b)

Figure 5.5. (a) Guide walls built in advance for the trench excavation. (b) Trench excavation using a clamshell in guide walls under bentonite slurry. (c) A crane operated trench cutter. (d) Details of the trench cutter.

(c)

(d)

Figure 5.5. (*Continued*)

After the first panel has been cast, the adjacent panel next to the cast panel would be excavated 12 hours later, starting from the outside bite, away from the stop-end joint. The soil directly next to the stop-end joint would only be excavated 24 hours later. Stop-end tubes of various shapes are available. Figure 5.6(b) shows the circular type while Figure 5.6(c) and (d) show the interlocking type with single and double water stops respectively. The stop-end joints are installed prior to the installation of the reinforcement steel cage. After the concrete is cast, while excavating the adjacent panel, the stop-end tube is extracted leaving behind the rubber water stop, which would eventually be cast with the new concrete to form an effective water barrier over the vertical joint.

The depth of the excavation is checked using a weight and rope. The width of the trench over the depth of excavation is checked by lowering plumb line, echo sounding sensors, etc. Further excavation is needed in depths where the trench width is insufficient. For depths with larger trench widths, the excess concrete cast will be hacked off at a later stage.

Check the purity of the bentonite regularly. A submersible pump is lowered to pump up the contaminated bentonite for recycling while fresh and recycled bentonite are fed back continuously as shown in Figure 5.4(a). Reinforcement steel cage with openings for tremie pipes is then lowered with a crane. For deeper trenches where more than one set of cage is needed, hold the cage temporarily with steel bars across the top of the trench and weld the next section to the first and repeat for subsequent cages. Lower the whole cage down, insert inclinometer pipes, services, etc as required. Place concrete through the tremie pipes and pump off the displaced bentonite slurry.

5.2.3. *Construction in Excavations Supported by a Contiguous Bored Pile or Secant Pile Wall*

Contiguous bored pile wall is a line of bored piles installed close together or touching. Smaller diameter micro-piles may be installed in between each adjacent pile to close the gaps between the main piles. The gaps between the micro-piles and main piles are grouted. Grouting may be pumped in through perforated pipes inserted into holes drilled in between the piles.

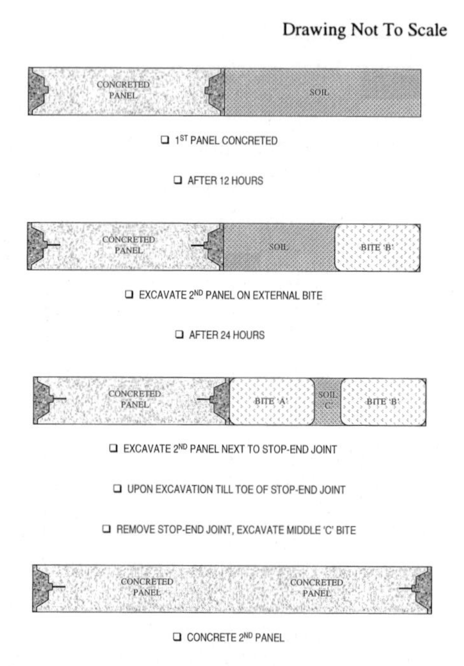

(a)

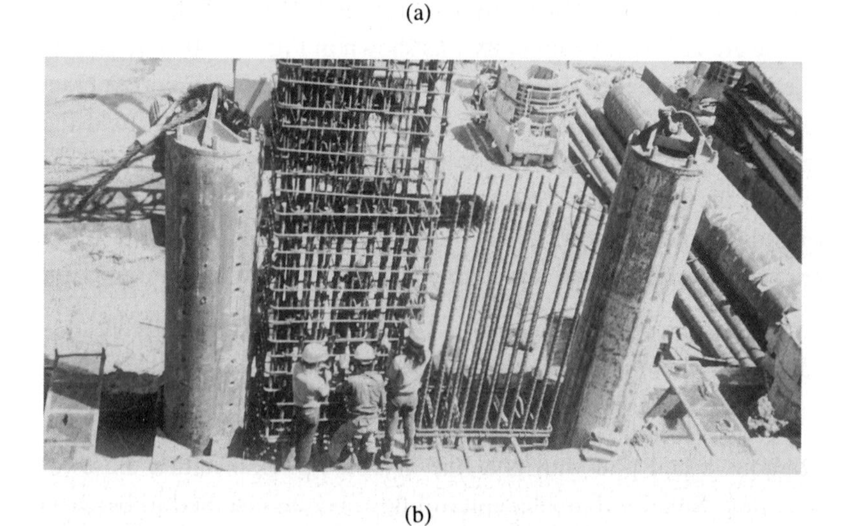

(b)

Figure 5.6. (a) Typical excavation sequence of diaphragm walls. (b) Circular stop-end tubes inserted at ends of each cast section, reinforcing cage being lowered. (c) Stop-end joint with single water stop. (d) Stop-end joint with double water stop.

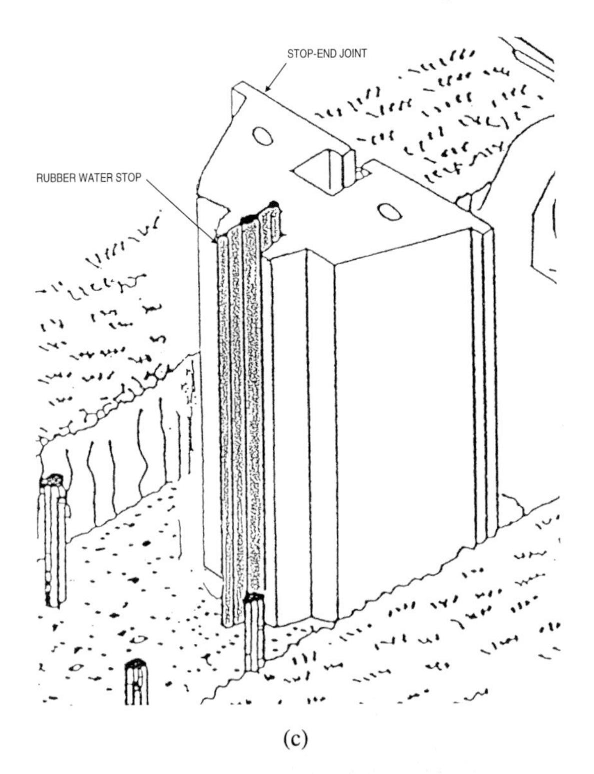

(c)

(d)

Figure 5.6. (*Continued*)

The typical sequence of construction is such that the next pile is to be constructed more than 3 m away from the previous one. Contiguous piling may be covered with mesh reinforcement or fabric faced with rendering or sprayed concrete/shotcreting/guniting. This method is useful in:

— built-up areas where noise and vibration should be limited.
— in industrial complexes where access, headroom and/or restriction on vibration may make other methods such as steel sheet piling or diaphragm walling less suitable.

Figure 5.7 shows contiguous bored piles laid in two stages, as an alternative for long piles in deep basement construction.

Secant piles are similar to contiguous bored piles except that they are constructed such that the two adjacent piles, i.e. male (hard) and female (soft) piles are interlocked into each other (Figure 5.8 and Figure 5.9). The male piles are reinforced while the female piles are not. In cases where lateral pressure from the soil is excessively high, male-male secant piles may be used. The advantages of secant pile walls over other forms of barriers such as contiguous pile wall and sheet pile walls are that they provide for a higher degree of watertightness and stronger resistance to lateral pressure, although their installation cost may be higher.

The construction sequence of a typical secant pile wall is shown below:

- Set up the exact location of piles.
- Install the steel guide frame along the proposed wall for augering (Figure 5.8).
- Boreholes with flight rock augers.
- When the bottom of the hole is reached, in the case of a female pile, discharge cement grout from the auger head during the withdrawal of the auger.
- Adjust the pumping rate of the grout during the withdrawal of the auger such that no void is created.
- Repeat the augering process according to the sequence as in Figure 5.8.
- The inner curvature of the secant pile wall may be smoothened to facilitate the placement of a membrane.

(a)

(b)

Figure 5.7. (a) Contiguous bored piles laid out in two stages to minimise the pile length. (b) Micropiles between adjacent piles and gap sealed by grouting.

Drawing Not To Scale

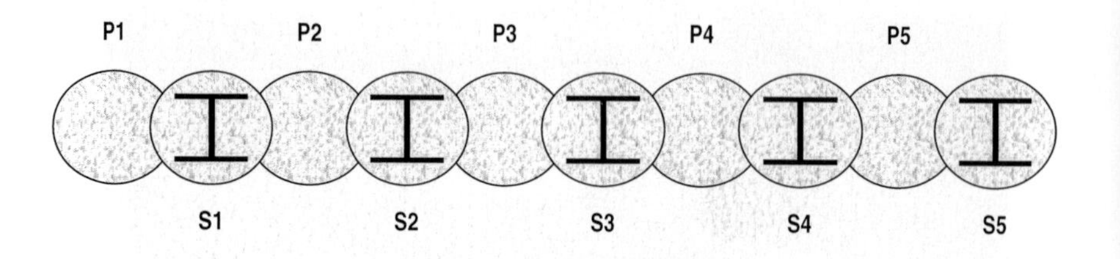

- ❑ DAY 1 – **P1 P3 P5 P7 P9** & SO ON.
- ❑ DAY 2 – **P2 P4 P6 P8 P10** & SO ON.
- ❑ DAY 3 – **P11 P13 P15** & SO ON.
- ❑ DAY 4 – **S1 S3 S5 S7 S9** & SO ON.
- ❑ DAY 5 – **S2 S4 S6 S8 S10** & SO ON.

Figure 5.8(a). Layout and sequence of work of secant piles.

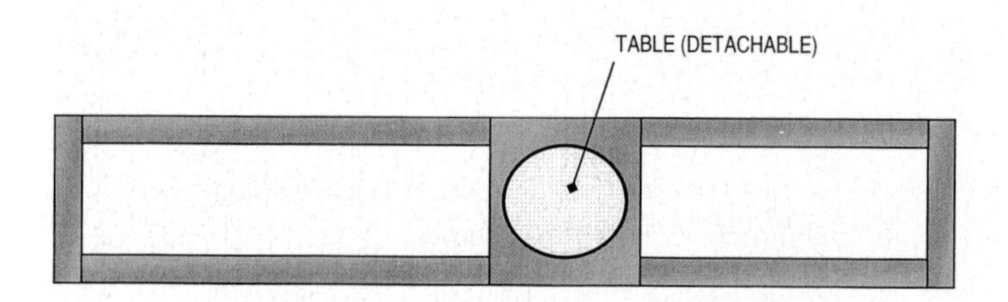

Figure 5.8(b). Guide frame for secant piles.

Figure 5.9(a). Secant piles on the left and sheet piles on the right.

Figure 5.9(b). Secant piles with alternate interlocking between male and female piles.

5.3. Control of Groundwater

Groundwater refers to the residual water that percolates downwards to the watertable after runoff, evaporation and evapotranspiration. The hydrologic cycle of ground water is shown in Figure 5.10. A part of the precipitation falling on the surface runs off toward the river, where some is evaporated and returned to the atmosphere. Of that part filtering into the ground, some is removed by the vegetation as evapotranspiration. Some part seeps down through the zone of aeration to the watertable. Below the watertable, the water moves slowly toward the stream, where it reappears as surface water. Water in a confined aquifer can exist at pressures as high as its source, hence the flowing well. Water trapped above the upper clay layer can become perched, and reappear as a small seep along the riverbank [18–31].

An aquifer is a zone of soil or rock through which groundwater moves:

Confined aquifer — a permeable zone between two aquicludes, which are confining beds of clay, silt or other impermeable materials.

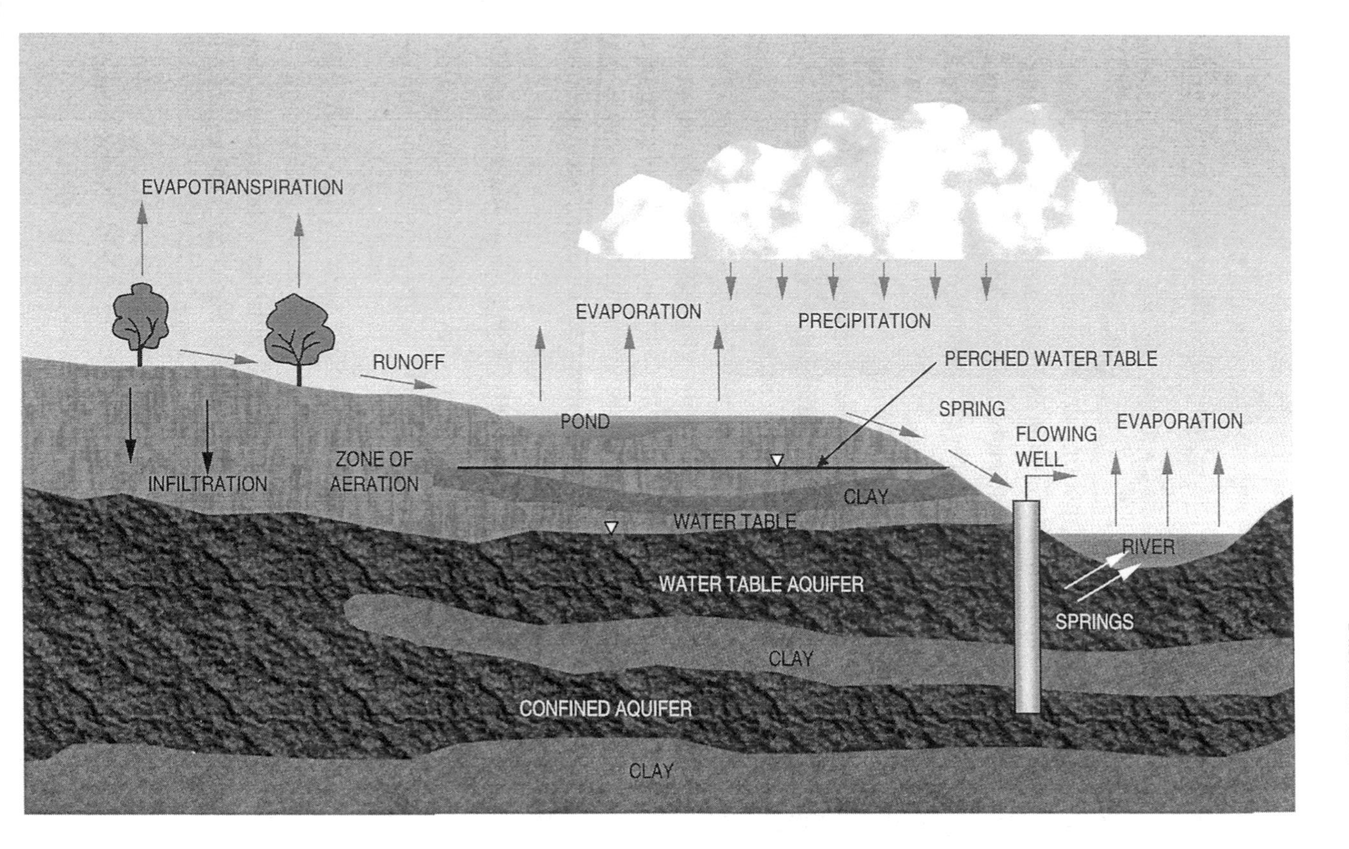

Figure 5.10. The hydrologic cycle of groundwater.

Watertable aquifer — this zone has no upper confining bed. The watertable rises and falls with changing flow conditions in the aquifer. The amount of water stored in the aquifer changes radically with watertable movement.

In aquifers adjacent to estuaries or to the sea, the water head fluctuates with the tide.

Groundwater is constantly in motion but its velocity is low in comparison to surface streams. There are continuing additions to the ground water by infiltration from the ground surface and by recharge from rivers. Continuing subtractions of groundwater occur through evaporation, evapotranspiration, seepage into rivers and by pumping from wells. Such pumping disturbs the groundwater flow patterns and their velocities increase sharply, especially in the immediate vicinity of the wells. Effects in dewatering however only induce temporary modification in groundwater patterns. It is the permanent deep basement structures that result in a permanent change.

Dewatering is the process of removing water from an excavation. Dewatering may be accomplished by lowering the groundwater table before the excavation work. This method of dewatering is often used by placing pipelines in areas with high groundwater levels. Alternatively, excavation may be accomplished first and the water simply pumped out of the excavation site as work proceeds.

There are four basic methods for controlling ground water:

Open pumping — Water is permitted to flow into an excavation and collected in ditches and sumps before being pumped away.

Predrainage — Lower groundwater table before excavation using pumped wells, wellpoints, ejectors and drains.

Cutoff — Water entry is cut off with steel sheet piling, diaphragm walls, contiguous bored pile and secant pile walls, tremie seals or grout.

Exclusion — Water is excluded with compressed air, a slurry shield, or an earth pressure shield. These methods are frequently used for tunnelling work.

The principal factors which affect the choice of the appropriate dewatering techniques are:

- The soil within which the excavation is to take place.
- The size of the excavation (and available space).
- The depth of groundwater lowering.
- The flow into the excavation.
- Proposed method of excavation.
- Proximity of existing structures and their depths and type of foundation.
- Economic considerations.

When the groundwater table lies below the excavation bottom, water may enter the excavation only during rainstorms, or by seepage through side slopes or through or under cofferdams. In many small excavations, or where there are dense or cemented soils, water may be collected in ditches or sumps (Figure 5.11) at the excavation bottom and pumped out. This is the most economical dewatering method.

Where seepage from the excavation sides may be considerable, it may be cut off with a sheetpile cofferdam, grout curtains or concrete-pile or slurry-trench walls. For sheetpile cofferdams in pervious soils, water should be intercepted before it reaches the enclosure, to avoid high pressures on the sheetpiles. Deep wells or well points may be placed around the perimeter of the excavation to intercept seepage or to lower the water table (Figure 5.12). Water collecting in the wells is removed with centrifugal or turbine pumps at the well bottoms. The pumps are enclosed in protective well screens and a sand-gravel filter [9].

Wellpoints are used for lowering the water table in pervious soils or for intercepting seepage. Wellpoints are metal well screens that are placed below the bottom of the excavation and around the perimeter. A riser connects each wellpoint to a collection pipe, or header, above ground. A combined vacuum and centrifugal pump removes the water from the header (Figure 5.13).

The well-point system is limited in its suction lift depending on the pump (usually < 6 m). Lifting water above its height limit would result in loss of pumping efficiency, owing to air being drawn into the system through the joints in the pipe. In deep excavation, the multi-stage well-point system may be practical (Figure 5.14).

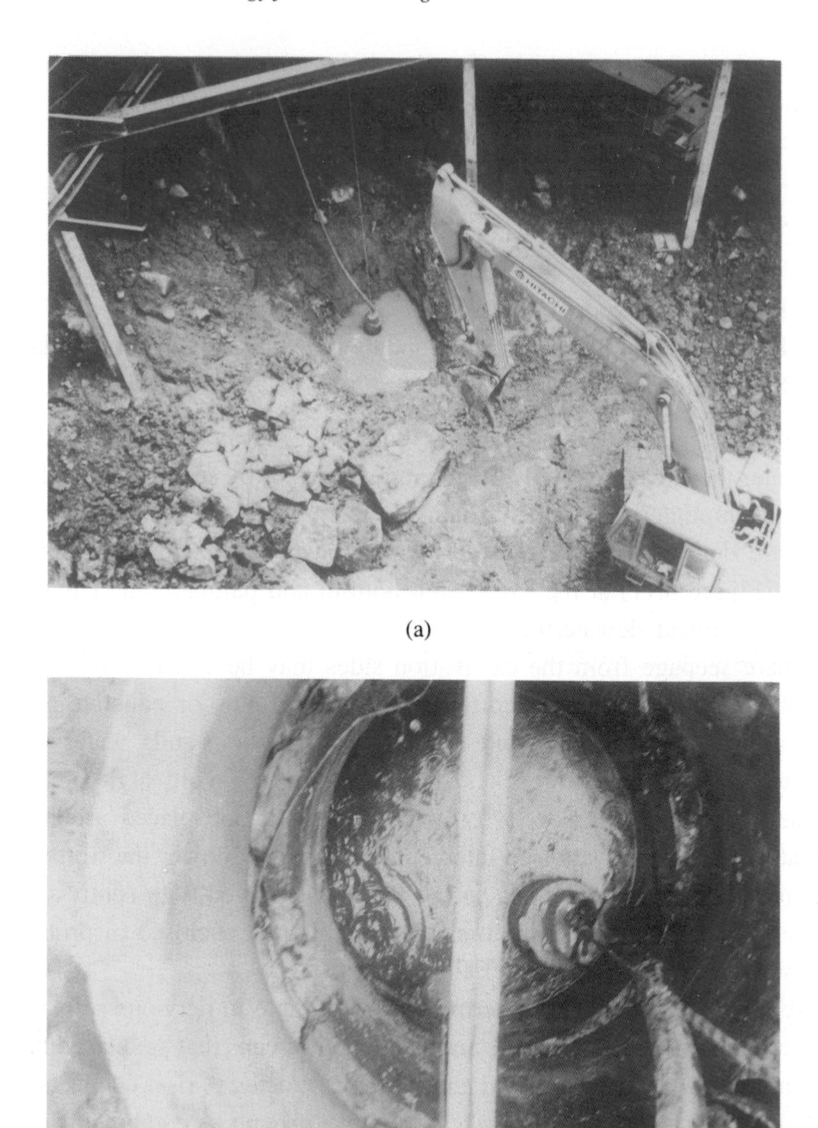

(a)

(b)

Figure 5.11. (a) Pumping from sumps. (b) A closer view of a sump.

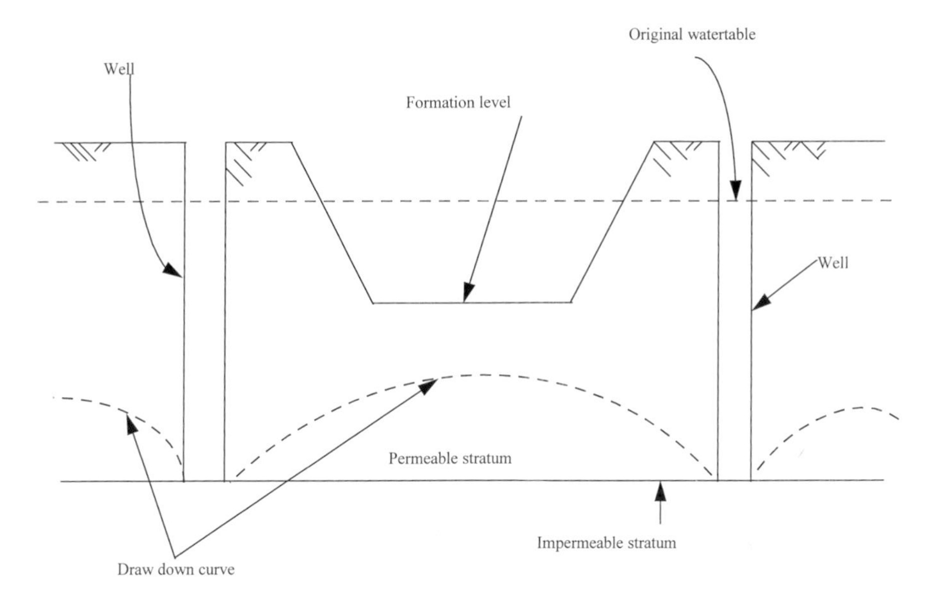

Figure 5.12. Wells used to lower water table to a level below the formation level.

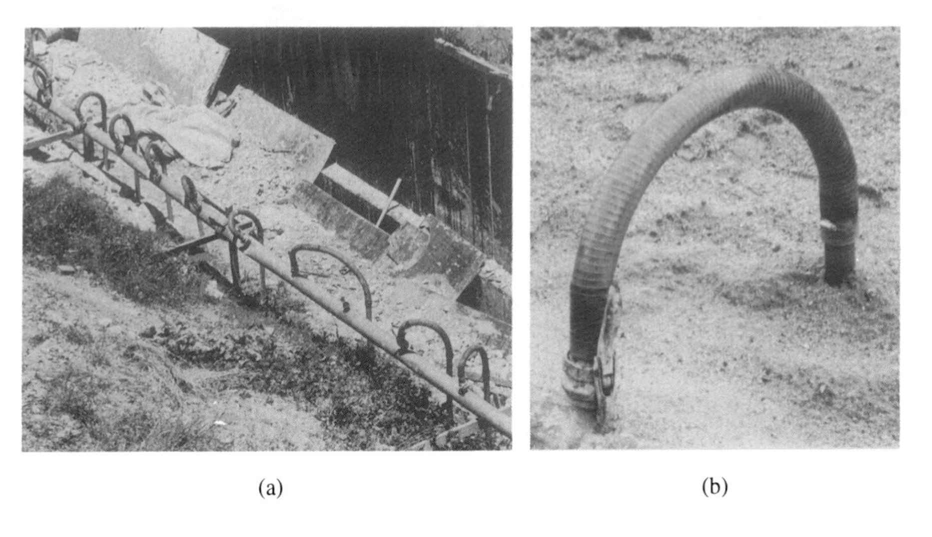

(a) (b)

Figure 5.13. (a) Pumping from wellpoints. (b) The connection of a riser to the collection pipe.

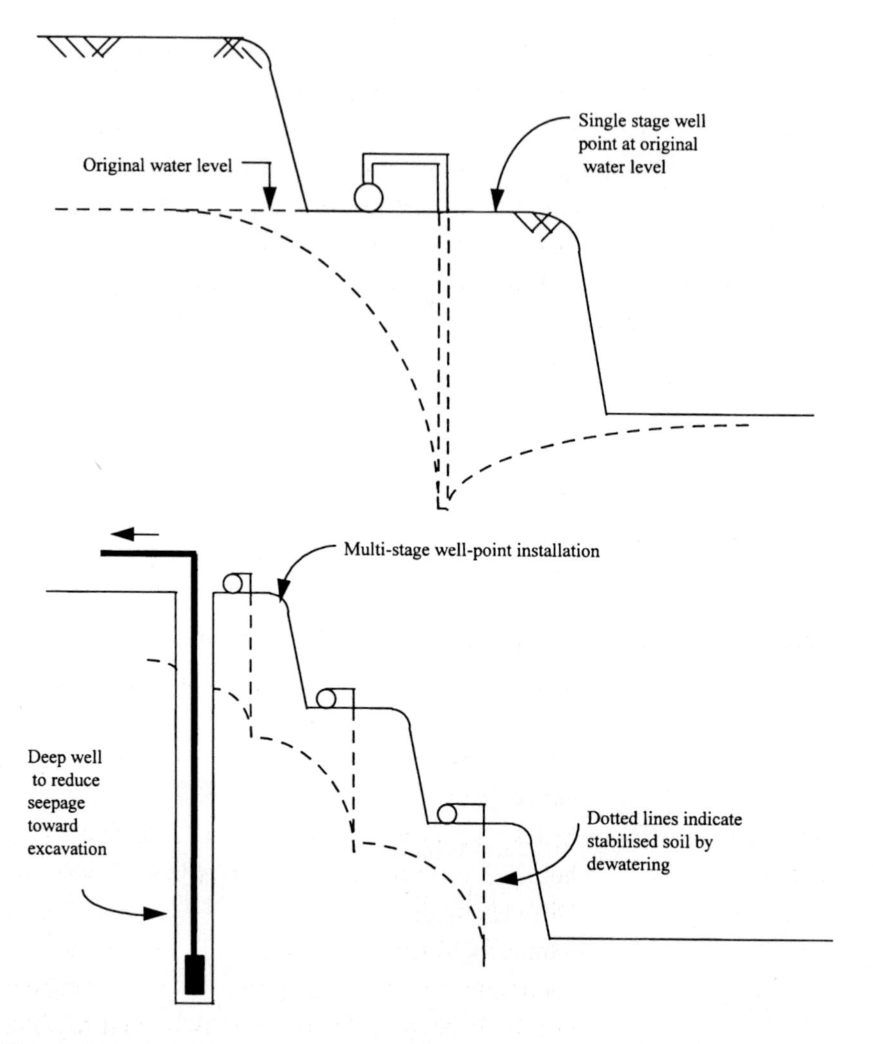

Figure 5.14. Top: single-stage wellpoint system. Bottom: multi-stage wellpoint system.

Jet grouting may be used when the water table is higher than the level of excavation. Holes are drilled throughout the whole site close to each other so that the grout overlap and form a layer of water barrier preventing water from rising up to the surface.

Pressure relief valves may be installed at regular interval on the basement floor to allow ground water to enter the excavation in controlled locations. The water can then be channelled to ejection pits or sumps. The relief valves help to control the water pressure preventing the effect of floatation at the base. This is particularly significant when constructing adjacent to estuaries or to the sea where the water head fluctuates with the tide.

The advantages and disadvantages of the various common temporary and permanent groundwater exclusion methods are described:

Temporary Extraction of Groundwater

(a) *Sump pumping*

Soil	Clean gravels and coarse sands.
Uses	Open shallow excavation.
Advantages	Simplest pumping equipment.
Disadvantages	Fine sand easily removed from ground. Encourages instability of formation.

(b) *Wellpoint systems with suction pumps*

Soil	Sand gravel down to fine sand (with proper control can also be used in silty sand).
Uses	Open excavations including rolling pipe trench excavations.
Advantages	Quick and easy to install in suitable soils. Economical for short pumping periods of a few weeks.
Disadvantages	Difficult to install in open gravel or ground containing cobbles and boulders. Pumping must be continuous and noise of pump may be a problem in a built up area. Suction lift is limited to about 5.25–5.5 m. If greater lowering is needed multi-stage installation is necessary.

(c) *Bored shallow wells with suction pumps*

Soil	Sandy gravel to silty fine sand and water bearing rocks.
Uses	Similar to (b).
Advantages	Costs less to run than (b).

	Filtering can be better controlled.
Disadvantages	Same as (b).

Permanent Exclusion of Groundwater

(a) *Sheet piling*

Soil	Suited for all types of soil except boulder beds.
Uses	Practically unrestricted.
Advantages	— Rapid installation.
	— Steel can be either permanent or recovered.
Disadvantages	— Difficult to drive and maintain seal in boulders.
	— Vibration and noise.
	— High capital investment if re-usage is restricted.
	— Seal may not be perfect.

(b) *Diaphragm wall*

Soil	Suited for all types of soil including boulder beds.
Uses	— Deep basements.
	— Underground carparks.
	— Underground pumping stations.
	— Shafts.
	— Dry docks.
Advantages	— Can be designed to form part of a permanent foundation.
	— Minimum vibration and noise.
	— Treatment is permanent.
	— Can be used in restricted space.
Disadvantages	— High cost often makes it uneconomical unless it can be incorporated into permanent structure.

(c) *Contiguous bored pile and secant pile walls*

Soil	Suited for all types of soil but penetration through boulders may be difficult and costly.
Uses	As for (b).
Advantages	— Can be used on small and confined sites.
	— Can be put down very close to existing foundations.

	— Minimum noise and vibration.
	— Treatment is permanent.
Disadvantages	— Ensuring complete contact of all piles over their full length may be difficult in practice.
	— Joints may be sealed by grouting externally. Efficiency of reinforcing steel not as high as for (b).
(d) *Cement grout*	
Soil	Fissured and jointed rocks.
Uses	Filling fissures to stop water flow.
Advantages	— Equipment is simple and can be used in confined spaces.
	— Treatment is permanent.
Disadvantages	Treatment needs to be extensive to be effective.
(e) *Clay/cement grout*	
Soil	Sands and gravels.
Uses	— Filling voids to exclude water.
	— To form relatively impermeable barriers, vertical and horizontal.
Advantages	Same as (d).
Disadvantages	— A comparatively thick barrier is needed to ensure continuity.
	— At least 4 m of natural cover is necessary.
(f) *Resin grout*	
Soil	Silty fine sand.
Uses	As for (e) but only some flexibility.
Advantages	Can be used in conjunction with clay/cement grout for treating finer strata.
Disadvantages	— Higher costs, so usually economical only on larger civil engineering works.
	— Required strict site control.

Recharging well may be located surrounding the perimeter of a site and/or at critical locations where settlement is likely to occur. The main function of recharging wells is to compensate for the water loss and thus maintaining the water pressure so as to reduce the settlement of soil. Perforations at the sides of the well introduce water into the area to maintain the required water pressure. Recharge wells of diameter ranging from 100 mm to 200 mm spaced between 5 to 10 m apart are common. Flow meters and pressure gauges are used to monitor the flow rate and pressure applied to the water into the recharge wells (Figure 5.15).

Drawing Not To Scale

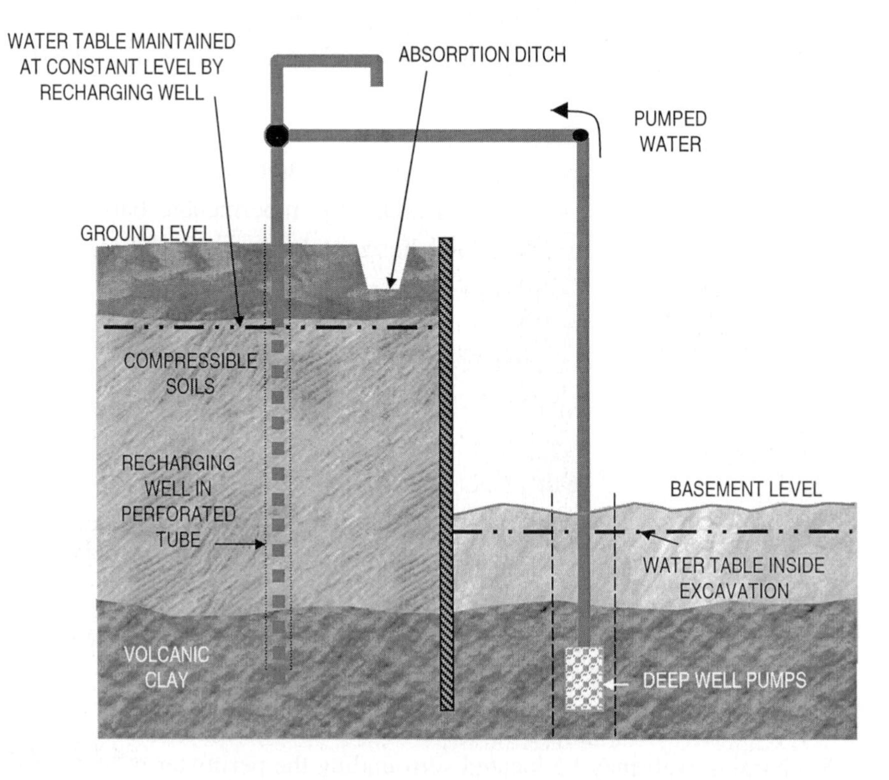

Figure 5.15(a). System of recharging wells.

Figure 5.15(b). Recharge wells.

Figure 5.15(c). Pressure meter to monitor pressure of water pumped into ground.

5.4. Open Cut Technique

This is the simplest and most straight forward technique of providing an excavation to the required depth. The sides of the excavation are sloped to provide stability, with possible slope protection to maximise the angle of the slope. Upon excavating to the required depth, the basement is constructed from bottom upwards. After the completion of the basement, the remaining excavated areas between the basement and the side slope are backfilled (Figure 5.16).

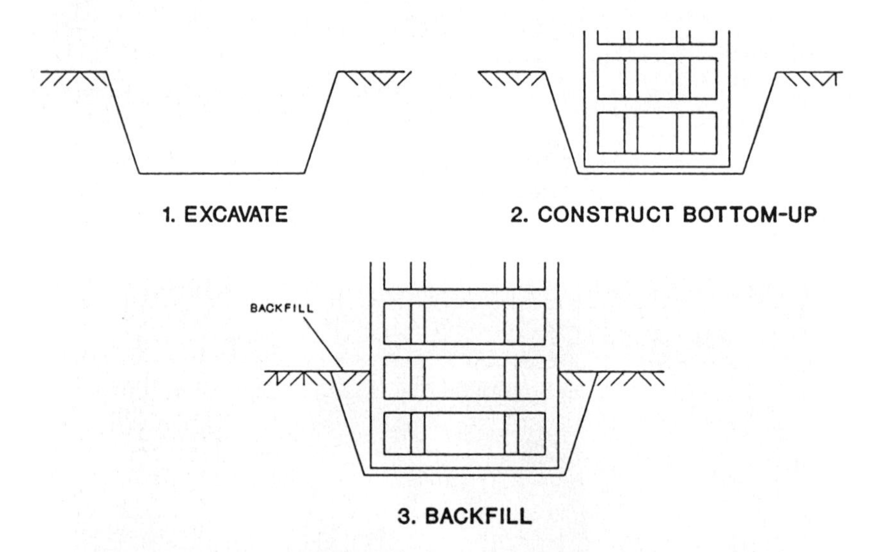

Figure 5.16. Construction sequence of the open cut technique.

Figure 5.17 shows an example of a protected slope. Rods are inserted into the soil alongside the slope, nets are cast over the slope and concrete are sprayed onto them, to prevent the soil from eroding or sliding during rainy days. In some cases, pile sheeting may be provided to reduce differential settlement to the adjacent buildings.

(a)

(b)

Figure 5.17. (a) An overall view of a typical excavation using the open-cut technique. (b) A protected slope with sheet piles and an inclinometer. The inclinometer is to measure deflection of sheet piles and soil movements below the sheet piles.

In built-up urban areas, such a technique is often impractical in view of site constraints and the need to restrict ground movements adjacent to the excavation. The main criteria to consider for an open cut technique is the geological condition of the site, as this has a direct effect on the earth slope. The main limitation of this technique is that the site is exposed to the weather. Flooding usually occurs after a downpour. Provision of dewatering and temporary drainage system are necessary.

5.5. Cut and Cover Technique

This technique is usually employed in constrained sites where ground movements to the adjacent surrounding has to be kept to the minimum. Retaining walls are required to support the excavation with the provision of bracing as the excavation proceeds downward until the deepest basement level. The basement is then constructed in the conventional way, bottom upwards in sequence with removal of the temporary struts (Figure 5.18).

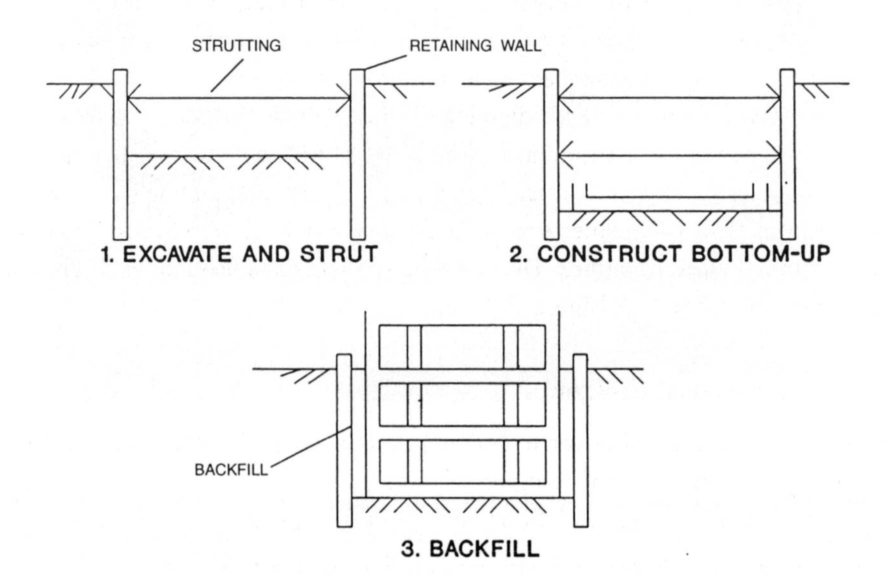

Figure 5.18. The construction sequence of the cut and cover technique.

Figure 5.19 shows the construction of a five storey basement with diaphragm wall around the perimeter in a congested site. The diaphragm wall is supported with wall bracing and heavy strutting. A working stage is erected to provide access in and out the site, and to provide the platform for mechanical plants to operate on. The excavation is carried out mainly by two large excavators with boom arms that are capable of reaching to basement level 4. Excavated earth is carted away by lorries (Figure 5.20). As the excavation proceeds to a deeper level, smaller excavators are mobilised into the basement for excavation under the stage. The sides of the excavation is supported using heavy lateral bracing (strutting), installed at various depths with the subsequent progress of excavation, and the intermediate vertical king posts and bracings (Figure 5.21). The stresses in the struts are monitored to ensure proper load transfer using jacks and adjustable connectors (Figure 5.22). After the excavation, the basement floor slabs are constructed with connection to the starter bars pre-formed with the reinforcement cage of the diaphragm wall (Figure 5.23). The struts are subsequently removed with the construction and gaining of strength of the floors.

In situations where a clear space may be needed for access, or the shape of the site makes supporting by horizontal strutting not economical (Figure 5.24), ground anchors/soil nails may be used.

Ground anchors are small diameter bored piles drilled at any inclination for the purpose of withstanding thrusts from soil of external loading. A ground anchor consists of a tendon, which is fixed to the retained structure at one end whilst the other end is firmly anchored into the ground beyond the potential place of failure. The construction sequence and details of ground anchors are shown in Figure 5.25 and Figure 5.26.

5.6. Top Down Technique

Similar to the cut and cover technique, permanent perimeter walls are first constructed. Prefounded columns are then constructed, followed by the construction of the ground floor slab. Prefounded columns are structural columns/piles formed before basement excavation. Boreholes are formed to the hard strata. Steel stanchion/H-sections are inserted and concrete

(a)

(b)

Figure 5.19. (a) Diaphragm walls supported by horizontal struttings. (b) A closer view showing the heavy horizontal strutting.

(a)

(b)

Figure 5.20. (a) Excavation proceeds from the working platform with large excavators until it reaches the depth where small excavators need to be mobilised. (b) Excavated earth carted away by lorries.

(a)

(b)

Figure 5.21. Excavation supported by (a) heavy strutting and (b) king posts and bracing as excavation progresses.

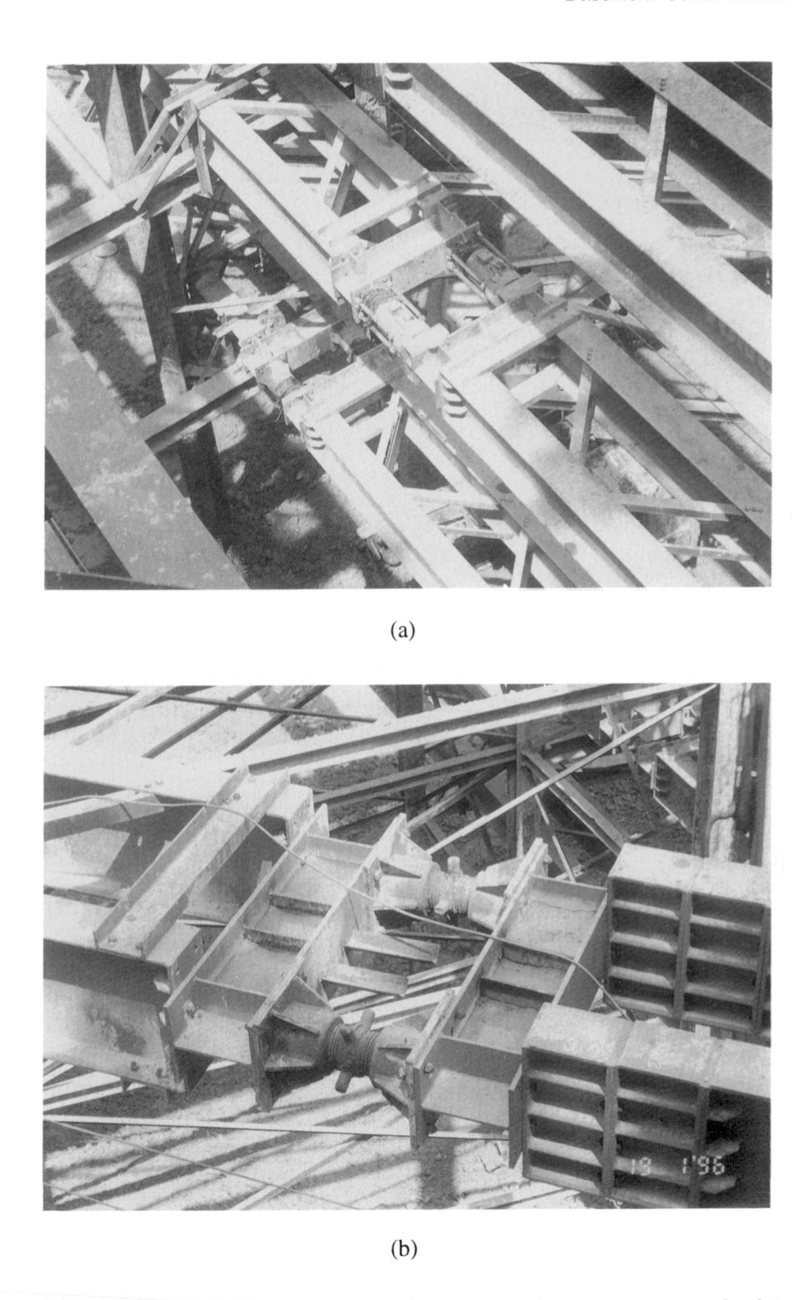

(a)

(b)

Figure 5.22. Jacks, adjustable connectors and gauges used to ensure proper load transfer through the struttings.

Figure 5.23. Exposed starter bars on a diaphragm wall to be connected to the reinforcement of the respective basement slab.

Drawing Not To Scale

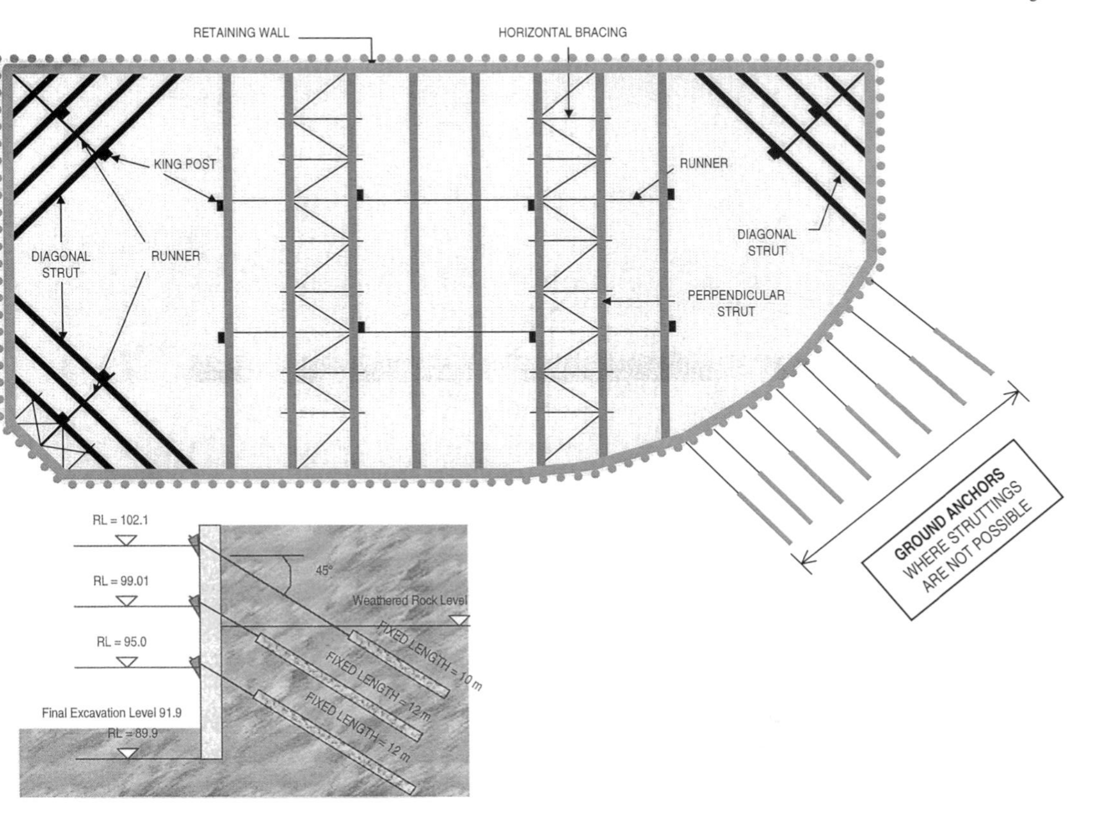

Figure 5.24. The use of ground anchors/soil nails and struttings.

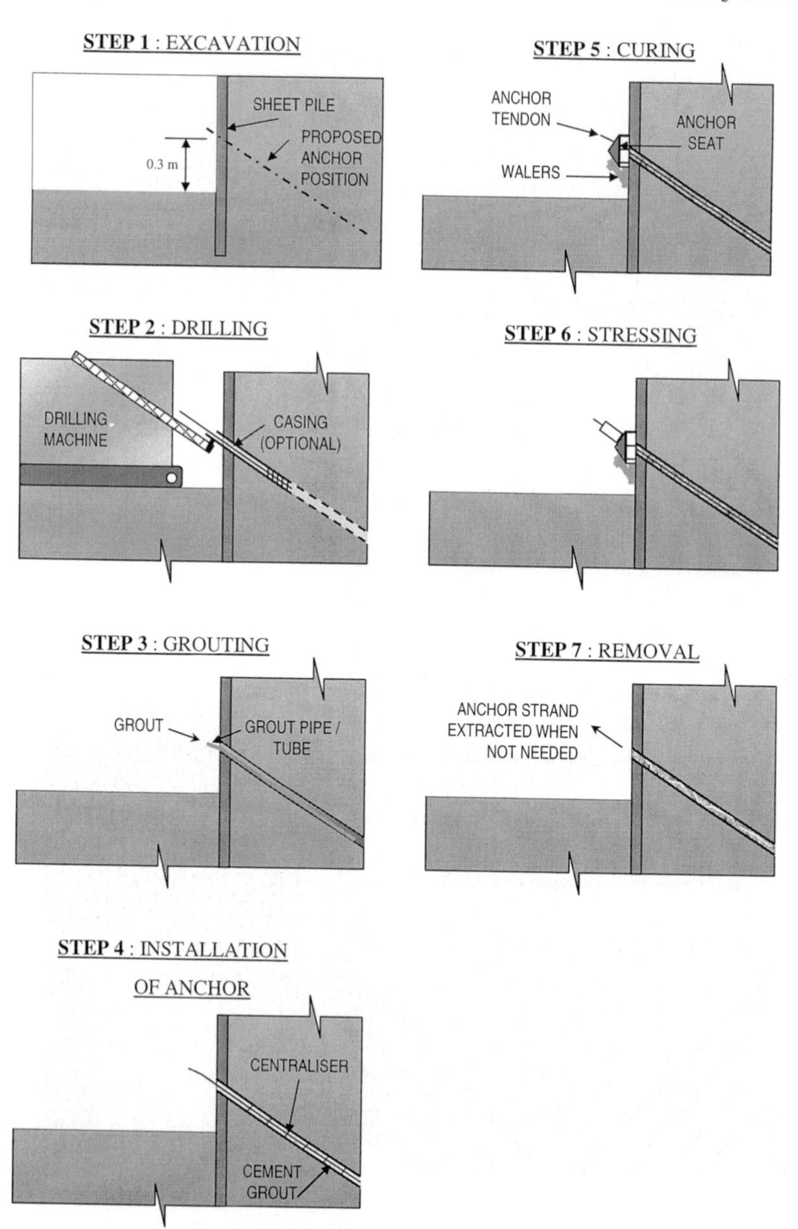

Figure 5.25. The construction sequence of ground anchors/soil nails.

Figure 5.26(a). Casings of various diameters for drilling.

Figure 5.26(b). Inclined drilling with water as lubricant.

Figure 5.26(c). Drilled holes with grout tubes inserted spaced at 1 m apart.

pumped in to slightly over the lowest basement slab level. The holes are then backfilled with soil. This lower part of the stanchions embedded in concrete will later form the integral part of the foundation for the structure. The upper part of the stanchions not embedded in concrete serves as the column supports for the subsequent basement floors. Excavation then proceeds downward and basement slabs constructed while construction of the superstructure proceeds simultaneously (Figure 5.27). Temporary openings are provided at various strategic locations on the basement floors to provide access for removal of excavated earth as well as delivery of excavation machinery and materials for the construction of the substructure. When the formation level is reached, construct pile caps, ground beams etc. Excavation proceeds without the need for strutting to support the excavation as the slabs act as the permanent horizontal supports.

Figure 5.28 shows the excavation in process under restricted headroom. The right hand side of the figure shows the reinforcing cage for the enlargement of the prefounded column.

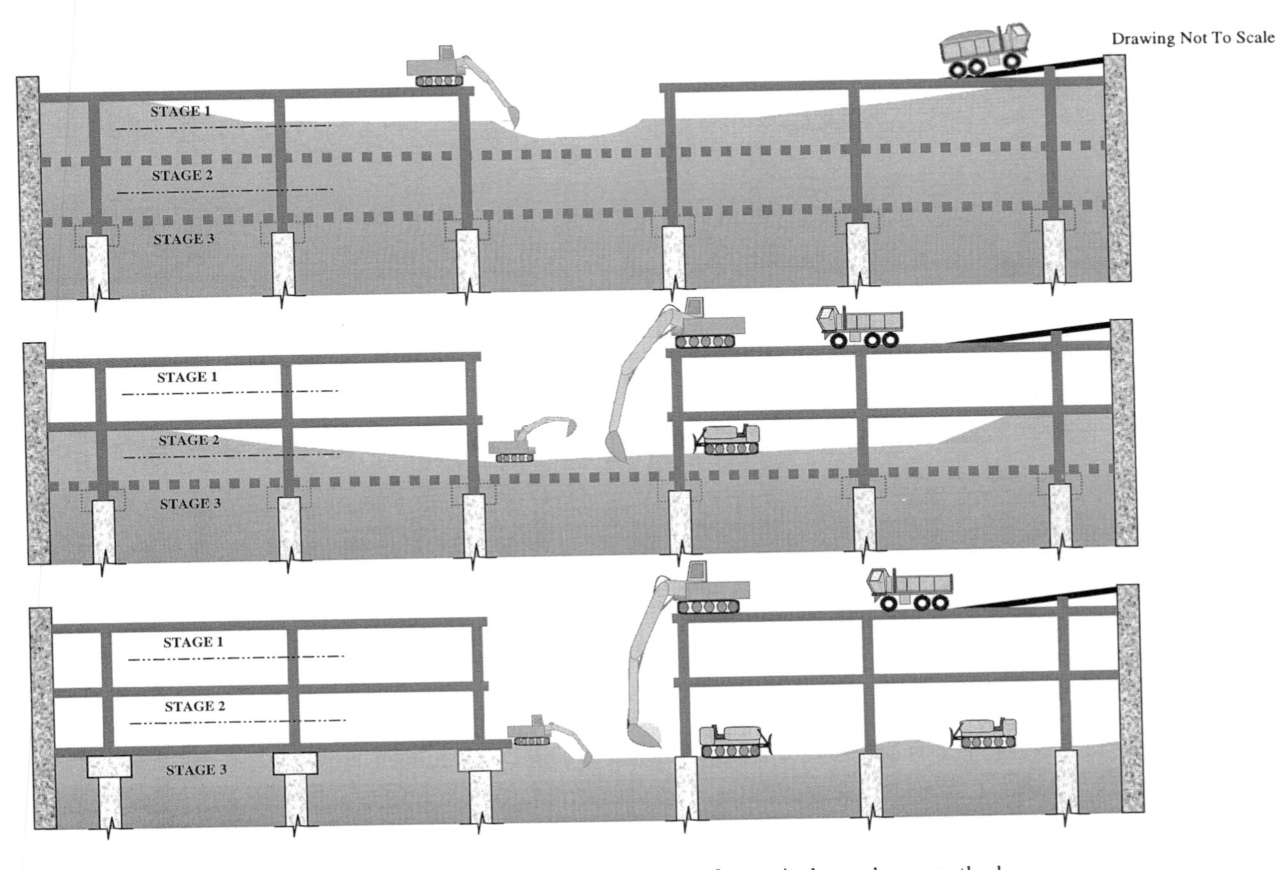

Figure 5.27. Construction sequence of a typical top-down method.

Figure 5.28. The use of small machinery due to the limited headroom.

The obvious advantage of utilising top down construction technique is that the superstructure can proceed upwards from ground level simultaneously with the excavation downwards. Strutting is not necessary. It also allows early enclosure of the excavation which would permit work to be carried out even in adverse weather condition. Ground movement to the adjacent area is minimised as excavation is always strutted during construction.

The main difficulties are the limited headroom for excavation, restricted access for material handling, dust and noise problems. Figure 5.29 shows the temporary opening in a floor slab to provide access for the material handling. Cost of construction is high as it involves installation of more sophisticated temporary support such as prefounded columns. Provision of mechanical ventilation and artificial lighting is necessary during construction (Figure 5.30). In cases where prefounded columns are to be removed or enlarged, great care must be taken to ensure effective load transfer. The constraints with headroom, access etc. can also make the removal of unwanted soil, e.g. boulders, a very tedious job (Figure 5.31).

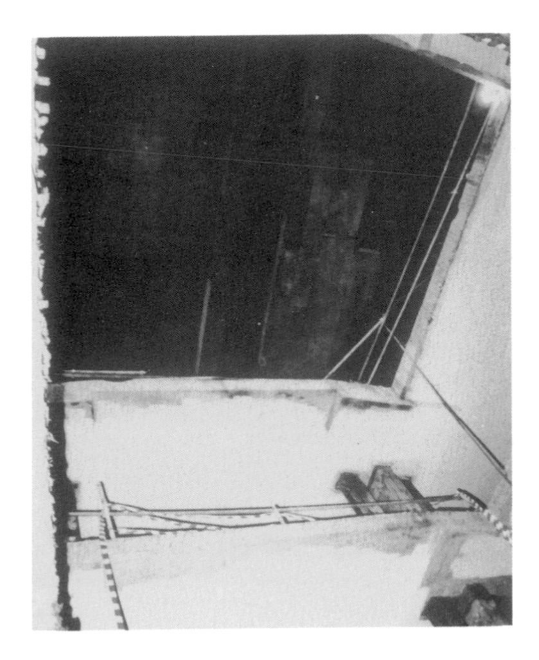

Figure 5.29. Temporary opening for material handling.

Figure 5.30. Mechanical ventilation, lighting are essential.

Figure 5.31. The removal of boulders using jack hammers.

5.7. Soil Movement Monitoring

Excavation works in areas with adjacent existing buildings, railways and roads require careful soil movement monitoring. Field conditions such as surface movements, subsurface deformations, in-situ earth and pore pressure, water table level, etc. are measured regularly. This is particularly important when dealing with viaducts and underground tunnels [32].

The most common instrumentation installed around the perimeter of an excavation are groups of inclinometers, water standpipes and pneumatic piezometers (Figure 5.32). Tilt meters, settlement markers to detect differential settlement and detectors for noise and vibration are also commonly used.

Inclinometers — to monitor lateral movements in embankments and landslide areas, deflection of retaining walls and piles, and deformation of excavation walls, tunnels and shafts (Figure 5.33) [33]. Inclinometer casing is typically installed in a near vertical borehole that passes through suspected zones of movement into stable ground (Figure 5.34). The inclinometer probe, containing a gravity-actuated transducer fitted with wheels, is lowered on an electrical cable to survey the casing (Figure 5.35). The first survey establishes the initial profile of the casing. Subsequent surveys reveal changes in the profile if ground movement occurs. The cable is connected to a readout unit and data can be recorded manually or automatically. During a survey, the probe is drawn upwards from the bottom of the casing to the top, halted in its travel at half-meter intervals for inclination measurements. The inclination of the probe body is measured by two force-balanced servo-accelerometers. One accelerometer measures inclination from vertical in the plane of the inclinometer wheels, which track the longitudinal grooves of the casing. The other accelerometer measures inclination from vertical in a plane perpendicular to the wheels. Inclination measurements are converted to lateral deviations. Changes in lateral deviation, determined by comparing data from current and initial surveys, indicate ground movements (Figure 5.36).

Figure 5.32. A group of pneumatic piezometer, water standpipe and inclinometer among many groups located around the perimeter of an excavation.

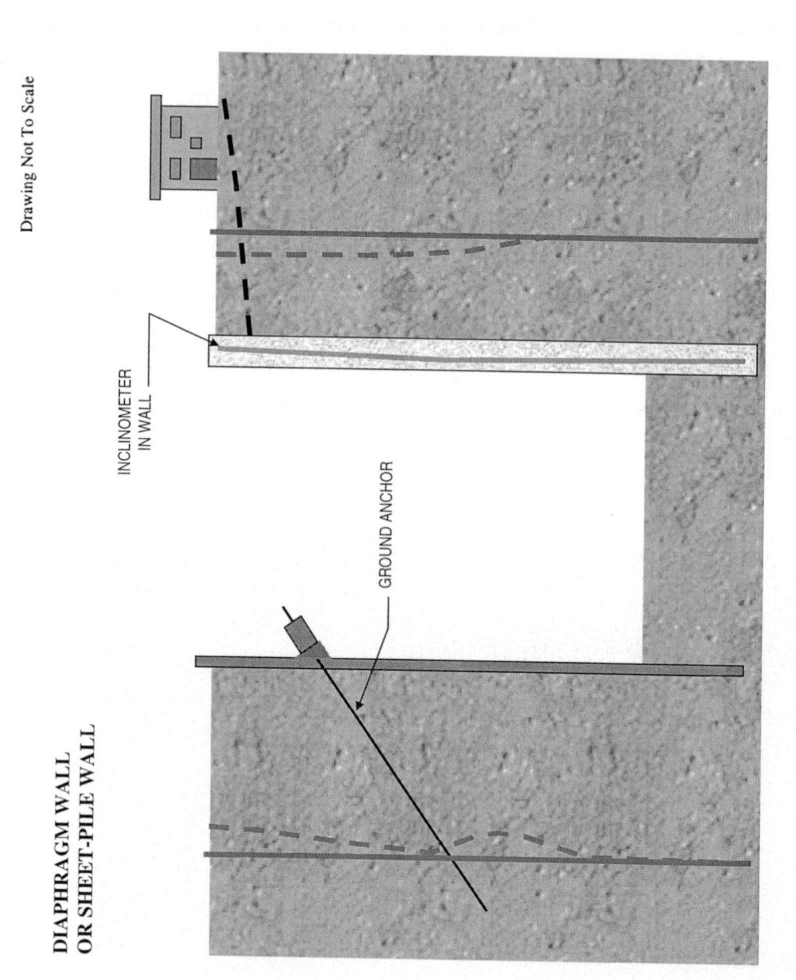

Figure 5.33. The use of inclinometers to check for (a) stability and deflections of retaining wall, (b) ground movement that may affect adjacent buildings, (c) performance of struts and ground anchors [33].

Drawing Not To Scale

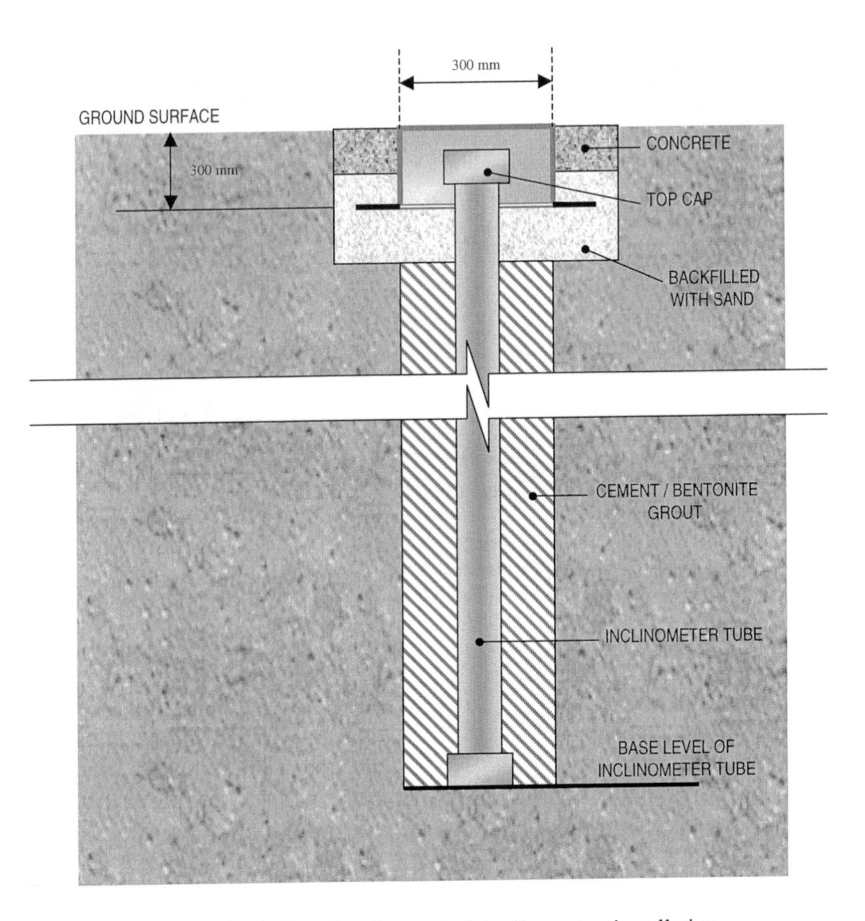

Figure 5.34. Details of a typical inclinometer installation.

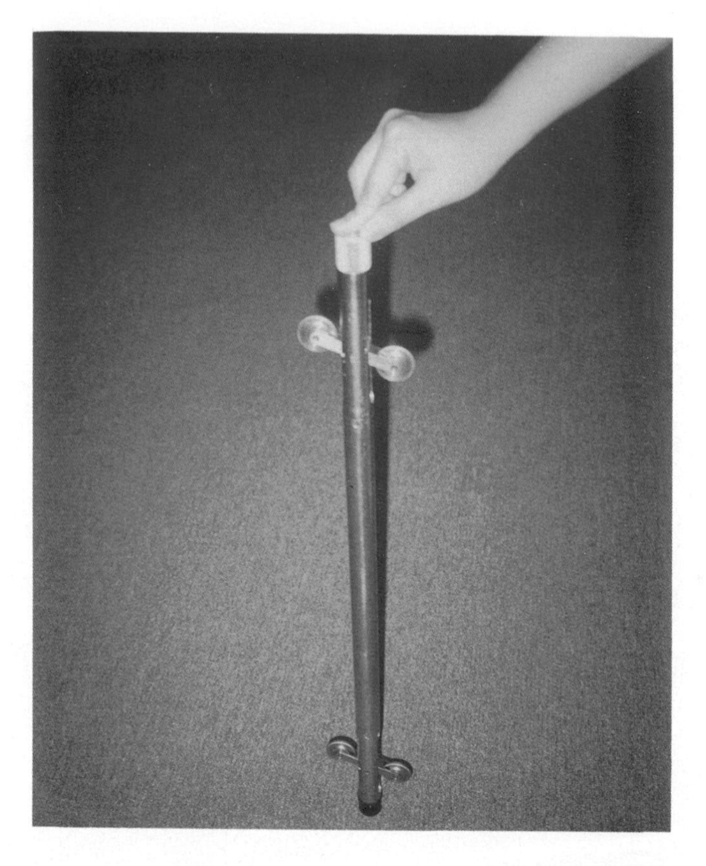

Figure 5.35. An inclinometer probe with wheels fitted with transducers.

Water stand pipes — to monitor the ground water level, control the rate of de-watering in excavation work, monitor seepage and to verify models of flow. It involves drilling a 150 mm borehole to the required depth, lowering the 50 mm standpipe into the borehole, backfill with sand, terminate the tubing at the surface and place a protective cap at the top of the tube (Figure 5.37). A water level indicator will be used to measure the water level. The water level indicator is lowered down the standpipe until a light and buzzer sounds indicate contact with water. Depth markings on the cable show the water level (Figure 5.38).

Drawing Not To Scale

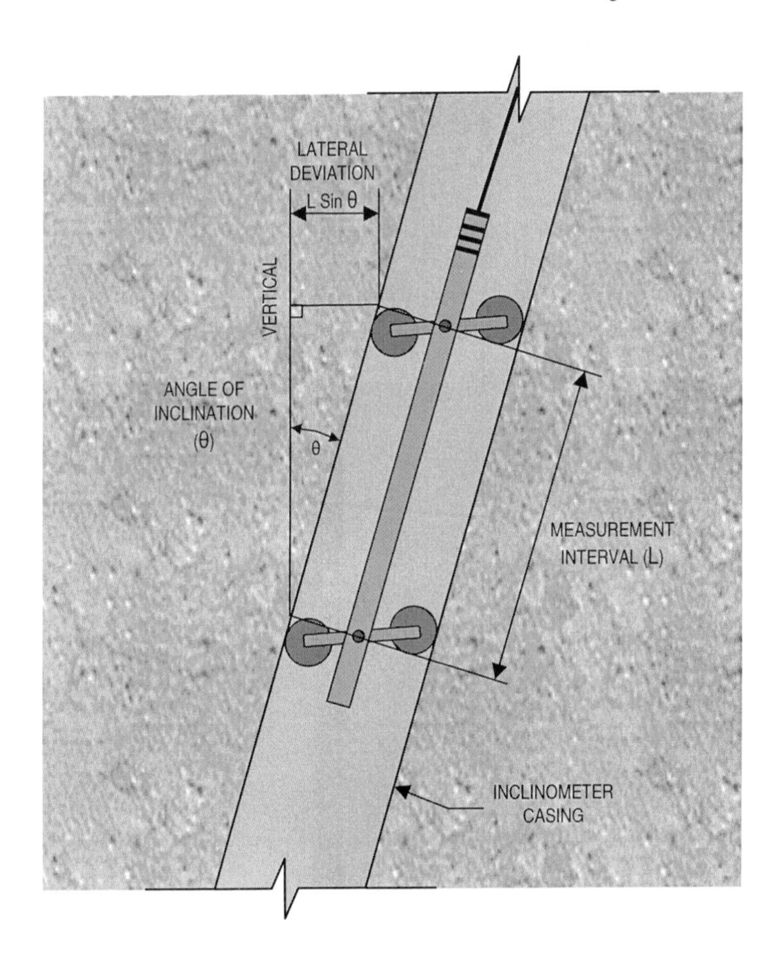

Figure 5.36. Angle of inclination and lateral deviation [33].

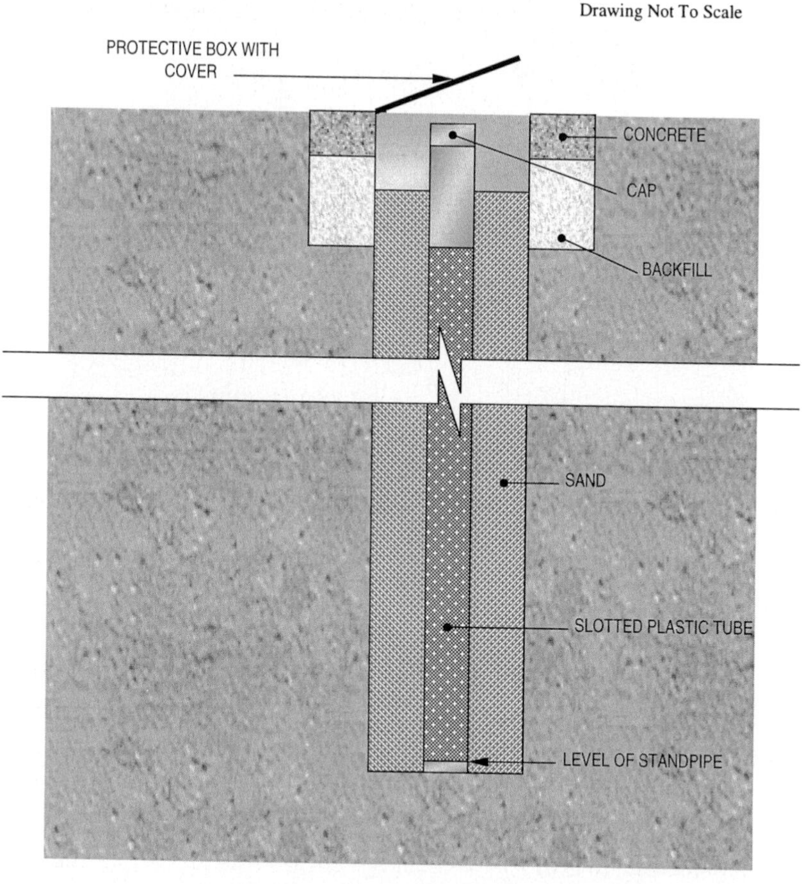

Drawing Not To Scale

PROTECTIVE BOX WITH COVER

CONCRETE

CAP

BACKFILL

SAND

SLOTTED PLASTIC TUBE

LEVEL OF STANDPIPE

Figure 5.37. Details of a typical water standpipe installation.

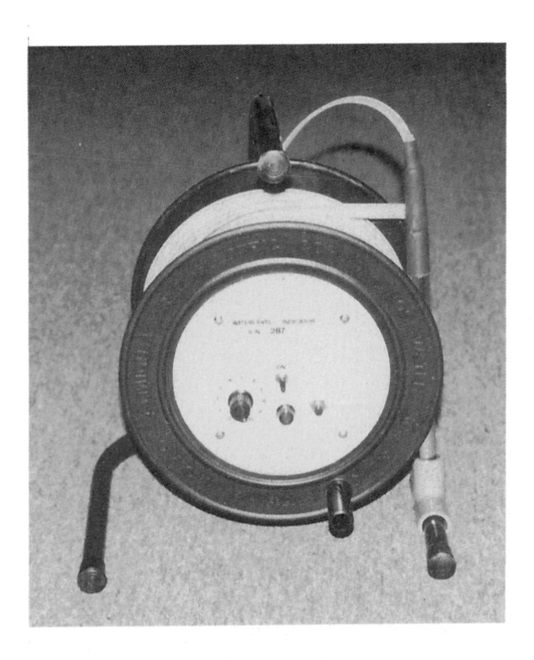

Figure 5.38. Water level indicator with the sensitivity knob on the reel panel to select the desired sensitivity on a scale from 1 to 10 [33].

Pneumatic Piezometers — to monitor pore-water pressure to determine the stability of slopes, embankments and ground movement. It is useful for monitoring de-watering schemes for excavations and underground openings, seepage and ground water movement, water drawdown during pumping tests, etc. (Figure 5.39). The piezometer consists of a filter tip placed in a sand pocket and lined to a riser pipe that communicate with the surface (Figure 5.40). A bentonite seal is placed above the sand pocket to isolate the ground water pressure at the tip. The annular space between the riser pipe and the borehole is backfilled to the surface with a bentonite grout to prevent unwanted vertical migration of water. The rise pipe is terminated above surface level with a vented cap. As pore-water pressure increases or decreases, the water level inside the standpipe rises or falls. The height of the water above the filter tip is equal to the pore-water pressure.

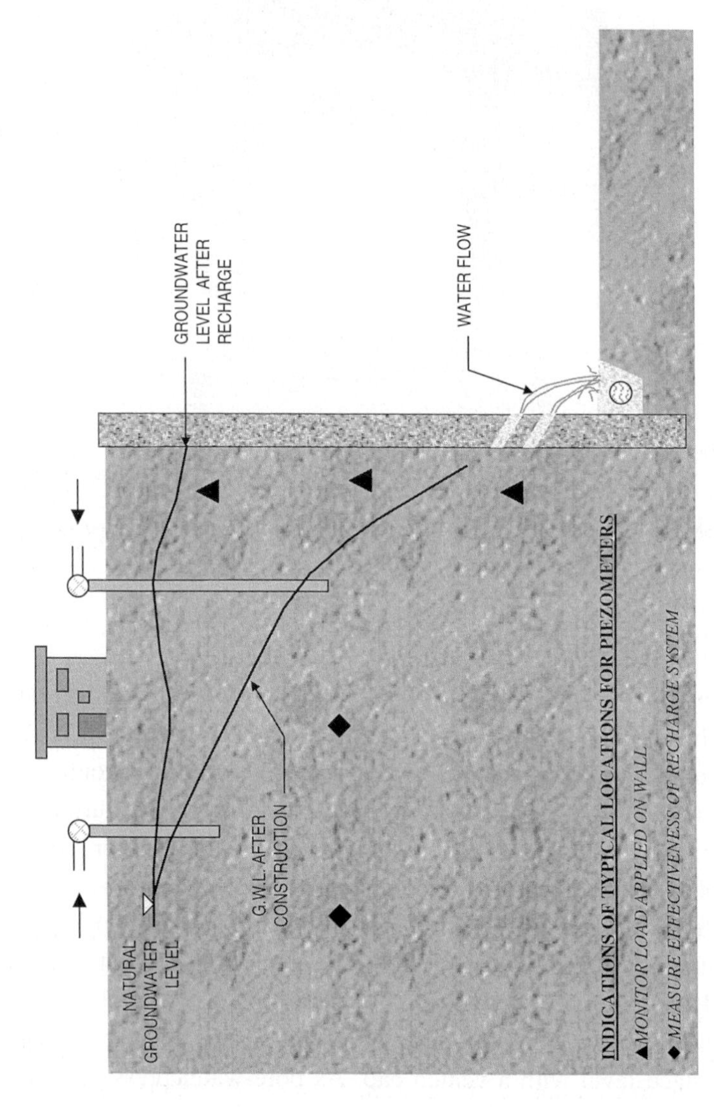

Figure 5.39. The use of piezometers to measure the effectiveness of a recharge system and load applied to a wall [33].

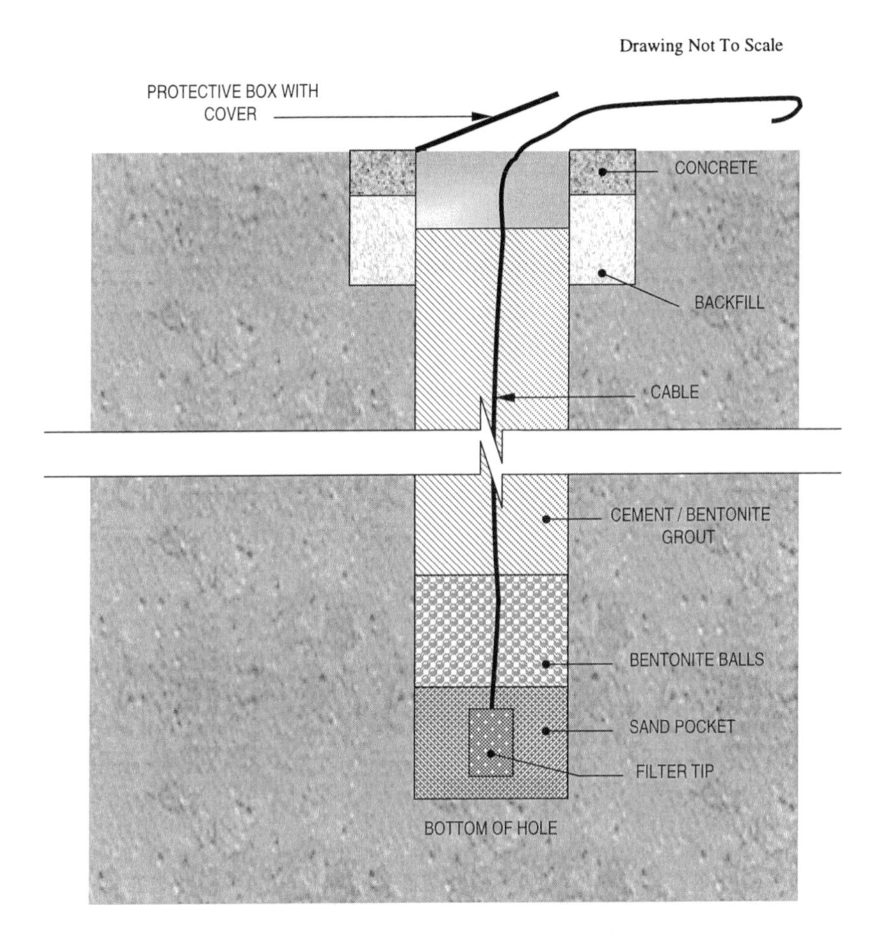

Figure 5.40. Details of a typical piezometer installation.

Tiltmeter system — to monitor changes in the inclination of a structure, to provide an accurate history of movement of a structure and early warning of potential structural damage. The tiltmeter system includes a number of tilt plates, the portable tiltmeter and a readout unit (Figure 5.41). Tilt plates are available in ceramic or bronze. They are either bonded or screwed onto the structure in specified locations (Figure 5.42). The portable tiltmeter uses a force-balanced servo-accelerometer to measure inclination. The accelerometer is housed in a rugged frame with machined surfaces that facilitate accurate positioning on the tilt plate. The bottom surface is used with horizontally-mounted tilt plates and the side surfaces are used with vertically-mounted tilt plates (Figure 5.43). To obtain tilt readings, the tiltmeter is connected to the readout unit and is positioned on the tilt plate, with the (+) marking on the sensor base plate aligned with peg 1. The displayed reading is noted, the tiltmeter is rotated 180°, and a second reading is taken. The two readings are later averaged to cancer sensor offset. Changes in tilt are found by comparing the current reading to the initial reading. A positive value indicates tilt in the direction of peg 1 (peg 1 down, peg 3 up). A negative value indicates tilt in the direction of peg 3 (Figure 5.44).

Figure 5.41. Portable tiltmeter with tilt plates and a readout unit [34].

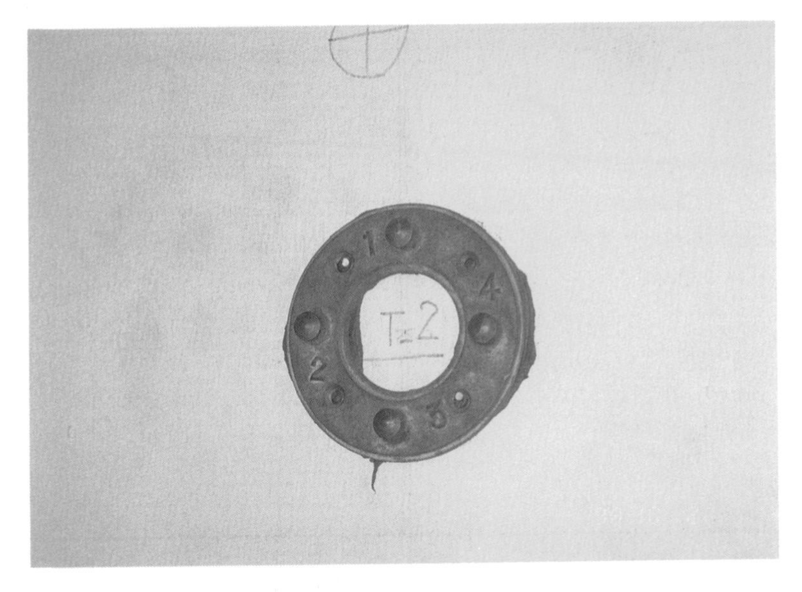

(a)

(b)

Figure 5.42. (a) and (b) Bronze and coated ceramic tilt plates bonded on vertical walls.

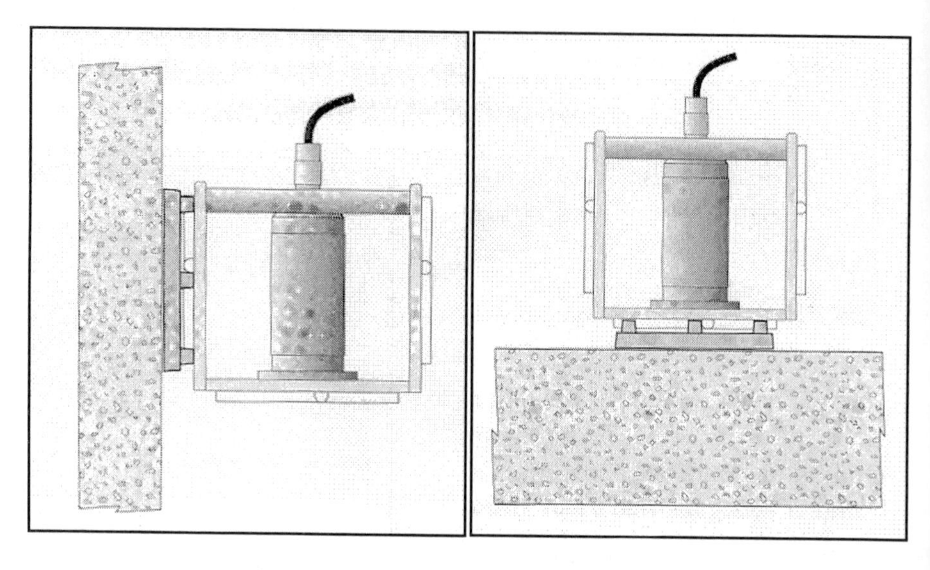

Figure 5.43. Tiltmeters positioned onto tilt plates on vertical and horizontal surfaces [33].

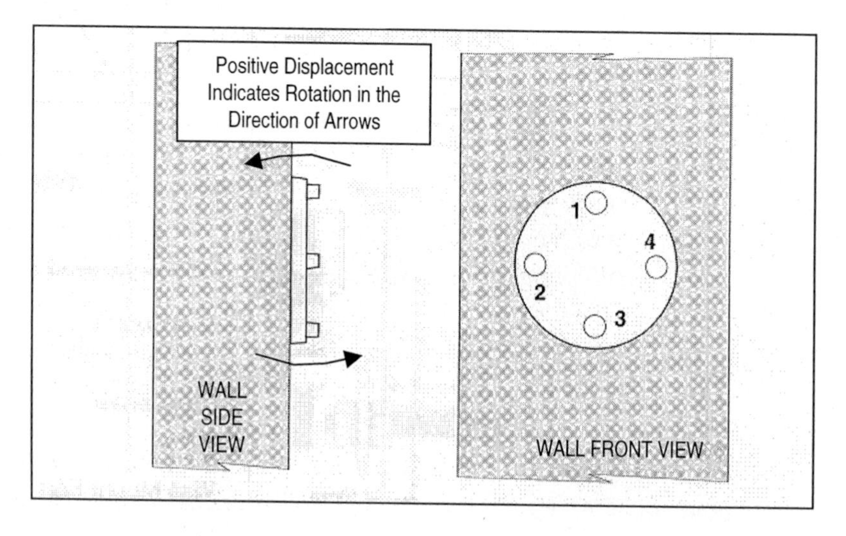

Figure 5.44. Orientation for vertical mounting [33].

Settlement Markers — to monitor vertical ground movement when measured against a known fixed datum (Figure 5.45). When the displacement of a retaining wall or structure indicated by the above instrumentation falls between the design and allowable values (red light), contingency plan will be activated. If the displacement exceeds that of the allowable values, site work will be stopped immediately until the problem is resolved. Ground treatments involving silicate/cement/bentonite grouting or other chemical injection may be used. Strengthening involving additional shoring/propping and tie bars may be used. Underpinning may be used as an effective tool to prevent the settlement of adjacent building during the course of excavation.

5.8. Shoring

Shoring is the means to provide temporary support to structures that are in an unsafe condition till such time as they have been made more stable. BOWEC specifies that no person shall be permitted to enter into any excavation unless sheet piling, shoring or other safeguards had been provided. It also specifies that where any person is exposed to the hazard of falling or sliding material from any side of an excavation that exceeds 1.5 m, adequate shoring shall be provided.

Raking shores: consist of inclined members called rakers placed with one end resting against the vertical structure and the other upon the ground. The angle of the shores is generally 60°–75° from the ground. Figure 5.46(a) shows the details of a traditional timber raking shores. Figure 5.46(b) shows the use of a traditional raking shore supporting the wall of an old shophouse. The raker is supported on a "wall-piece" fixed to the wall using wall-hooks driven into the brick joints. The wall-piece receives the heads of the rakers and distributes their thrust over a larger area of the wall. The top of each shore terminate at a short "needle" set into the wall below each floor level. The function of the needle is to resist the thrust of the raker and prevent it from slipping on the wall-piece. The tendency of the shore to move upwards is resisted by a "cleat" fixed to a wall plate set immediately above the needle. The feet of the raker rest upon an inclined "sole plate" embedded in the ground. When the shore has been tightened it is secured to the sole

(a)

(b)

Figure 5.45. (a) and (b) Settlement markers.

plate by iron dogs and a "cleat" is nailed to the sole plate. Figure 5.47 shows the use of H-section steel raking shores to support the side of an excavation.

Flying shores: consist of horizontal members with details similar to that of raking shores (Figure 5.48). Strutting used for an excavation is also often referred as flying shores (see Section 5.5).

Dead shores: consist of columns of framed structures shored up vertically with or without needle beams (Figure 5.49). There are used extensively for underpinning works (Section 5.9).

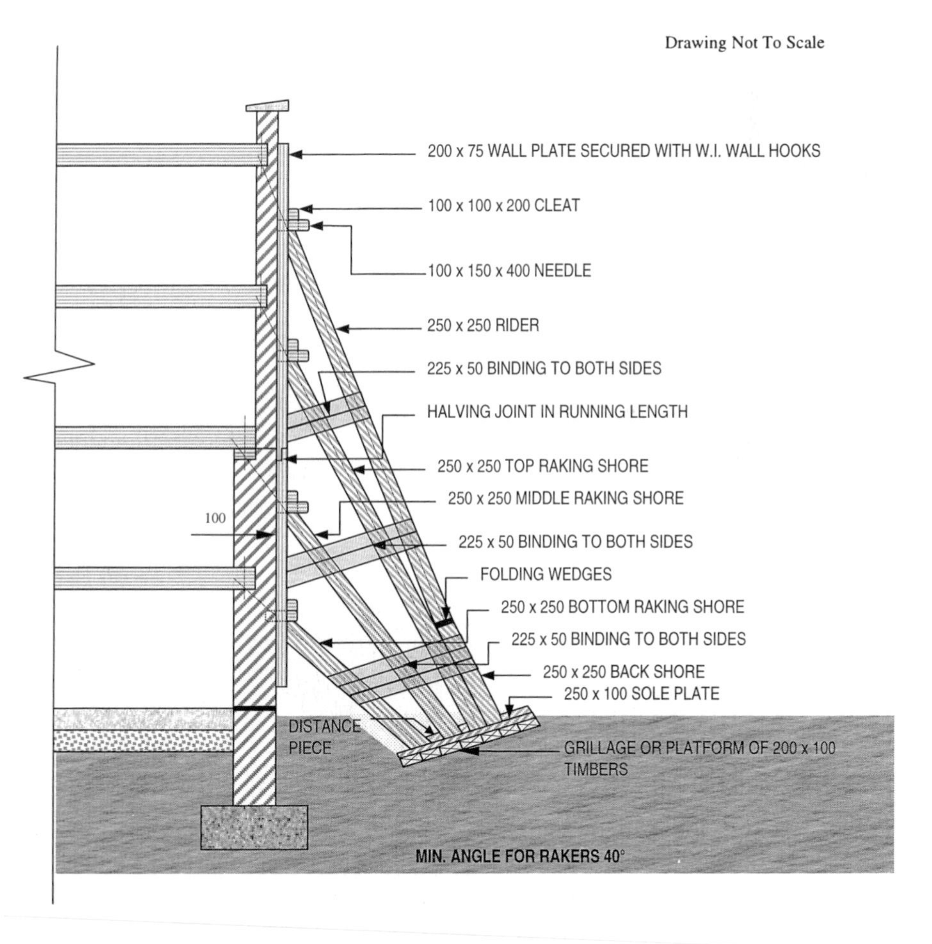

Figure 5.46(a). Details of a traditional timber raking shore.

Figure 5.46(b). A traditional timber raking shore on the external wall of an old shophouse.

Figure 5.47. The use of H-section steel raking shores to support the side of an excavation.

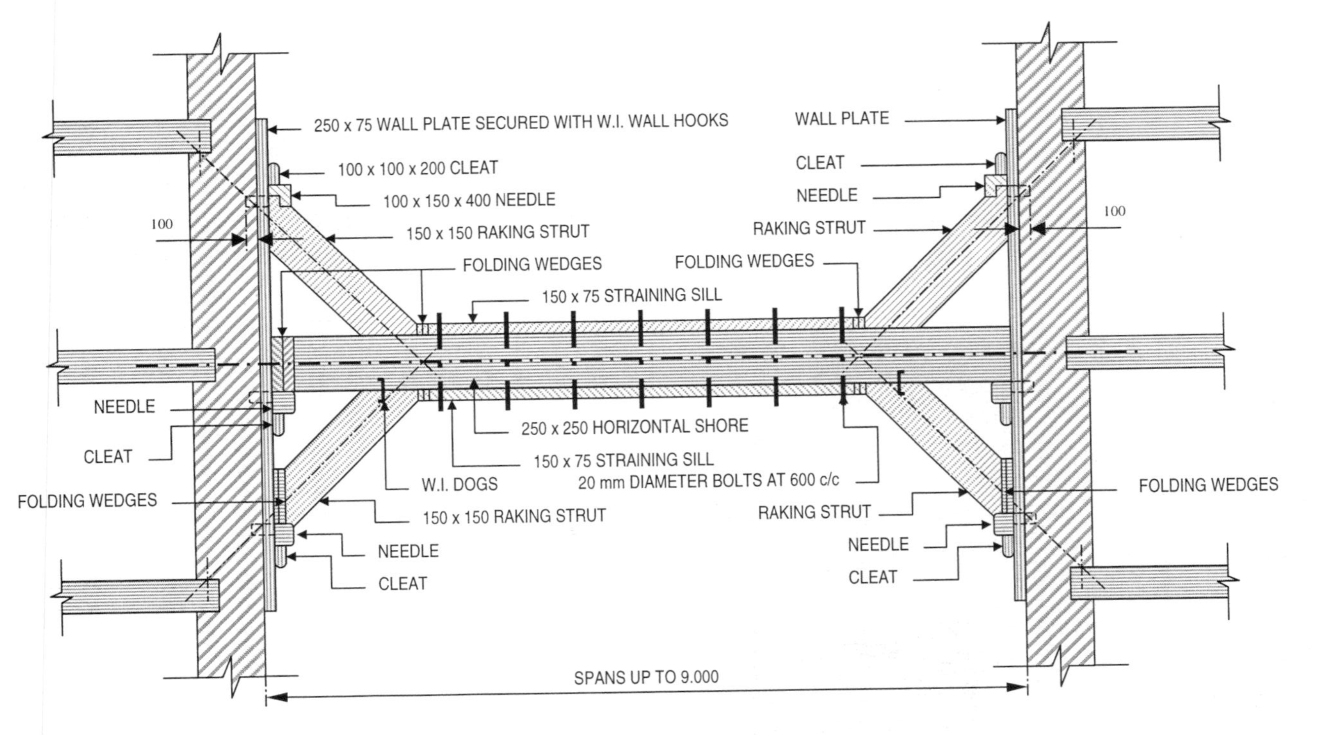

Figure 5.48(a). Details of a traditional flying shore.

Figure 5.48(b). The use of a traditional flying shore supported on 2 vertical walls.

5.9. Underpinning

Underpinning is the process of strengthening and stabilising the foundation of an existing building. It involves excavating under an existing foundation and building up a new supporting structure from a lower level to the underside of the existing foundation, the object being to transfer the load from the foundation to a new bearing at a lower level [35–41]. Examples of buildings that may require underpinning include:

- Buildings with existing foundation not large enough to carry their loads, leading to excessive settlement.
- Buildings overloaded due to change of use or partial reconstruction.
- Buildings affected by external works such as adjacent excavations or piling which lead to ground movements and vibrations.
- Buildings going for new extensions, e.g. higher storeys, new basements, etc.
- Buildings with new buried structures, e.g. service tunnels and pipelines in close proximity.

Drawing Not To Scale

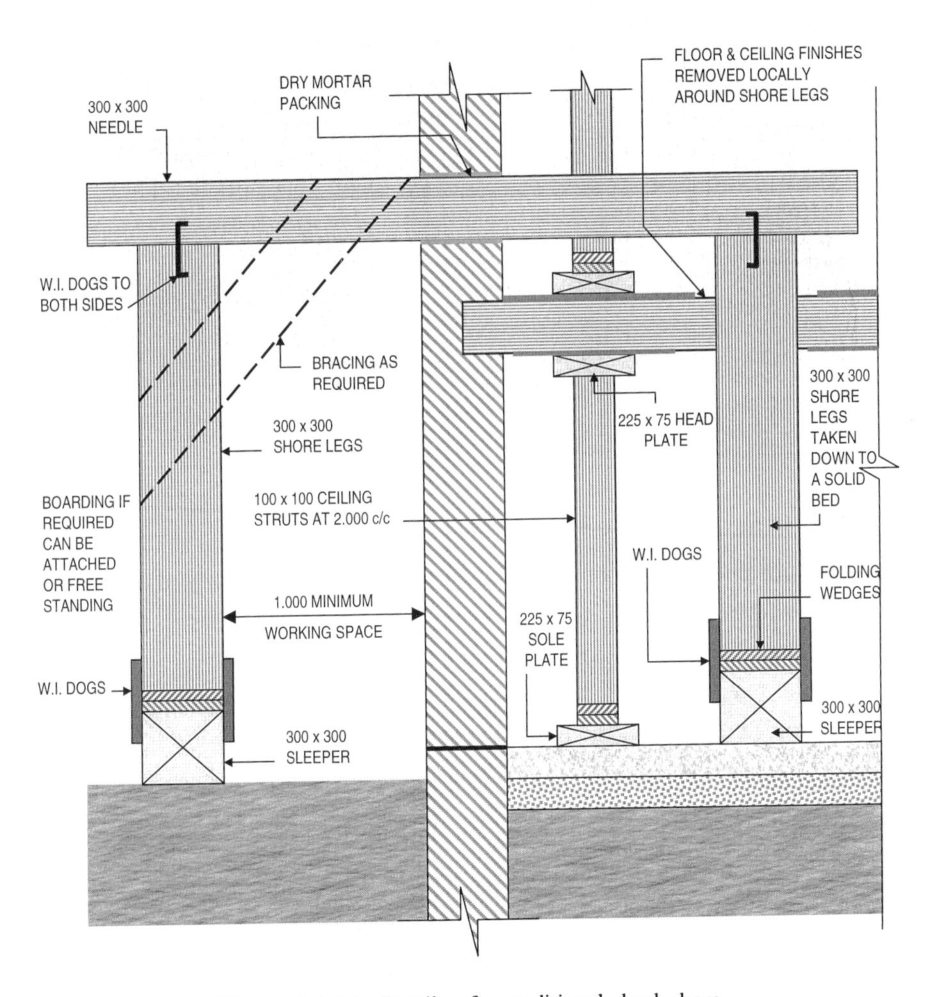

Figure 5.49(a). Details of a traditional dead shore.

Figure 5.49(b). The use of a traditional dead shore in refurbishment works.

There are three fundamental methods to increase the capacity of an existing foundation:

- The foundations may be enlarged (Figure 5.50).
- New deep foundations may be inserted under shallow ones to carry the load to a deeper, stronger stratum of soil (Figure 5.51).
- The soil itself may be strengthened by grouting or by chemical treatment (Figure 5.52).

The two common methods of supporting a building while carrying out underpinning work beneath its foundation are by:

(a) *progressive method* — where the work is carried out in "discrete" or alternate bays so as to maintain sufficient support as works proceeds. Normally no more than 20% of the total wall length should be left unsupported. The bays are then joined together to form a continuous beam (Figure 5.53).

(b) *needling method* — where the foundation of an entire wall is exposed at once by needling, in which the wall is supported temporarily on needle beams threaded through holes cut in the wall. After underpinning has been accomplished, the jacks and needle beams are removed and the trench backfilled (Figure 5.54).

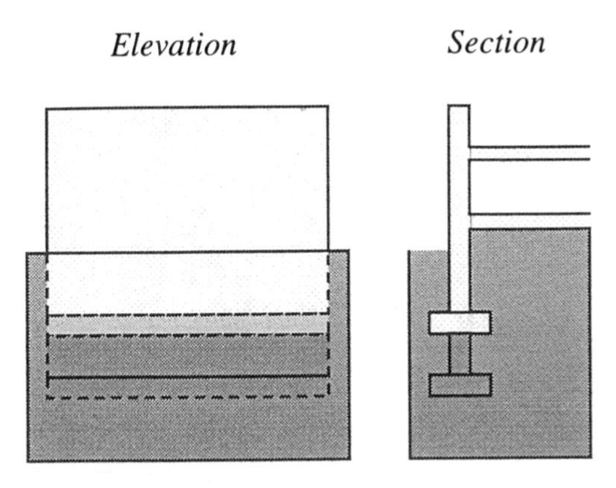

Figure 5.50. A new foundation wall and footing are constructed beneath the existing foundation.

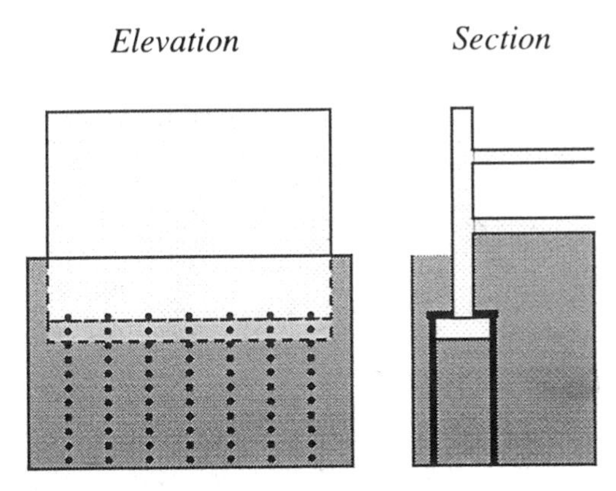

Figure 5.51. New piles or caissons are constructed on either side of the existing foundation.

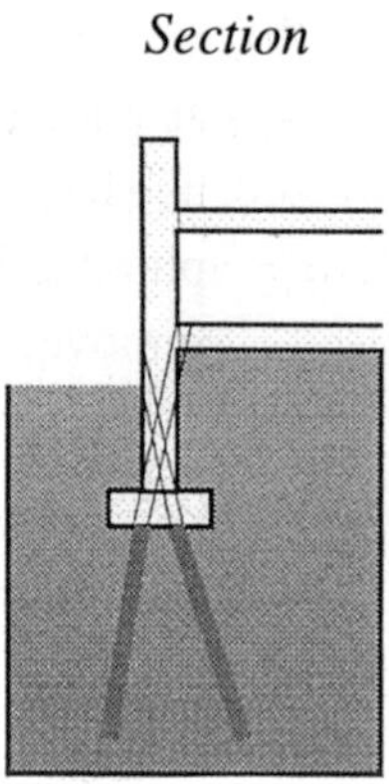

Section

Figure 5.52. Cast in-situ concrete minipiles cast into holes drilled diagonally through the existing foundation.

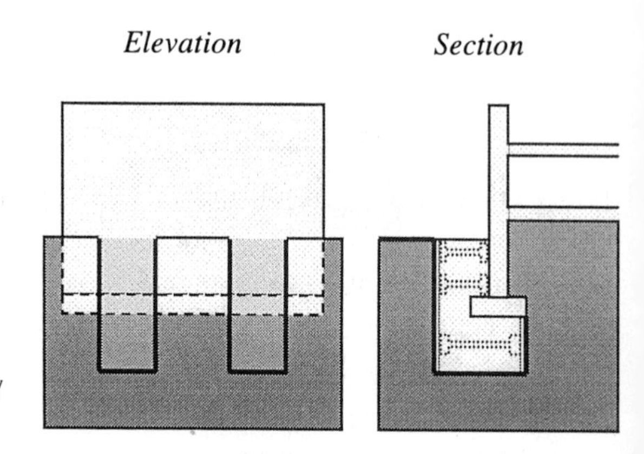

Figure 5.53. Underpinning by the progressive method.

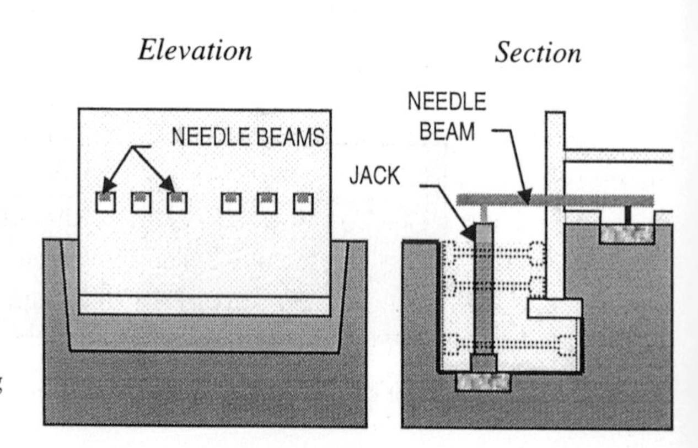

Figure 5.54. Underpinning by the needling method.

The use of micropiles for underpinning is suitable for locations with limited headroom, e.g. in a basement. Micropiles are small diameter (up to 300 mm) piles come in different lengths. An example of underpinning using micropiles around an existing pad foundation is as follows:

- Expose the pad foundation by hacking away the slab, etc. (Figure 5.55a).
- Drive micropiles around the pad foundation (Figure 5.55b).
- Join lengths of micropiles by welding or chemical methods.
- When the required depth is reached, grout the piles through a tremie pipe.
- Insert new bars into the pad foundation and connect to the micropiles (Figure 5.55c).
- Cast concrete around the micropiles and the pad foundation as a new integrated foundation (Figure 5.55d).

5.10. Basement Waterproofing

It is important that the basement is kept free from water not only to ensure that it can carry on with its commercial activities without interruption, but also for its long term structural integrity.

There are generally 3 categories of basement waterproofing systems [27]:

- Membrane system
- Integral system
- Cavity/drainage system

5.10.1. *Membrane System*

It provides a physical barrier forming a tanking system using either sheet membranes or liquid membranes or both to the flow of water [42–43].

Sheet membranes are preformed, factory-made in rolls, which are bonded or cast against the substrate to form a continuous membrane by lapping. Side laps of 100 mm and end laps of 150 mm are common. Lapping may be achieved by torching/flaming, use of bonding compound or self-adhesive membranes. The various generic types of sheet membranes are shown in Figure 5.56.

Figure 5.55(a). Expose the existing foundation by hacking.

Figure 5.55(b). Driving micropiles adjacent to the existing foundation.

Figure 5.55(c). Inserting into the existing foundation new bars which will be connected to the micropiles.

Figure 5.55(d). Concrete cast around the new micropiles and the existing foundation to form a new integrated foundation.

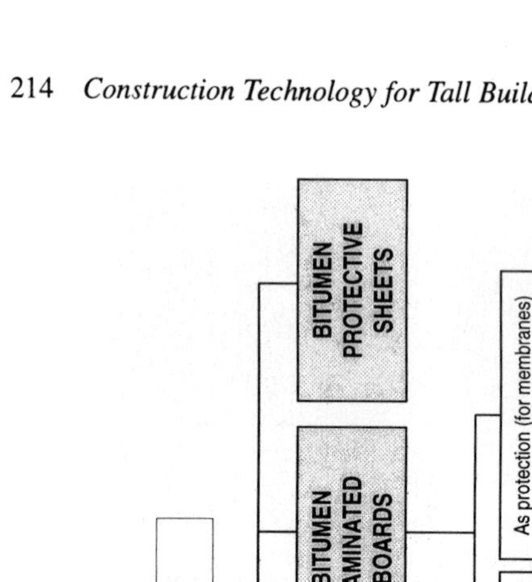

Figure 5.56. Generic types of sheet membrane.

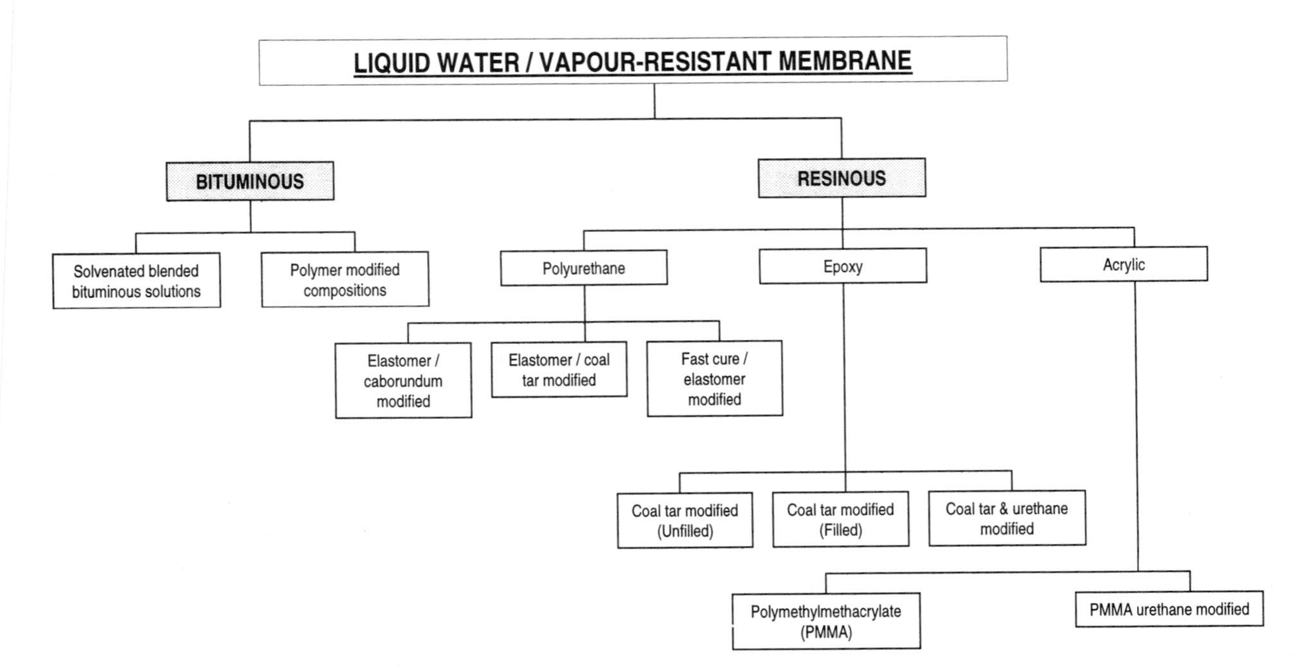

Figure 5.57. Generic types of liquid membrane system.

Liquid membranes come in either one or two-component in liquid or gel form. They are applied either by brush, rollers or spray. Liquid membranes give the flexibility for works on uneven surface and complicated details. The various generic types of liquid membranes are shown in Figure 5.57. Typical details for basement waterproofing using a membrane system is shown in Figure 5.58.

5.10.2. *Integral System*

It provides protection against water penetration based on the use of admixtures with waterproofing properties in the concrete mix to form concrete with surfaces that are repellent to water, and to fill the capillary pores hence reduces the permeability of the concrete. Common admixtures used are the reactive hydrophobic pre-blocking ingredient and silica fume.

5.10.3. *Cavity/Drainage System*

A cavity system allows water to enter the structure, contain and direct it to sumps from where it is removed by drainage or pumping. Cavity system is suitable for cases where tanking system, i.e. physically stopping the water could result in higher water table and/or unacceptable stresses behind the structure. The inner wall is generally non-load bearing and may need to be designed to be free standing to prevent moisture paths occurring across ties. Cavities under floor may be formed from no-fines concrete or proprietary systems such as profiled drainage sheets or purpose-made tiles. Wider cavities may be formed using precast concrete planks to give a raised floor, which may be useful where access is required for maintenance of drainage channels or for servicing pumps (Figure 5.59).

A drainage system consists of high-density polyethylene sheets with dimples formed in the sheets, attached on the external face of the basement wall. The sheets create a cavity between the basement wall and the sheet to allow water to pass through which is then directed through a pipe to sumps and subsequently removed by pumping [44]. Unlike the cavity system, the groundwater is drained off at the external face of the basement wall.

Drawing Not To Scale

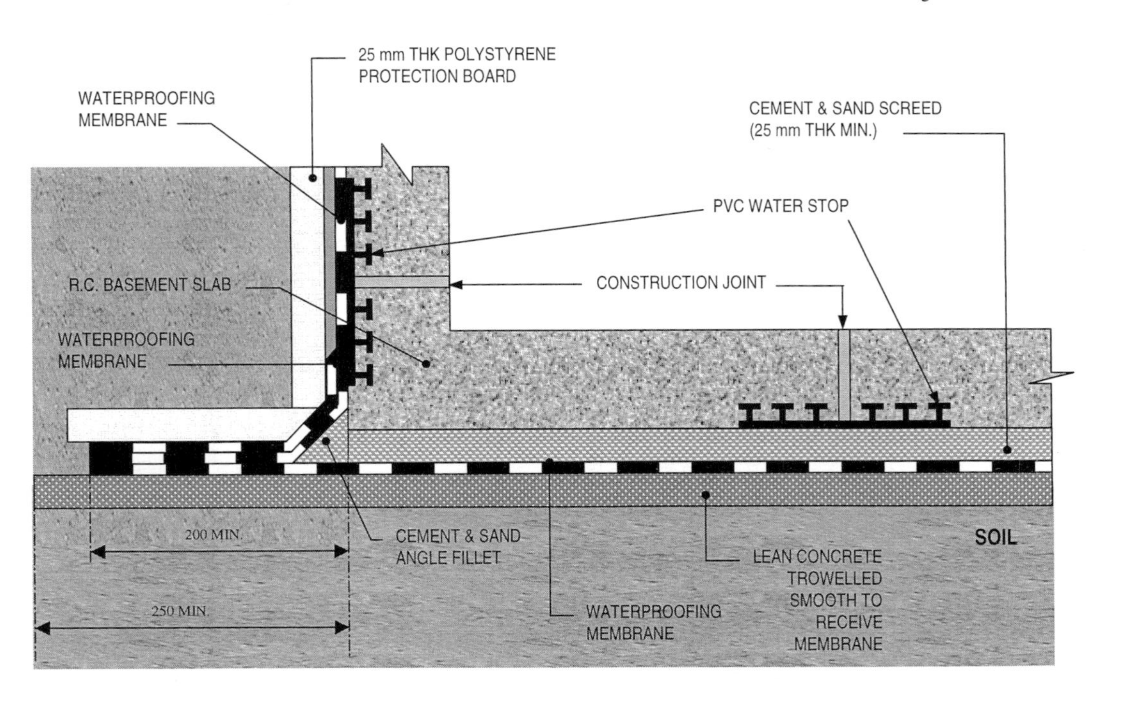

Figure 5.58. Typical details for basement waterproofing using membrane system.

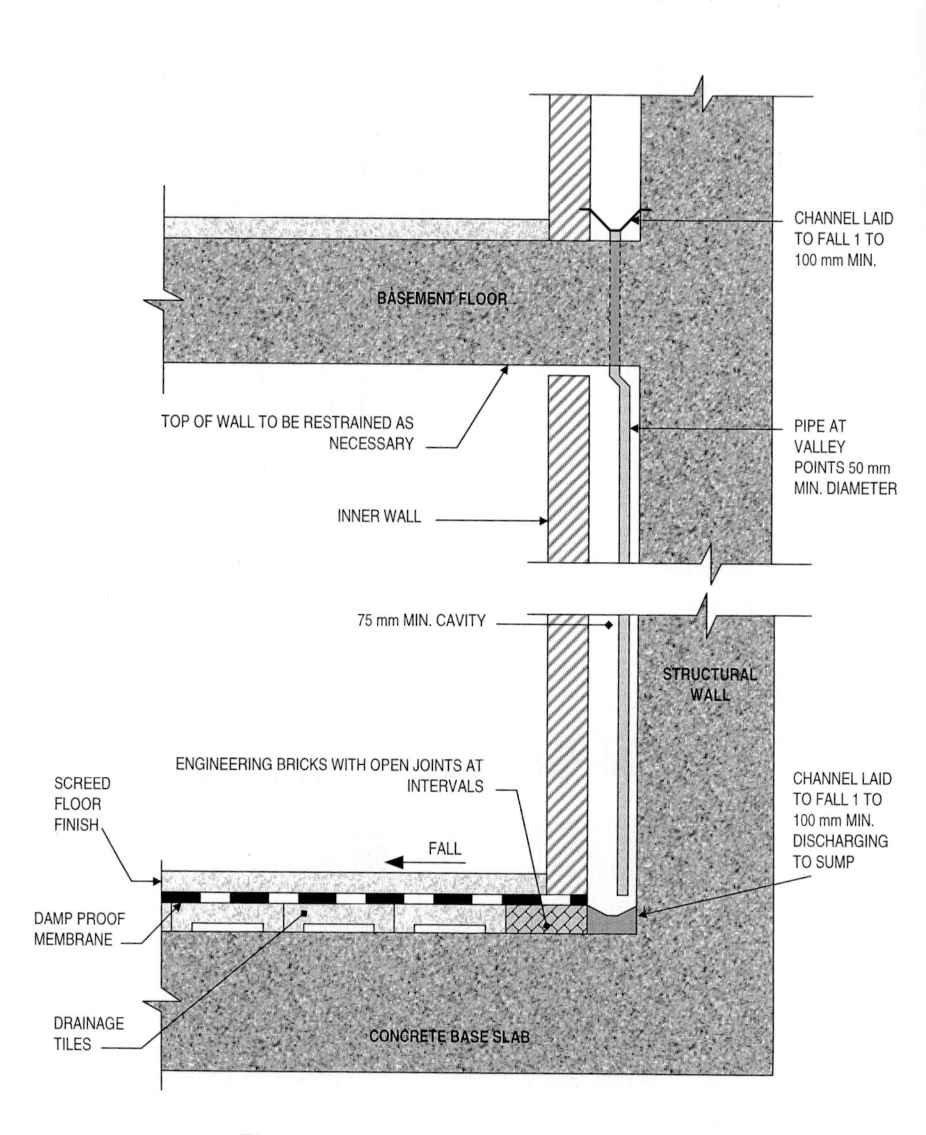

Figure 5.59. An example of a cavity system.

Drawing Not To Scale

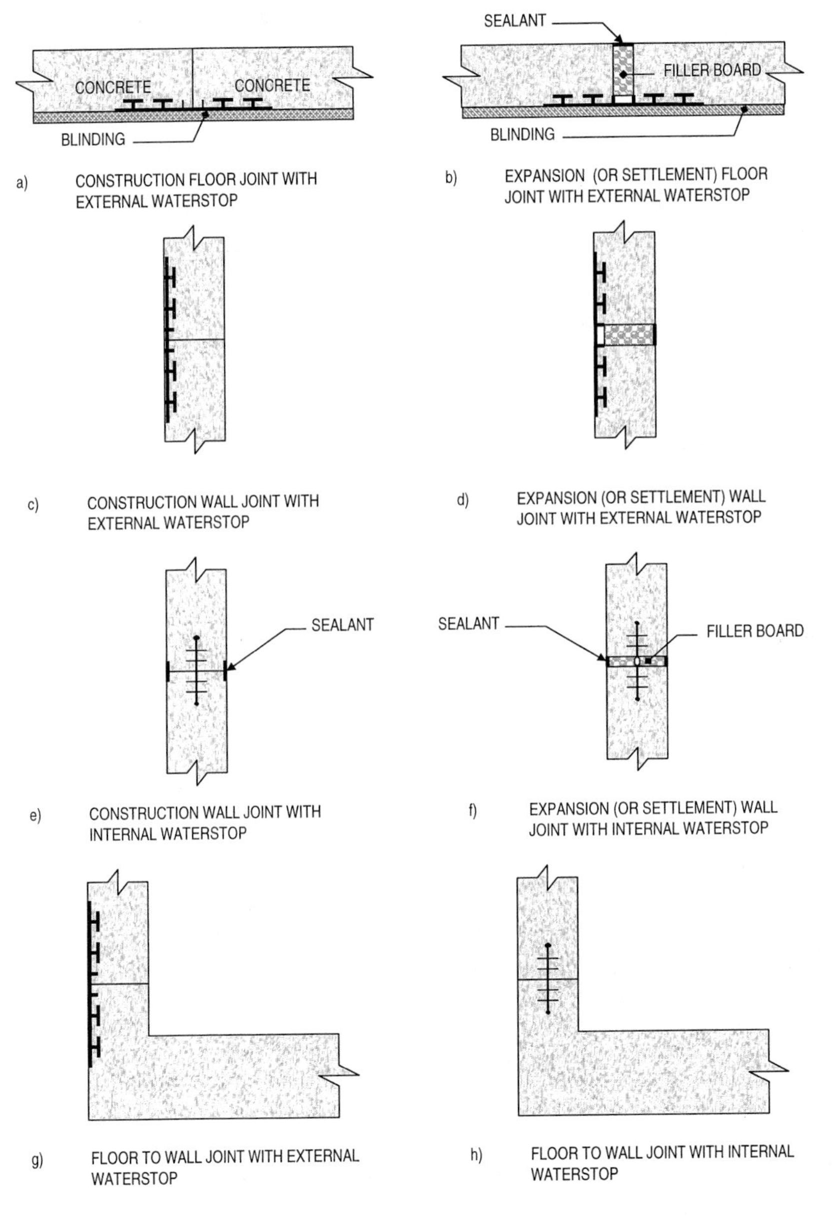

Figure 5.60. Typical applications of waterstops [27].

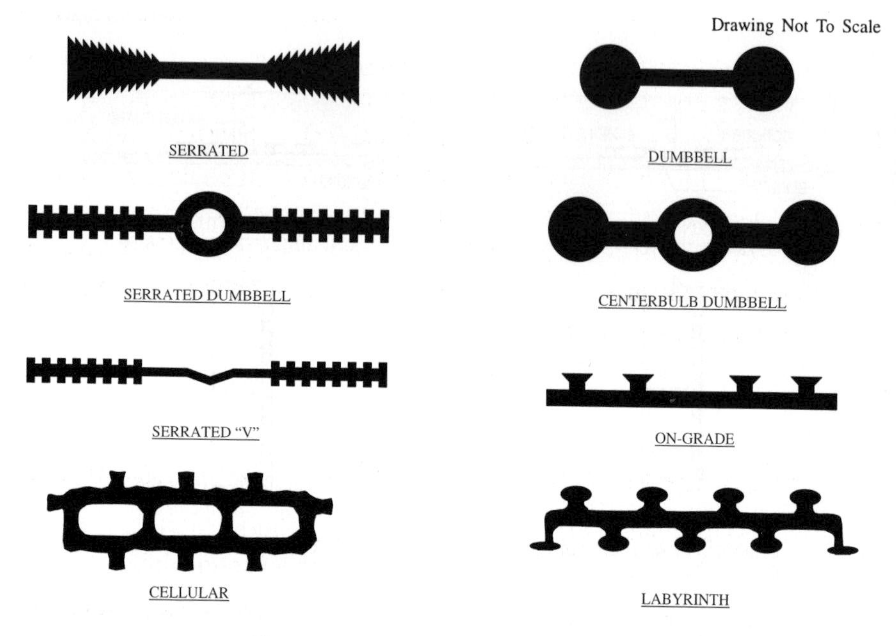

Figure 5.61. Basic types of waterstops.

5.10.4. *Joints and Penetration*

Joints in basement structures are vulnerable to water ingress. The waterproofing membrane bridging all construction joints, expansion joints and any other movement joints shall form a continuous watertight layer. A double layer of waterproofing membrane shall be placed at these locations. This strip shall be at least 200 mm wide, with the edges of the overlaps welded to the first layer and sealed with a liquid mastic compatible with the membrane and/or other methods such as flaming/torching. To allow for movements, the central 100 mm along this strip shall not be bonded to the substrate and shall form a loop of 50 mm. No construction joints or expansion joints shall be placed along the lappings of the waterproofing membrane. Waterstops, i.e. materials crossing joints to provide a barrier, or longer water path, to the transmission of water, shall be used (Figure 5.60). The basic types of waterstop are shown in Figure 5.61.

When embedded fixtures, pipes or cast holes for fixing are required through the waterproofing membrane, temporary plugs shall be inserted to allow the fixing to be located through the membrane. The membrane shall be cut using a cylindrical stamp cutter and not by drilling. Horizontal waterproofing membrane shall be protected with a minimum of 50 mm cement/sand screed placed within 24 hours after the waterproofing membrane is laid. Vertical waterproofing membrane shall be protected with a layer of protective material, such as polystyrene foam, placed within 24 hours after the waterproofing membrane is laid.

References

[1] British Standard Institution, "Code of Practice for Waterproofing", BS8000: Part 4, 1989, UK, 1989.

[2] M. J. Tomlinson, *Foundation Design and Construction*, 5th Edition, Longman, 1986.

[3] M. J. Puller, "Economics of basement construction", in *Diaphragm Walls and Anchorages*, Institute of Civil Engineers, London, 1975.

[4] P. Gray, "Deep basement construction and facade retention", Architect and Surveyor, Vol. 64, No. 10, Nov 1989, pp. 22–25.

[5] Institute of Structural Engineers, "Design and Construction of Deep Basements", London, 1975.

[6] C. Liu and J. B. Evett, *Soil Properties*, 2nd Edition, Prentice Hall, 1990.

[7] F. E. S. Alexander, "Shear strength characteristics of the recent marine clays in Southeast Asia", *Journal S.E.A.S.S.E.*, Vol. 1, No. 1, p. 1.

[8] J. S. Foster, *Structure and Fabric*, Longman, 1994.

[9] R. Holmes, "Introduction to Civil Engineering Construction", College of Estate Management, 1983.

[10] R. Y. H. Tan, "Evaluation of Deep Basement Construction Techniques in Singapore", School of Building & Estate Management, National University of Singapore, 1991/92.

[11] Committee for Waterfront Structures, "Recommendations of the Committee for Waterfront Structures Harbours and Waterways", EAU, 1990, the Committee for Waterfront Structures of the Society for Harbor Engineering and the German Society for Soil Mechanics and Foundation Engineering, 6th English Edition, Berlin, Ernst & Sohn, 1993.

[12] G. C. McRostie, "Performance of Tied-Back Sheet Piling in Clay", Ottawa, National Research Council of Canada, 1972.

[13] British Steel Corporation, BSC Sections, "Steel Piling Products", Middlesbrough, Cleveland, The Corporation, 1978.

[14] D. H. Lee, *An Introduction to Deep Foundations and Sheet Piling*, Concrete Publications, London, 1961.

[15] P. A. Dupeuble, "A new diaphragm wall technology providing mechanical and watertightness continuity", Proceedings of the International Conference on Deep Foundation, Vol. 1, China Building Industry Press, 1986.

[16] R. E. Nex, "Diaphragm walls — Answers to difficult excavations", 2nd Conference on Diaphragm Walls and Ground Anchors, Singapore, CI Premier, 1983.

[17] R. Davies, "Some special considerations associated with the construction of diaphragm walls in marine deposits and residual soils found in South East Asia", Diaphragm Walling Conference, Singapore, 1984.

[18] Drainage Department, Ministry of the Environment, "Code of Practice on Surface Water Drainage", 4th Edition, Singapore, 1991.

[19] M. D. Nalpon, "Basement waterproofing application of performance specifications on Suntec City project", The Professional Builder, Vol. 7, Issue 1, 1st Quarter 1993.

[20] CIRIA — Construction Industry Research and Information Association, "Protection of Reinforced Concrete by Surface Treatment", CIRIA Technical Note 130, CIRIA, London, 1987.

[21] P. C. Hewlett, *Cement Admixtures — Uses and Application*, 2nd Edition, Cement Admixture Association, Longman, UK, 1988.

[22] P. H. Perkins, *Repair, Protection and Waterproofing of Concrete Structure*, Elsevier Applied Science Publishers, London, 1986.

[23] C. H. Lee, "Basement Waterproofing — High Performance Systems", School of Building & Estate Management, National University of Singapore, 1993/94.

[24] G. F. Sowers, "Excavation bracing technology", Proceedings of the International Conference on Deep Foundation, Vol. 2, China Building Industry Press, 1986.

[25] R. Y. H. Tan, "Evaluation of deep basement construction techniques", Proceedings of the International Conference on Deep Foundation, Vol. 1, China Building Industry Press, 1986.

[26] B. P. Williams, "The Design and Construction of Sheet-Piled Cofferdams", Construction Industry Research and Information Association, T. Telford Publications, London, 1993.

[27] British Standard 8102, "Code of Practice for Protection of Structures Against Water from the Ground", British Standard Institute, UK, 1990.

[28] CIRIA Report 139, "Water Resisting Basement Construction: A Guide", Construction Industry Research and Information Association, UK, 1995.

[29] S. W. Nunnally, *Construction Methods and Management*, 3rd Edition, Regents/Prentice Hall, USA, 1993.

[30] J. P. Powers, *Construction Dewatering: New Methods and Applications*, 2nd Edition, Wiley, New York, 1992.

[31] US Department of the Army, "Maintenance of Waterfront Facilities", US Government Print Office, Washington, 1978.

[32] Mass Rapid Transit Corporation, "Code of Practice for Railway Protection", MRTC, Singapore, May 1990.

[33] Slope Indicator, "Applications Guide — Geotechnical, Environmental, and Structural Instrumentation", 2nd Edition, Slope Indicator Company, 1994.

[34] Slope Indicator, "Geotechnical and Structural Instrumentation", Slope Indicator Company, 1998.

[35] F. Harris, *Modern Construction and Ground Engineering Equipment and Methods*, Longman Scientific & Technical, England; Wiley, New York, 2nd Edition, 1994.

[36] R. A. Bullivant, *Underpinning: A Practical Guide*, Blackwell Science, Cambridge, MA, 1996.

[37] J. Smith, "Underpinning and Repair of Subsidence Damage", Camden Consultancy, London, 1994.

[38] S. Thorburn and G. S. Littlejohn (eds.), *Underpinning and Retention*, 2nd Edition, Blackie Academic & Professional, London, 1993.

[39] G. S. Littlejohn (ed.), "Ground anchorages and anchored structures", Proceedings of the International Conference, Institution of Civil Engineers, Thomas Telford, London, UK, 20–21 March 1997.

[40] C. Vipulanandan, (ed.), "Geo-Logan '97 Conference — Grouting: Compaction, remediation, and testing", American Society of Civil Engineers, New York, 1997.

[41] P. P. Xanthakos, *Ground Control and Improvement*, Wiley, New York, 1994.

[42] MACEF, "Tanking Handbook", Mastic Asphalt Council & Employers Federation/Mastic Asphalt Producers Association, UK, 1990.

[43] CIRIA, Report FR/CP/17, "The Design and Construction of Joints in Concrete Structures", Construction Industry Research and Information Association, UK, 1994.
[44] H. Phil, "Waterproofing Concrete", The Concrete Society, UK, 1997.

CHAPTER 6

MATERIALS HANDLING & MECHANISATION

6.1. Materials Handling

Materials handling may be defined as the study of all aspects of the movement of materials on site, with a view to reducing effort and maximising site efficiency. It may be considered as an integral part of the construction method. It can have a very substantial effect on the cost and time of completion of construction.

Materials handling involves lifting, movement and placing. These different activities are coordinated to form one operation. The objective will be to determine the most efficient method of lifting, movement and placing according to the types of load and movement involved. This is adjusted from phase to phase especially for congested site [1–3].

6.1.1. *Selection of Equipment*

The selection of the types of equipment to be acquired, whether to hire or to purchase, etc., requires consideration of the following:

- — capital investment
- — duration of actual utilisation
- — effect on the speed with which the work is completed
- — re-usability of equipment on the same job as well as on other jobs
- — erection and dismantling
- — transportation, mobilisation required

— site constraints
— safety
— maintenance
— environmental issues

6.1.2. *Classification of Materials*

Materials can be classified according to [4]:

— Characteristics of materials
  • bulk (such as gravel, sand)
  • packaged and individual items (bags or drums of cement, sinks, window/door units, curtain walling panels, etc.).
— Physical state
  • effect of temperature, humidity, abrasion, etc.
— Possible chemical action
  • corrosion, deterioration, decomposition, ignition, explosion, etc.
  • Volume to be handled.

6.1.3. *Storage*

— Shelter against condensation for cement storage, corrosion for metal roofing, security for expensive items such as cabinets, door locks, etc.
— Support off the ground, shelves, racks, bins, pallets.
  • Temporary or permanent. In many cases, the storage system is adjusted with the progress of a building.
  • Capacity of structure or building — Floor load and distribution of loading, accessibility through loading docks, hoists, etc.

6.1.4. *Protection*

— Physical
— Chemical

As an example, curtain walling panels may be protected by plastic framing to give physical and chemical protection, or wrapped in plastic film to protect against dirt, moisture, etc. Marble may be protected by covering with sawdust after laying, etc.

6.1.5. *Packaging*

The packaging depends on both the protection and handling requirements. The common packaging methods are crates, drums, barrels, bulk handling equipment (cement, aggregate, etc.), pallets and special packaging (e.g. precast prefabricated items such as hollow core slabs), etc.

6.1.6. *Handling*

— Equipment availability and control in relation to demand, e.g. crane allocation.
— Special handling equipment in relation to the type of packaging, e.g. grabs, vacuum and magnetic lifting gear, etc.
— Temporary support and bracing, e.g. bracing during lifting of precast concrete units, spreader, lifting beams, etc.
— Multiple handling to be minimised.
— Progressive breakdown of bulk and package items for allocation to final position, e.g. bricks, tiles, doors, etc.

Table 6.1 describes the handling, storage, protection and wastage of different materials of a typical site.

6.2. Mechanisation

Mechanisation is the use of machines or mechanisms in place of traditional labour intensive methods. It has been regarded as another form of industrialisation [5]. The shortage and high cost of labour have made the selection of an efficient mechanisation critical. Mechanisation is most suited to

Table 6.1. Schedule of material controls.

Material	Handling	Storage	Protection	Loss or Waste
Topsoil & hardcore	• Bulldozer or scraper box to storage areas. • Strip with mechanical plant. • Load into trucks. • Remove to storage areas.	• In mounds as required.	• Avoid storing or discharging chemical or deleterious matter adjacent to topsoil mounds.	• Strip taken too deep/too shallow. • Storing in excessive mounds. • Indiscriminate handling. • Double handling.
Fine and coarse aggregates	• Delivered by high-sided truck. • Check quality, grade and quantity. • Tip into prepared bays or selected areas. • Transfer by grab, mechanical shovel or power-assisted equipment on weigh batcher. • Transport by bucket, hopper, dumper and barrow.	• In prepared bays adjacent to mixer in selected grading and particle size. • On concrete base laid to falls.	• Cover against rain. • Avoid contamination with mud, clay or oil. • Check calibrations on weighing equipment.	• Indiscriminate handling. • Contamination of any kind. • Failure to trim stocks. • Used as site dressing or to fill site voids.
Cement	• Delivered by tanker and pumped into silo. • Offload by forklift or crane. • Use bagged cement in order of delivery.	• In special silo adjacent to mixer. • On raised platforms in shed or in the open. • Cover completely with polythene sheet.	• Avoid accidental bursting of sacks. • Restrict rising moisture affecting cement in store. • Cover against rain.	• Humidity causing materials to lump together. • Dampness initiating setting. • Indiscriminate handling causing bags to burst. • Failure to use deliveries in rotation. • Pilfering. • Leaving stocks unused.

Table 6.1. *(Continued).*

Material	Handling	Storage	Protection	Loss or Waste
Concrete	• By crane, skip, hopper, dumper or barrow. • Special site truck fitted with paddle agitator. • Transfer by concrete pump from delivery to placing point. • Ready-mixed concrete. • Ensure good access roads to discharge point or hoisting area.	• In special hoppers or skips when transferring to upper levels. • On suitably sized timber or metal sheets with side supports to restrict concrete spreading.	• Avoid deleterious matter affecting concrete. • Cover where necessary to avoid rain.	• Overloading containers. • Overestimating quantity required. • Late delivery of ready-mixed concrete. • Failure to plan work in hand. • Mix not to specification. • Careless placing. • Unspecified use of concrete.
Formwork	• Clean formwork between operations. • Remove all nails and screws. • Keep formwork together including bolts and fixings. • Hoist with care. • Do not drop formwork when stripping from concrete.	• Arrange formwork in sections for identification. • Lay flat where possible. • Stack vertically where necessary. • Protect edges.	• Avoid damage from site traffic. • Oil and clean formwork.	• Mishandling formwork when stripping from concrete. • Dropping bolts or fixings. • Using formwork for other purposes.

Table 6.1. (*Continued*).

Material	Handling	Storage	Protection	Loss or Waste
Reinforce-ment	• Offload as close to fixing point as possible. • Lay on timber skids to keep steel above ground. • Check type and quantity delivered. • Cut with proper tools. • Use correct size mandrels for bending bars. • Hoist bundles of bars or sheet reinforcement with care.	• Mat reinforcement to be laid flat. • Bar reinforcement to be stored in racks. • Separate bent bars into types according to bending schedule. • Retain all off-cuts.	• Guard against oil, mud or loose scale on steel. • Avoid distortion of bars by site traffic or mishandling.	• Careless handling. • Leaving bars behind in fixing areas. • Failing to fix steel as specified. • Cutting bars incorrectly.
DPCs & Polythene Sheeting	• Unroll quantity to be used only. • Replace surplus amounts in safe storage. • Use proper knife for cutting to length. • Use correct width as specified.	• Store in cool, dry place. • Keep different widths in separate groups.	• Store away from site traffic. • Do not drop rolls.	• Leaving unused rolls lying around. • Using incorrect widths or cutting unnecessarily.

Table 6.1. (*Continued*).

Material	Handling	Storage	Protection	Loss or Waste
Bricks	• Delivered in packs or on pallets. • Offload by vehicle-crane, forklift or mobile crane. • Transfer on site by forklift, dumper, crane, hoist or elevator. • Sort all chipped or damaged bricks and set aside. • Do not tip facing bricks. • Cut bands holding packs with proper cutters.	• In selected stock-piles adjacent to work in progress. • On prepared base of hardcore, concrete or ash. • Store according to type. • Keep stacks within reasonable height.	• Guard against indiscriminate handling methods. • Avoid chipping. • Cover stacks to protect against rain. • Contamination from soluble salts or sulphates.	• Using facings for common work or as supports and packings. • Faulty workmanship. • Double handling.
Precast items	• Offload and transport with care using correct equipment and any lifting points provided in the casting. • Protect edges where slings or cables are used for fitting. • Lifting equipment to be fully capable of raising the load.	• Stack on level base according to size and type. • Use softwood bearers placed at equally spaced centres. • Wall panels stacked vertically in special frames. • Store close to fixing area where possible.	• Cover stocks to avoid impact damage. • Cover to avoid weathering from rain, etc. • Projecting reinforcement should be covered to avoid rust stains to units.	• Fragile nature — damage implies replacement of the whole unit.

Table 6.1. *(Continued)*.

Material	Handling	Storage	Protection	Loss or Waste
Door and window frames	• Delivered by manufacturer. • Offload by contractor. • Stack carefully in compound. • Transfer to working area required for fixing.	• Frames to be laid flat where possible and fully supported, particularly at joints. • Placed on bearers to raise off ground. • Stack in groups according to type.	• Primed as soon as possible after manufacture. • Where frames have been pressure-treated they should be allowed good air-circulation before priming.	• Damaged frames.
Doors	• Delivered in large loads. • Check quantities. • Offload by contractor. • Door and doorset may be on pallets. • Use forklift for offloading. • Site trailer and hoist for distribution.	• Store in area where temperature not likely to produce sudden changes. • Doors laid flat on timber platform or stacked vertically. • Arrange in lots according to size and thickness. • Doorsets laid flat should have "pull" side uppermost. • Stack under cover at all times.	• Remove polythene covers before application of door finish. • Prime hardboard doors on delivery. • Store in locked premises.	• Poor stacking causes bowing or twisting. • Damp conditions can affect joints and glue. • Theft likely. • Damage generally results in total loss.
Nails, bolts and screws	• Delivered in boxes and sacks. • Issue in plastic bags in correct quantities and sizes. • Return surplus to store.	• Use boxes and bins to retain different sizes, etc.	• Retain in dry storage to avoid corrosion.	• Indiscriminate use on site.

Table 6.1. (*Continued*).

Material	Handling	Storage	Protection	Loss or Waste
Glass	• Delivered on specially equipped vehicles. • Offload by glaziers. • Size of pane and type of glass affect handling procedures. • Large panes carried by slings or special suction pads with handles. • Edges of glass should be covered.	• Store must be clean and dry. • Lay glass on edge in racks with felt or rubber covered base. • Glass to be fully supported with bearers at right-angles to glass.	• Full support required to avoid sheet bending. • Guard against any impact. • Avoid splashes of cement or plaster, cleaning down may create surface scratches. • Mark glazed areas with whitening or plastic stickers.	• Scratches. • Breakages. • Leaching. • Inaccurate cutting. • Faulty glazing.
Putty and mastics	• Delivered by supplier in drums. • Do not use turps in putty to make workable. • Keep different materials separate.	• Store on racks on staging. • Avoid extreme temperatures. • Keep lids firmly closed. • Issue materials in correct quantities. • Keep separate drum for surplus returned.	• Follow manufacturers' directions for use. • Guard against rain. • Pilfering.	• Use for specified purposes only. • Leaving drums open. • Using too much material for glazing purposes.
Sanitary-ware and fittings	• Delivered by supplier. • Offload by contractor directly into store. • Check quantities delivered. • Deliver items to working area and store inside locked buildings.	• Retain in racks or on platforms. • Keep smaller items in cartons. • Do not stack units too high. • Avoid items projecting into route of site traffic.	• Maintain protective covering. • Avoid chipping & impact to glazed surfaces. • Avoid film of plaster or cement on glazed surfaces.	• Damage to these items inevitably means they are written off. • Check carefully on completion to avoid subsequent claim for breakages.

Table 6.1. (*Continued*).

Material	Handling	Storage	Protection	Loss or Waste
Gutters and rainwater pipes	• Fix as work proceeds to avoid damage to structure. • Check angles, offsets and bends. • Finish each section once work has commenced.	• Store in racks in compound. • Keep plastic gutters etc. away from heat sources. • Form components in separate stacks according to type. • Keep fittings in boxes or crates.	• Avoid impact damage. • Do not leave work partly fixed. • Provide light waterproof covers to stack. • Plastic components must be stored with support to length of pipe gutter.	• Pilfering. • Damage by site traffic. • Vandalism. • Failure to collect surplus lengths and fittings. • Plastic materials brittle in low temperature.
Paints	• Offload directly into store. • Replace lids securely after use. • Apply from kettles not the tin. • Do not mix different brands.	• Place tins on racks in well ventilated dry store. • Keep different paints in specific grouping. • Display NO SMOKING signs. • Keep store clean.	• Avoid extreme temperature. • Seal cans and tins after use.	• Poor storage. • Poor distribution. • Leaving surplus in kettle.

excavating and earthmoving operations, operations related to the structure of the building (e.g. concrete mixing, transport and placing) and the handling of materials and components. Mechanisation can reduce building costs and speed up the construction process. In particular, off-site production of materials and components is usually highly mechanised. However, complete mechanisation of the construction process is not possible. The plant has to be moved from job to job and has to remain versatile; different jobs give different problems, and site conditions vary. Machines usually should be operated for a reasonably high proportion of their time on site and close to full capacity for their use to be really economical [6, 7].

In Singapore, mechanisation involving a low capital outlay has been as fast as in most other places, although the previous low cost of labour had meant that some equipment and techniques available had not been fully utilised. However mechanisation involving a high capital outlay, e.g. tower cranes, has been slower due to some instability of the market, contract conditions, and the general structure of the construction industry. Recent government incentives over the last few years have encouraged higher capital investment in plants but the downturn in the industry has left plants idle and contractors have generally over invested in relation to plant returns. Capital investment tends to be more critical to contractors than other forms of investment and has to be considered very carefully. Contractors have been quite willing to invest in non-mechanical equipment such as formwork or scaffold systems (although even these required heavy capital commitment), but less keen to invest in plants which require skilled maintenance. Investment in equipment requires a stable market, providing some continuity of construction activity, or some incentive. The use of mechanical plants makes possible early completion; and consequently early recovery and re-investment of capital, resulting in better rates of return. It also makes the contractor more vulnerable when delays occur (equipment lying idle does not bring any returns but interest is lost on the money invested in the equipment).

6.3. Earthmoving

Earthmoving is the process of moving soil or rock from one location to another and processing it so that it meets construction requirements of location, elevation, density, moisture content and so on. Activities involved in this process include excavating, loading, hauling, placing (dumping and spreading), compacting, grading and finishing. Efficient management of the earthmoving requires accurate estimating of work quantities and job conditions, proper selection of equipment and competent job management.

6.3.1. *Earthmoving Fundamentals*

The efficiency and profitability of a construction site are to a large extent affected by the selection of equipment. The selection requires a good understanding of the fundamentals of earthmoving [8].

Loadability is the term used to define the ease with which material may be dug and loaded; the choice of the type of earthmoving equipment depends to a great degree on the loading characteristics of the material.

Weight or volume determine the carrying capacity of equipment; dense material may give a full load to the equipment before the volume is filled. Weight also affects the manner of loading of a scraper, pushing by a bulldozer or casting by a grader, etc. Weight affects the ability of equipment to turn, manoeuvre and haul.

Swell is the volume increase in material when removed from its natural state, expressed as a percentage of the increase in volume.

Load Factor is the percentage decrease in density of a material from its natural to a loose state.

Compactability is the compression of soil from its loose to its final state at the point of placing after moving.

Power Required is the power required to do work (e.g. pulling or pushing a load). The amount of power available compared to the required power can determine how fast a machine can operate.

Rolling Resistance is the retarding force of the ground against the wheels of a vehicle. This force must be overcome before the vehicle will move.

It is affected by internal friction of a tyre, tyre flexing, penetration of the tyre into the soil and load being carried by the tyre. Track type tractors are not affected by rolling resistance.

Grade Resistance is the retarding force of gravity which must be overcome to move a vehicle uphill. When a vehicle travels downhill it receives grade assistance due to gravity. It is most frequently measured as percent slope (ratio between vertical and horizontal distance). More power is required to overcome grade resistance.

Power Available — horsepower, watts, newtons metres per second or joules per second are the primary measure of the power available. These must be considered in terms of rate of doing work and consequently the speed at which the machine moves and exerts a given force.

Cycle Time is the amount of time it takes a machine to complete a set of operations, e.g., loading, hauling, dumping and returning to a position ready to start loading again. The time taken to get a job done is determined from the time taken for a machine to make one round trip or cycle.

Production is governed by the number of trips per hour and the payload per trip. Profitable production means moving large quantities of material at the lowest possible cost.

Efficiency Factor — An equipment which operates effectively for 40 to 50 minutes/hour has an efficiency factor of 0.67 to 0.83.

There are 3 principal conditions or states in which earthmoving materials may exist:

(a) Bank: Material in its natural state before disturbance. Often referred to as 'in-place' or '*in situ*'. A unit volume is identified as a bank cubic meter (Bm^3).
(b) Loose: Material that has been excavated or loaded. A unit volume is identified as a loose cubic meter (Lm^3).
(c) Compacted: Materials after compaction. A unit volume is identified as a compacted cubic meter (Cm^3).

Table 6.2 and Figure 6.1 show the volume change during earthmoving [8].

Table 6.2. An example of weight and volume change in earthmoving.

	Loose (kg/m^3)	Bank (kg/m^3)	Compacted (kg/m^3)	Swell (%)	Shrinkage (%)	Load factor	Shrinkage factor
Clay	1370	1780	2225	30	20	0.77	0.80
Rock	1815	2729	2106	50	−30*	0.67	1.30*
Sand	1697	1899	2166	12	12	0.89	0.88

*Compacted rock is less dense than natural rock
Swell (%) = [(Bank/Loose) − 1] × 100
Shrinkage (%) = [1− (Bank/Compact)] × 100
Load Factor = Loose/Bank
Shrinkage Factor = Bank/Compact

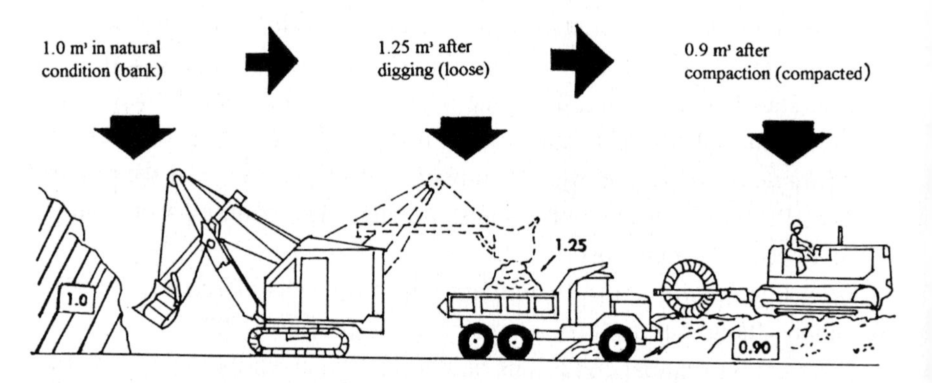

Figure 6.1. Typical soil volume change during earthmoving [8].

6.3.2. *Earthmoving and Excavation Equipment*

The process of excavation includes both surface and deep excavation and consists of three basic steps — excavation, transportation and placing. These steps may be carried out by one machine only or by two, three or more pieces of equipment, depending on a number of factors: type of material, quantity, distance to be traveled, lift, the condition in which it is to be left, nature of excavation (size, accessibility), condition and gradient of site, cost of available equipment, etc. Machinery is expensive and so should not be

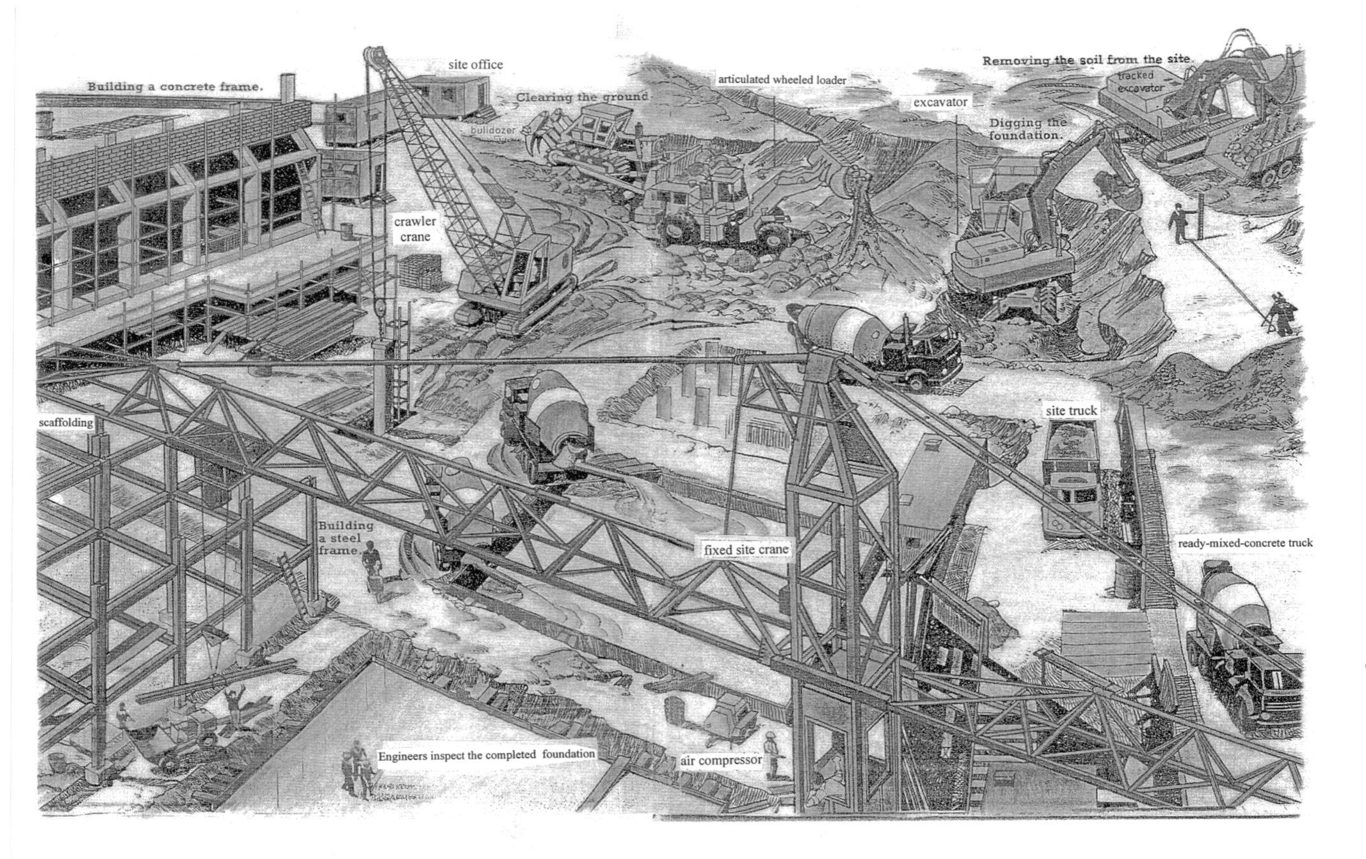

Figure 6.2. The common machinery on a building site.

left idle for long and must be carefully selected for the size of the task and maximum return. Some equipment have interchangeable parts using the same basic machine while others are more specialised and non-adaptable. Machines may be wheel or crawler mounted — a factor that can have a considerable effect on the cost and efficiency of the machine [9, 10].

There are many types of equipment available for excavation, cutting, etc. Unlike civil engineering works, normally only a few major equipment are used on a building site due to site constraints, nature and the extent of the work. Figure 6.2 shows the common equipment employed on a typical building site.

Shovel: This is used to excavate against a face or bank. It may have an open top bucket or dipper with bottom opening door fixed to an arm which slides and pivots on the jib of the crane. It is suitable for excavating clay, chalk, friable materials and loosened/broken rock and stone (Figure 6.3).

Backacter (*back hoe, trench hoe*): This is similar to the face shovel except that the dipper stick pivots at the end of the jib and the dipper or bucket works towards the chassis and normally has no bottom door but is emptied by swinging the dipper stick up and so inverting the bucket. It can excavate to more than 10 m below ground level (Figure 6.4).

Figure 6.3. A typical face shovel.

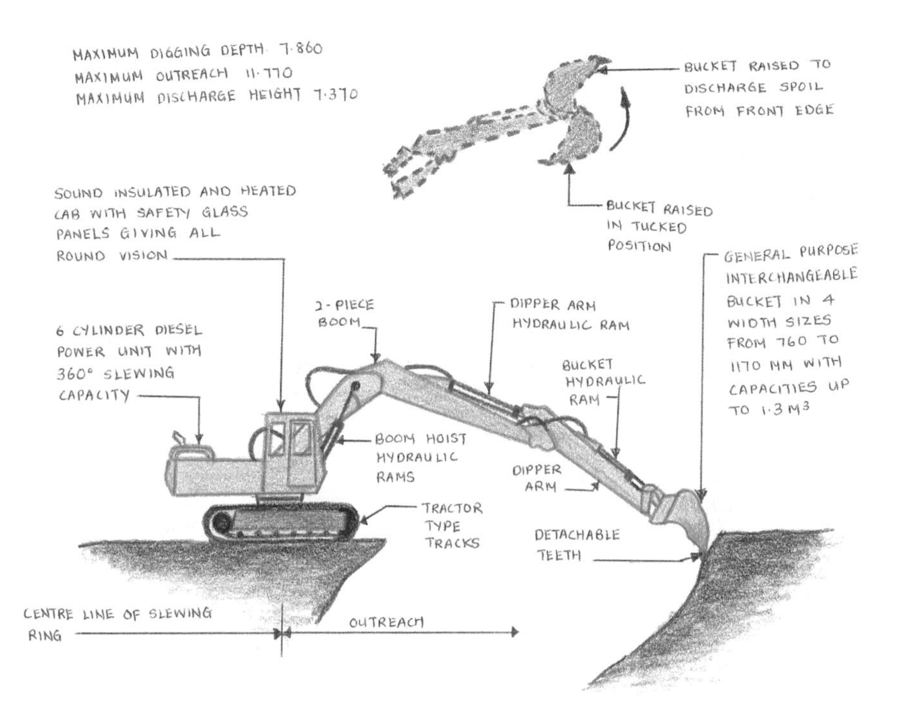
MAXIMUM DIGGING DEPTH 7.860
MAXIMUM OUTREACH 11.770
MAXIMUM DISCHARGE HEIGHT 7.370
BUCKET RAISED TO DISCHARGE SPOIL FROM FRONT EDGE
BUCKET RAISED IN TUCKED POSITION
SOUND INSULATED AND HEATED CAB WITH SAFETY GLASS PANELS GIVING ALL ROUND VISION
GENERAL PURPOSE INTERCHANGEABLE BUCKET IN 4 WIDTH SIZES FROM 760 TO 1170 MM WITH CAPACITIES UP TO 1.3 M3
2-PIECE BOOM
DIPPER ARM HYDRAULIC RAM
6 CYLINDER DIESEL POWER UNIT WITH 360° SLEWING CAPACITY
BUCKET HYDRAULIC RAM
BOOM HOIST HYDRAULIC RAMS
DIPPER ARM
TRACTOR TYPE TRACKS
DETACHABLE TEETH
CENTRE LINE OF SLEWING RING
OUTREACH

Figure 6.4. A typical hydraulic backactor.

Figure 6.5. Attachment of jack hammer on a backactor.

Crane and grab (*grabbing crane, clamshell etc.*): It uses buckets or grabs of different types for different materials or purposes, typically of two half buckets hinged and pivotting on a frame which is suspended by cable from the jib. The grab is closed when suspended but opens on dropping and closes under the soil as it lifts it. The jaws open for discharge by means of a trip cord or automatically as the grab is laid to rest. The grab is used for deep excavations of limited area for all soils except rock (Figure 5.5 of Chapter 5).

Attachments are available to equip a piece of machinery to perform other functions, e.g. a hydraulic breaker (Figure 6.5) from a backactor.

6.4. Horizontal and Vertical Movement

The common machineries employed in a typical site are discussed.

Dumper: This is used mainly for transporting excavated materials (Figure 6.6). It usually has a capacity of 0.3 m^3 to 5 m^3.

Figure 6.6. A typical dumper.

Lorry: The most general hauling vehicle on a site. It is usually side tipped or rear tipped for easy unloading (Figure 6.7).

Wheel barrow: (Figure 6.8) and *Fork Lift* (Figures 6.9 and 6.10) both are common basic means of horizontal transportation.

Hoists: These are intended for vertical movement only and so are only able to move in one direction (vertical). The maximum reachable height is virtually unrestricted in theory but depends on the particular hoist design. Figures 6.11 and 6.12 show passenger hoists with tracks attached to the main structure for stability.

Gondola/Swinging Stage: This provides vertical movement for workers working on the external finishes of a building (Figure 6.13). The gondola is powered by motors which are either situated at the top of the building or on the gondola itself.

Elevators: These move on tracks. They are more stable and have higher capacities as compared with a gondola (Figure 6.14).

Cranes: These are generally capable for moving objects in all directions. Various attachments are available for a crane to perform different functions.

Common types of cranes used on building sites are the truck mounted crane (Figure 6.15), the mobile crane (Figures 6.16 and 6.17), the tower crane (Figure 6.18) and the climbing crane (Figure 6.19).

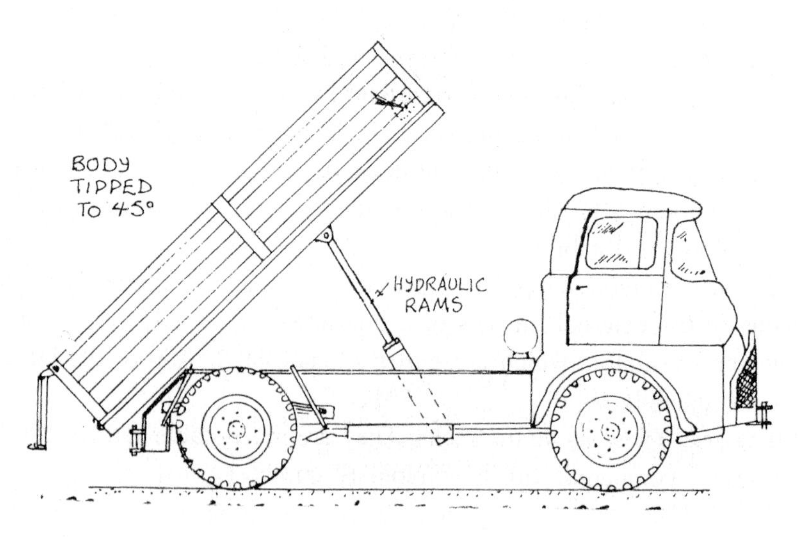

Figure 6.7. A typical lorry with hydraulic rams.

Figure 6.8. A typical wheel barrow.

Figure 6.9. A typical manual fork lift.

Figure 6.10. A typical hydraulic fork lift.

Figure 6.11. Passenger hoist attached to the structure.

Figure 6.12. Passenger hoist attached to the structure.

Figure 6.13. Gondola or swinging stage for works on external finishes.

Figure 6.14. Elevator with greater stability and surface coverage.

Figure 6.15. A typical lorry mounted crane.

Figure 6.16. A typical mobile crane with a flying jib.

Figure 6.17. A typical telescopic jib.

Figure 6.18. Tower crane with attachments to the building structure.

Figure 6.19. Climbing crane located within the building.

Figure 6.20. A typical concrete pump.

Concrete is normally pumped up from the ground to the required height using a mobile concrete pump (Figure 6.20). In cases where the height of a concrete pour exceeds the capacity of the mobile concrete pump, a crane is used to lift the concrete using a hopper to the required height. Figures 6.21 and 6.22 illustrate the transportation of concrete from a ready-mix truck to a hopper, which is in turn lifted to the required height by the use of a crane.

Figure 6.21. Ready-mix concrete truck discharging concrete into a hopper to be hoisted by the crane.

Figure 6.22. Concrete in hopper being lifted by a crane to the place of discharge.

In cases where the point of loading or delivery is on the intermediate floors rather than the topmost open floor, receiving platforms are used for the loading/unloading (Figure 6.23(a)). Unwanted materials and debris can be collected in a dumper placed on a receiving platform (Figure 6.23(b)) and disposed off using a crane. Alternatively, a rubbish chute can be used (Figure 6.24).

(a)

(b)

Figure 6.23. (a) Receiving platforms for loading/unloading of materials through the use of a crane. (b) A bin loaded with debris on the receiving dock waiting to be lifted away.

Figure 6.24. Rubbish chute to direct disposals from various floors to a bin on the ground floor.

The crane may be fitted with either a derricking (or luffing) jib or a fixed horizontal (saddle or hammerhead) jib. Counterweights are generally not adjustable with the angle of luffing or location of hook trolley, although some may be adjusted into the mast during the climb up. Tower cranes are tied to the main structure for stability (Figures 6.25 and 6.26).

Shorter tower cranes are often self-erecting whilst larger ones require the assistance of mobile cranes. Some smaller cranes have no provision for mast lengthening and simply fold down ready for towing. For others the mast and parts are erected by mobile cranes.

Figure 6.25. Tower crane attached to the structure using ties at various intervals.

Figure 6.26. Details of a tie connected to an anchor embedded in the concrete slab.

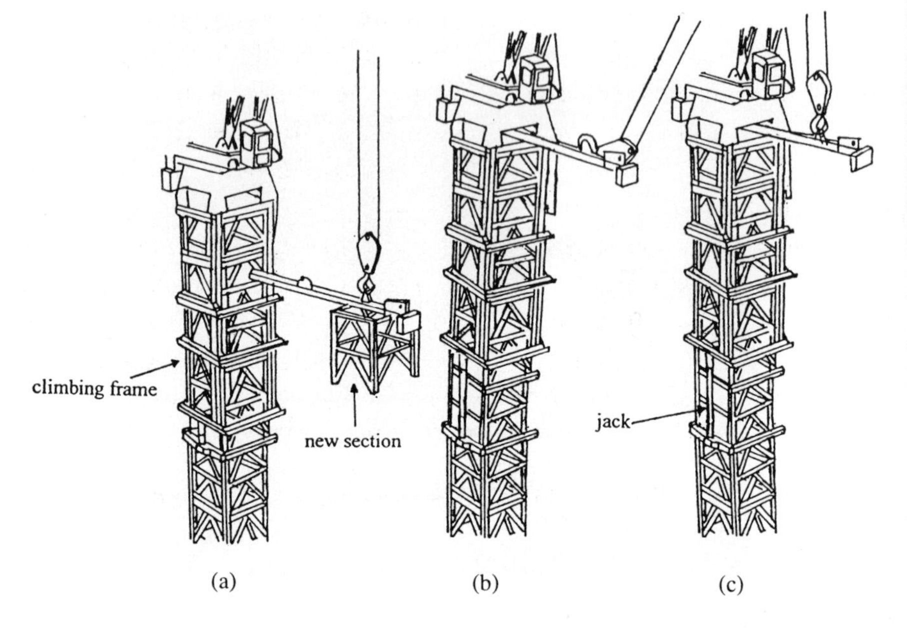

Figure 6.27. Self-raising tower crane mast.

Some, however, are self-erecting in the sense that they can lengthen their mast by either:

(a) inserting inner tower units through one side below the working platform after jacking the crane on an external sliding tower piece.
(b) inserting tower units through the top of the crane (in this case the platform and parts are attached to a frame that fits around the tower).
(c) jacking the platform and parts upon an inner sliding mast section and then attaching two external L-shaped sections to the main tower.
(d) mounting the platform and parts on an inner tower that fits into an outer tower fixed to the ground. Tower units are inserted through the side of the outer tower at ground level.

Cranes may also be climbing in the sense that they can climb up a shaft in the building using only a limited number of mast sections. The crane sits on a slewing ring on an outer frame that is braced on three sides but open on the fourth. The frame extends down over slightly more than two units of the tower. It picks up a tower unit with its jib, luffs in and hooks the unit to a monorail immediately adjacent to the open side. The outer frame is then jacked up hydraulically and the tower unit is slipped in.

Figure 6.27 shows an example of a self-climbing crane. The crane first lifts an additional tower section together with a monorail beam and trolley. The monorail beam is fastened to the crane's turntable base and the new section is trolleyed close to the tower. The turntable is unbolted from the tower. The climbing frame and the new section is inserted into the frame using the monorail beam trolley. The climbing frame is then lowered and the new section is bolted to the tower and the turntable base.

6.5. Factors Affecting the Selection and Location of Tower Cranes

6.5.1. *Selection Criterion for Tower Cranes*

Carrying capacity

In every construction project, it is important that sufficient load carrying capacity is available within the operating radius to pick up and deposit the loads. The engineer or the project manager will firstly have to determine the various loads to be lifted and the maximum load for the whole construction. Apart from this, the load carrying profile over the jib length should also be identified.

Maximum coverage

One of the major criteria is to ensure that the tower crane covers 100% of the required building area. The length of the jib should be chosen so as to reach the furthest point of delivery. Any excess length will result in the jib being underutilised and jeopardise the safety of the public.

Sufficient space for assembly, erection and dismantling of crane

The proposed tower crane position must be reachable by a mobile crane of an appropriate size. Apart from this, space for staging and assembly of components is also required. At the end of construction, access must also be provided for the mobile crane to disassemble the tower crane. If the tower crane is expected to climb down at the completion of its work, room must be left for it to clear away from any adjacent structure.

Ability to weathervane freely

As far as possible, the tower cranes must be able to turn 360 degrees freely without any obstruction. They must also be able to "luff" vertically, especially for congested sites surrounded by existing buildings. The luffing boom crane is also frequently the most viable choice when two or more tower cranes are employed at close proximity within the same site.

Building height

The height of the tower crane depends on the height of the building and its adjacent building over which the crane jib passes, or the height of the tallest crane over which the crane jib passes. The height at which a climbing crane is preferred to an external one is where either the extra cost of the mast sections is outweighed by the additional building structure, jacking up costs, and extra dismantling costs, or the building is higher than the supported height to which a crane can stand.

Cost of acquisition

Tower cranes, like any other mechanical plant or machinery incur a substantial capital investment. However, crane selection should not be based solely on acquisition price, because the true cost of a tower crane includes many other items that can dwarf differences in price. With this in mind, one should consider all the various factors such as the company's financial status, size and duration of the project, and also the types of projects handled.

Availability of crane

After the maximum radius and critical lift for the crane have been calculated, it is necessary to check whether tower crane with such a jib length and capacity is available. Following which the contractor will then be able to select the most appropriate crane to suit his project from the load diagram [11–15].

6.5.2. *Location Criterion For Tower Cranes*

Site area

The ideal location for a tower crane is outside the building footprint to which there is vehicular access to within at least 10 metres for the assembly and dismantling operations. Erection and dismantling costs are then at their minimum. It has been shown that most contractors will usually ensure that the delivery of all materials, plants and equipment to be lifted by the tower crane would be within 10 to 15 metres from the crane base [16].

In sites with limited space, such as in the case where the building footprint covers almost the whole area, then it would be better for the tower crane to be located within a building. This is because the positioning of the internal climbing crane is undoubtedly the most strategic as contractors can get by with a much shorter jib than with an external tower crane located at the perimeter of the building. It is important to note that the shorter the jib for a given task, the smaller the bending moment, and the lighter and cheaper the crane structure.

Building height

To erect an external fixed tower crane for a tall building is not considered cost efficient by contractors [16]. A climbing crane has similar advantages as that of a static model used within a building, the mast height of the internal climber can be just a fraction of the maximum height necessary. Thus there will be no extra cost incurred for the additional mast sections, regardless of the height of the project.

However, in the case of retrofitting and upgrading of an existing tall building, it would be justifiable to employ an external static tower crane rather than a climbing crane.

Maximum coverage

The primary consideration for a tower crane is to ensure that it can cover the whole plan area, and the pick up zone for materials. The tower crane must be located at such locations where it is possible to provide 100 percent lifting coverage over the plan area of the building.

Under most circumstances it is insufficient for a single crane to cover the whole site. Hence, it is common to see supplementary mobile cranes or additional tower cranes at most sites. When multiple cranes are required, the building should be sectioned off in accordance with the physical characteristics of the project, so as to establish the locations of multiple tower cranes.

Openings within the building for climbing crane

To accommodate the mast of the crane and to facilitate climbing, an opening has to be created for this purpose. A common practice by local contractors is to erect the tower crane in the lift shaft if it is spacious enough, thus leaving only one hole on the roof to be cast after the crane has been dismantled. However, this may cause an unacceptable delay in the installation of elevators, which in many cases are used to assist in the handling of materials and human traffic. In some other cases, when the client demands partial completion for early occupation, then the lift cars have to be operational on time.

Alternatively, a series of floor openings can be provided through the height of the building if existing openings such as lift shafts are not to be used. These temporary openings must not be too close to the building edge or to other large floor openings. In addition, the openings must not materially penetrate through major structural elements, and they must be repeated throughout the height of the whole construction regardless of changes in framing arrangement.

It is important to consider whether a structure can withstand the forces placed on it by the crane. A tower crane usually exerts two distinct types of loads to the structure namely, vertical forces that derive from crane and load weight, and lateral forces that derive from overturning moment and swing torsion. This implies that shoring on a number of floors below the crane is necessary to distribute the loads. As for lateral forces which are often of smaller magnitude, these are applied to the plane of the floor structure. These forces will be distributed without posing much of a problem because the floors are stronger in this direction [17–20].

Soil condition

A static tower crane is similar to a rail-mounted type without its wheeled undercarriage. The tower or mast is fixed to a ballasted framework, or to a specially designed reinforced concrete foundation that helps to transfer the load to the soil.

The maximum pressure exerted on the soil by a tower crane footing is the combined effect of the vertical loads and the moments. The soil condition will therefore have a direct effect on the location of a tower crane. This is especially true when a particular site uses track mounted tower crane for the construction of the project. On poor soil, track differential settlements can be a problem, as they may cause track elevations to go beyond permitted tolerances and endanger operations.

Location of existing structure and underground hazard

Temporary facilities for construction must be in position before the crane location is determined so that there would not be any restraints on the location of the tower crane. There is always the threat of locating a tower crane above buried pipes and mains. A lower crane is usually not positioned near existing buildings as this would restrict the slewing of the crane's boom.

References

[1] J. E. Johnston, *Site Control of Materials: Handling, Storage and Protection*, Butterworths, London, 1981.

[2] H. F. Edward, *Material Handling Systems and Terminology*, Lionheart Pub., Atlanta, GA, 1992.

[3] R. M. Eastman, *Materials Handling*, M. Dekker, New York, 1987.

[4] J. S. Foster and R. Harrington, *Structure and Fabric — Part 2*, Mitchell's Building Series, B.T. Batsford Limited, London, 1994.
[5] P. A. Stone, *Building Economy*, Pergamon Press, UK, 1976.
[6] F. Harris, *Modern Construction and Ground Engineering Equipment and Methods*, 2nd Edition, Longman Scientific & Technical, Essex, England, New York, 1994.
[7] F. Harris, *Construction Plant*, Granada, 1981.
[8] S. W. Nunnally, *Construction Methods and Management*, Prentice-Hall, 1987.
[9] R. Holmes, "Introduction to Civil Engineering Construction", College of Estate Management, 1975.
[10] R. Chudley, *Building Site Works, Substructure, and Plant*, Construction Press, New York, 1982.
[11] S. Furusaka and C. Gray, "A model for the selection of the optimum crane for construction sites", *Construction Management and Economics*, Vol. 2, 1984.
[12] C. Gray and J. Little, "A systematic approach to the selection of an appropriate crane for a construction site", Proceedings, *Construction Management and Economics*, Vol. 3, 1985.
[13] A. Warszawski, "Expert systems for crane selection", *Construction Management and Economics*, Vol. 8, 1990.
[14] C. Gray, "Crane location and selection by computer", Proceedings, 4th International Symposium on Robotics and Artificial Intelligence in Building Construction, 22–25 June 1987, Haifa, Israel.
[15] A. Warszawski and N. Peled, "An expert system for crane selection and location", Proceedings, 4th International Symposium on Robotics and Artificial Intelligence in Building Construction, 22–25 June 1987, Haifa, Israel.
[16] M. Y. L. Chew and S. K. Ang, "A Systematic Approach to the Site Location of Tower Cranes", The Professional Builders, 1995.
[17] D. V. MacCollum, "Crane Hazards and Their Prevention", American Society of Safety Engineers, 1993.
[18] C. D. Reese, *Handbook of OSHA Construction Safety and Health*, Lewis, Boca Raton, 1999.
[19] John Laing Construction Ltd, "Crane Stability on Site: An Introductory Guide", Construction Industry Research and Information Association, London, 1996.
[20] R. L. Peurifoy, *Construction Planning, Equipment, and Methods*, 5th Edition, McGraw-Hill, New York, 1996.

CHAPTER 7

WALL AND FLOOR CONSTRUCTION

7.1. Structural and M&E Systems

The structure of a tall building can be visualised as floor framing supported by columns and walls. The horizontal planes and the vertical planes are tied together forming a three-dimensional closed structure. The tubular, core interactive, and staggered truss buildings are typical examples of three-dimensional structures.

As buildings get taller, they need to be stiffer to resist lateral forces. These are commonly wind loads but also include serious earthquake forces in seismically active areas. Up to the 60's, almost all high-rises were designed as rigid frames. With the advancement in structural engineering, numerical modelling such as finite element analysis, accurate wind tunnel simulation, etc., more efficient structural systems have evolved, replacing the conventional rigid frame structures. Innovative design methods include the semi-rigid frame, the frame tube, and the composite structure combining the use of reinforced concrete and steel.

Figure 7.1 shows the various structural systems proven efficient for tall buildings of different heights. Table 7.1 shows the efficiency of various structural systems used in some famous buildings according to their slenderness (height/weight). For example, the 102-storey Empire State Building, with its rigid frame shear wall system, uses 2.02 kN/m^2 of structural steel. The 100-storey John Hancock Center in Chicago, with its trussed tube system, uses only 1.42 kN/m^2 of structural steel, utilising 30% less structural steel. In another case, the 60-storey Chase-Manhattan Bank Building in New York, with a braced long-span rigid frame structure, is inefficient with

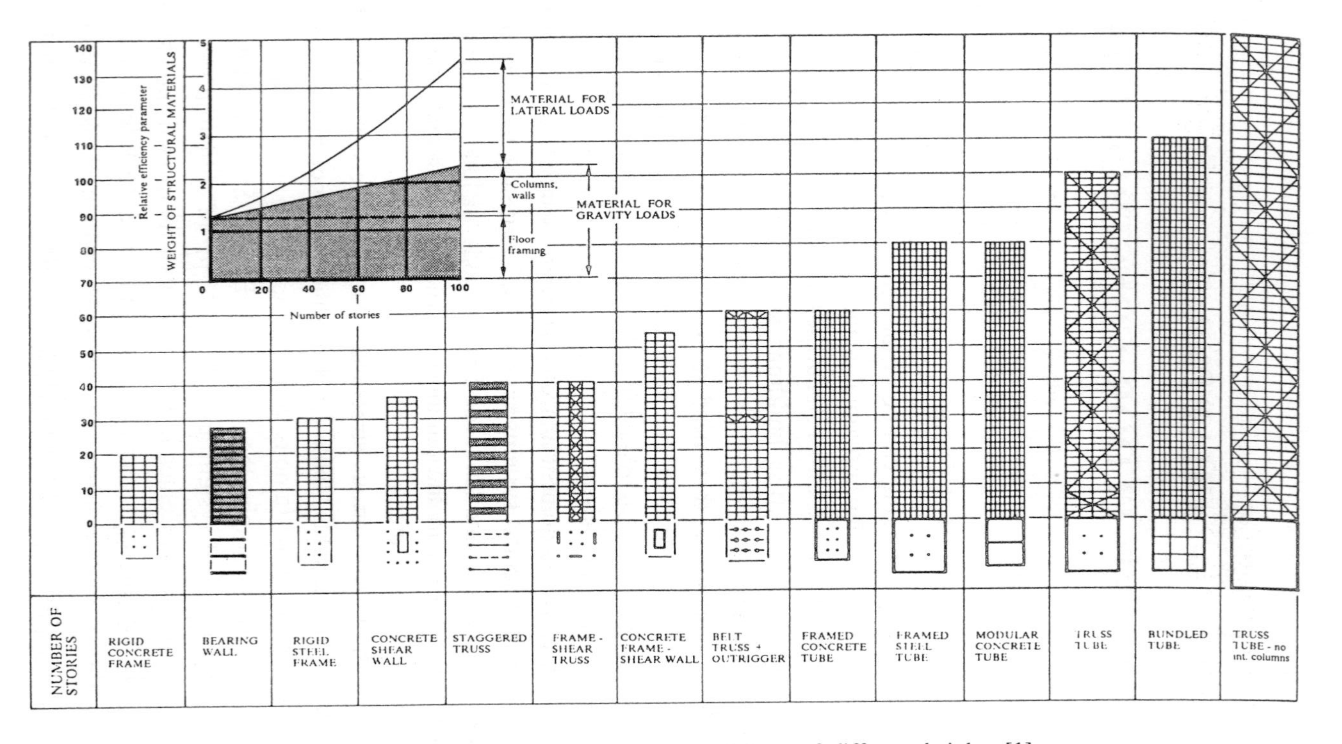

Figure 7.1. Structural systems for tall buildings of different heights [1].

Table 7.1. Efficiency of structural systems of tall buildings [1].

Building Cases	Year	Stories	Slender-	kN/m²	Structural
Empire State Building, New York	1931	102	9.3	2.02	Braced rigid frame
John Hancock Center, Chicago	1968	100	7.9	1.42	Trussed tube
World Trade Center, New York	1972	110	6.9	1.77	Framed tube
Sears Tower, Chicago	1974	109	6.4	1.58	Bundled tube
Chase Manhattan, New York	1963	60	7.3	2.64	Braced rigid frame
US Steel Building, Pittsburgh	1971	64	6.3	1.44	Shearwalls+outrigger
IDS Center, Minneapolis	1971	57	6.1	0.86	+belt trusses
Boston Co. Building, Boston	1970	41	4.1	1.01	K-braced tube
Alcoa Building, San Francisco	1969	26	4.0	1.24	Latticed tube

2.64 kN/m^2 structural steel when compared to the slightly lower 54-storey IDS Building in Minneapolis with only 0.86 kN/m^2, using shear walls with outriggers and belt trusses. It is apparent from these comparisons that each structural system is efficient within certain height limits. However, there are other factors to consider such as the building shape and size, building slenderness, the functional requirements, etc. Detailed descriptions of the various structural systems used for tall buildings can be found in Schueller [1].

The mechanical and electrical (M&E) system may represent between 20% to 45% of the total building cost of a tall building. It also constitutes a significant portion of the long-term operating costs [2]. It is important to understand the interaction of M&E systems with the construction of the superstructure. The typical M&E systems in tall buildings are:

- HVAC
- Cold and hot water systems
- Plumbing system (storm drainage and sanitary drainage)
- Fire protection and security
- Electrical distribution
- Lighting
- Transportation (e.g. lifts and elevators)

A mechanical system consists of various equipment in mechanical rooms and the fluids/gases/current are distributed to the various parts of the building via ducts, pipings and wiring networks. The flow of the distribution systems has close interaction with the construction of walls and floors. Good planning is needed for proper co-ordination of different trades and allowance for the passage of the distribution systems and minimising double handling. In buildings with fixed cellular subdivisions (e.g. apartments, hotels), a decentralised branching is common where each activity unit provides adjustable cooling services, water supply and plumbing stacks. In the case of commercial buildings, a more centralised branching of the mechanical services is more economical.

7.2. Construction

The construction industry has to respond to the swift changes in design by coming up with systematic methods of monitoring and controlling every phase of the construction activity. The advancement in construction technology saw in succession over the years the development and popularisation of such construction techniques such as jointing of precast, prefabricated components, slip forming, fly forming, etc. The continuous development of new concrete additives has increased the designability and buildability of structures which would have otherwise been impossible.

7.3. Core Wall Construction

7.3.1. *Slip Form*

Slipform is a method used for tall structures which are constant in cross-section, such as silos or the service cores of buildings. Figure 7.2 shows the arrangements of a conventional slip form. The construction sequence is generally as follows:

- Steel form of about 1.05 m high is built rigidly to the desired shape and clamped together by strategically placed adjustable yokes.
- Hydraulic climbing jacks are mounted in the yokes.

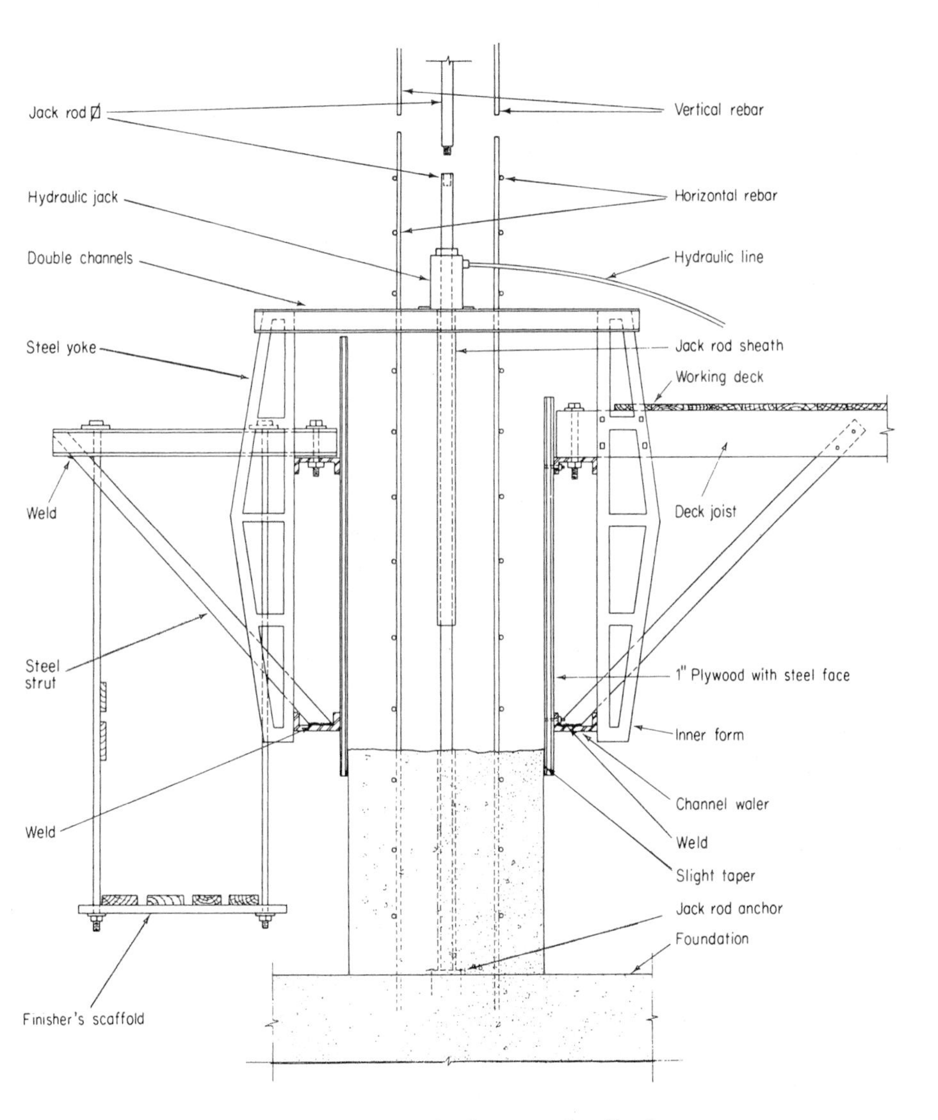

Figure 7.2. Schematic diagram of a slip form.

- Jacks climb on a jack rod embedded in the concrete.
- Horizontal rebars are placed in the wall as required to the full height of the forms.
- Vertical rebars are allowed to extend to a convenient height.
- The form is filled with concrete slowly.
- The forms are raised by the jacks after the concrete has partially set.
- All jacks are synchronised to operate equally and simultaneously.
- The verticality of the structure is monitored with plumbs, levels etc.

The system is not popular for buildings more than 30-storey as too many consumables are required as the operation is continuous. The continuous supply of concrete can also be difficult.

7.3.2. *Jump Form*

This is currently the most popular formwork system for core wall construction of tall buildings. It is constructed in successive lifts whereby the current portion being constructed is supported by the previously poured lift. This is a method of casting a wall in set vertical lift heights using the same forms in a repetitive fashion in order to obtain maximum usage from a minimum number of forms. After setting up the form, concrete is poured and allowed to cure. The forms are then removed, cleaned, and with the help of a crane, they "jump" to the next section and are fixed to the newly cast concrete, ready for the next pour (Figure 7.3). The construction sequence is generally as follows:

- Erect jumping form — waling with the use of soldiers with keys and wedges.
- Attach plywood or metal formwork panels.
- Hop up brackets fixed to the top and bottom of soldiers to create working platforms.
- Incorporate sockets and scaffold fittings to take a guard post and rail of standard scaffold tubes.
- Install toe board to complete working platform.

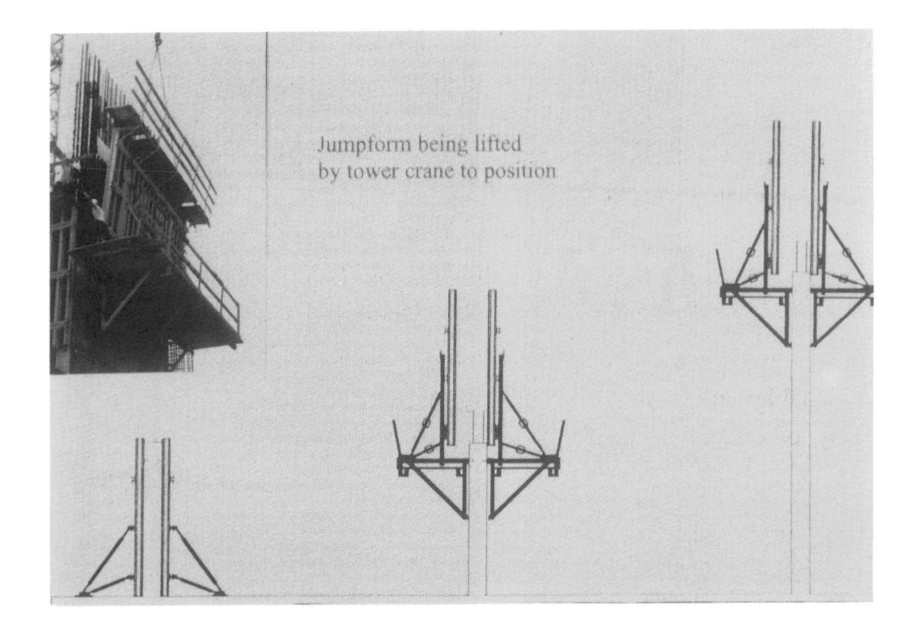

Figure 7.3. Construction sequence of a jumpform.

- Lift brackets bolted to top of soldiers at pre-selected points to facilitate lifting.
- Concrete poured and cured.
- Connect rebars for the next level.
- Lift jump forms using a crane to secure to previously cast wall portion.
- Repeat the process before the next pour.

Figure 7.4 shows a section through a typical jumpform.

7.3.3. *Climbing Form*

Similar to a jumpform except that jacks are used for the "climbing". The jacks are fitted through the lifting yoke and the whole external frame is lifted up to the next level.

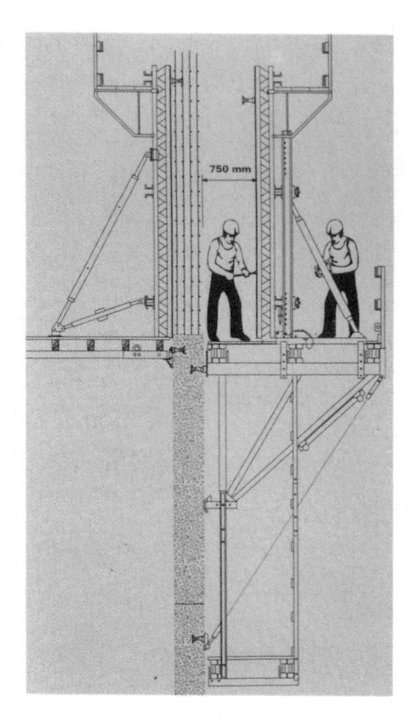

Figure 7.4. A section through a typical jumpform.

7.4. Floor Construction

7.4.1. *Flying Form*

Flying form, also known as table form is widely used in tall buildings where construction is identical from bay to bay and from floor to floor. It requires high regularity in structural bay width. A flying form consists of three principal components: (a) adjustable post shoring, (b) manufactured truss forms and (c) column supported forms. The construction sequence is generally as follows:

- Props or adjustable post shoring, normally scaffolding rests on a wood sill and blocking raised by jacks are installed to support the table form.
- The forms are moved horizontally by means of rollers. Manufactured truss forms use trusses raised by a series of uniformly distributed jacks. The levels are carefully checked.

- Reinforcements are placed and tied to columns and beams where necessary. Concreting proceeds.
- After concrete is set, the props are winded down with the table forms resting on rollers. The props not taking any load can then be removed (Figure 7.5).
- The forms are then fixed onto rollers and pushed to the slab edge (Figure 7.6).
- The first crane hook is attached to one end of the table form. The form is then slowly pulled out to allow for the second hook to be attached to the intermediate length of the form. Similar action for the third hook to be attached to the other end of the form (Figure 7.7).
- Lifting of the form to the next level proceeds (Figure 7.8).
- Positioning of the lifted form (Figure 7.9). Continuity of reinforcement with other structural member such as columns is critical.
- Positioned table forms ready for placement of reinforcement (Figure 7.10). Note the intersections between adjacent table forms, beams and columns.

Figure 7.5. Winding down of props after concrete is set.

Figure 7.6. Forms are fixed onto roller and pushed to the slab edge.

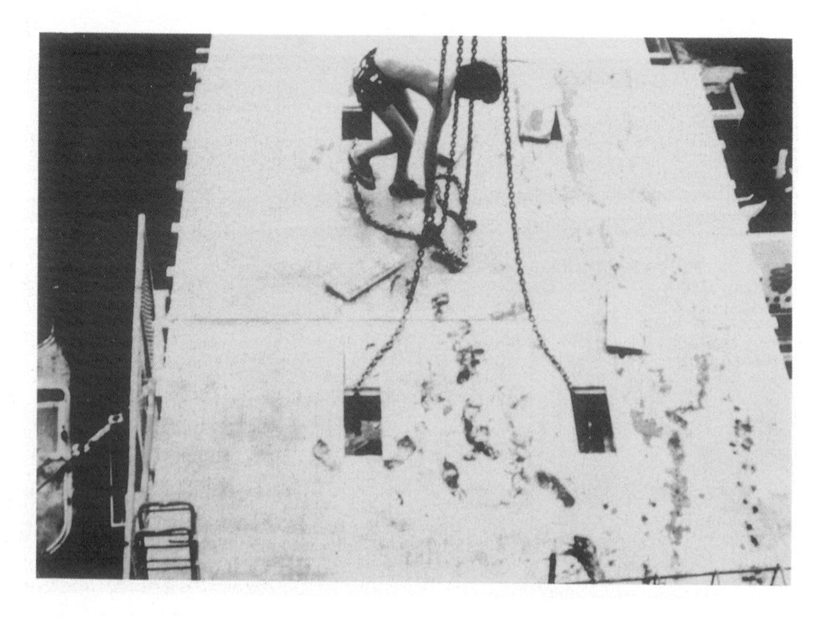

Figure 7.7. Attachment of crane to the table form.

Figure 7.8. Lifting of the form to the next level.

Figure 7.9. Positioning of the lifted form.

Figure 7.10. Table forms ready for the placement of reinforcement.

Usually the same work crew sets and strips the flying forms; half of the crew works below the deck level that has been cast, while the other half works above the previously cast area, setting the forms that have been removed.

7.4.2. *False Ceiling*

The most common types are suspended ceilings fixed to the framework hung from the main structure, creating a void or space between the two components. The basic functional requirements of false ceilings are:

- Easy to construct, repair, maintain and clean.
- Means of access to hidden services.
- Good sound and thermal insulation.
- Conform to fire regulations.

Figure 7.11 shows a typical suspended ceiling. The construction sequence is generally as follows:

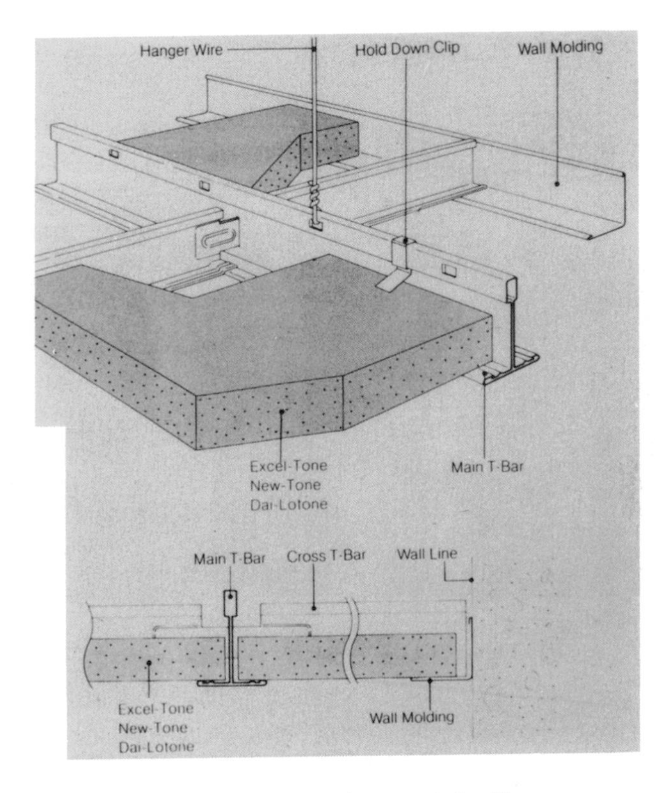

Figure 7.11. A typical suspended ceiling.

- Mark bracket locations for hanger rods.
- Confirm level and ceiling height, marking for wall angles, fix wall angles.
- Aligning and fixing the main and cross tees.
- Reconfirm and re-adjust suspension level with laser leveller.
- Laying in ceiling tiles for positioning of sprinkler heads, lightings, etc.
- Installation of all services/accessories.
- Fixed ceiling tiles (Figure 7.12).

Figure 7.13 shows the M&E ducts and pipings under a Bondek floor before the installation of a false ceiling.

Figure 7.12. The fixing of ceiling tiles.

Figure 7.13. M&E ducts and pipings behind the false ceiling.

7.4.3. *Raised Floor*

A raised floor provides a space above the structural floor level, allowing cables to be laid. The installation procedure is generally as follows:

- Inking is carried out with the help of datum lines to set out the grids and positions for the raised flooring.
- Levelling strings are used to mark and confirm the height of finished floor.
- Adhesives are used to secure the base plates used to support the raised floor panels to the sub-floor. The supports have to be lined up according to the inked lines (Figure 7.14).
- Laying of floor panels and finishes (Figure 7.15).

Figure 7.14. The use of adhesives to secure the base plates of the raised floor.

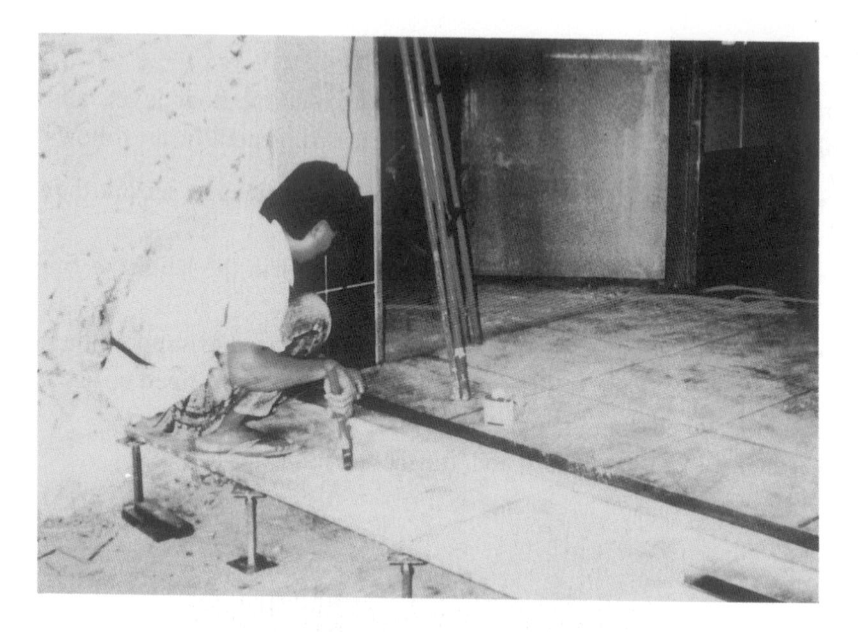

Figure 7.15. Laying of raised floor planels and finishes.

7.5. Precast Prefabricated Elements

The definition of precast prefabricated concrete is the manufacture of concrete elements, either plain, reinforced and/or prestressed whose elements require transporting to the site for final positioning. Figure 7.16 shows the laying of pipes on and through installed precast hollow core slabs. Figure 7.17 shows the positioning of a precast solid slab onto a building frame.

7.5.1. *Advantages*

— Precast concrete construction reduces total project time since units of components are cast and stockpiled while other phases of the project are carried out, thereby reducing the total required time. Accelerated curing methods further speed up the casting of precast components. Continuous uninterrupted erection of components makes the rapid formation of structural frame possible.

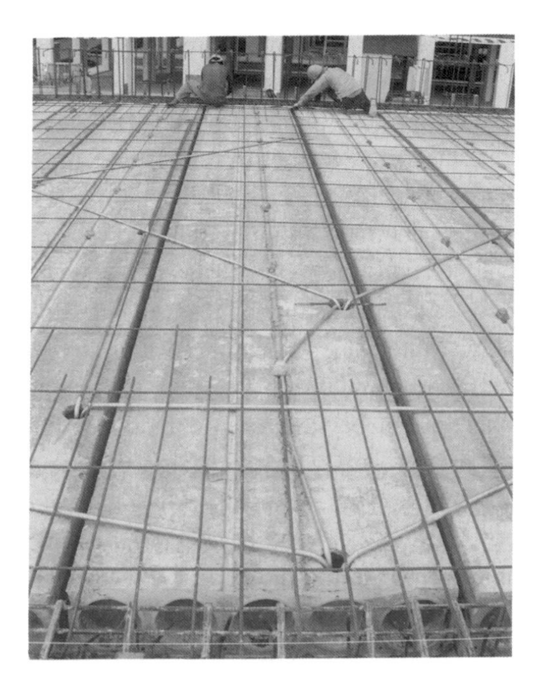

Figure 7.16. Laying of pipes in precast hollow core slabs.

Figure 7.17. Positioning of precast solid slab onto a building frame.

— The quality can be controlled as precast concrete components are fabricated under optimum conditions of forming, fabrication and placement of reinforcement and embedded fittings. Precast concrete has high resistance to weathering, abrasion, impact, corrosion and the general ravage of nature. They also exhibit smooth surfaces which resist moisture penetration, fungus and corrosion.

— Precast concrete construction is also economical as it reduces on-site labour cost. Faster construction leads to reduced general contractor overhead, lower interest paid on construction loan, earlier occupancy and quicker investment return.

— Precast concrete can be designed to facilitate future horizontal or vertical expansion. Thus with the design of precast concrete buildings on a modular system, necessary additions can be made with low removal costs. They may also be erected without falsework, maintaining uninterrupted traffic movement. Precast elements are quickly erected in buildings providing an instant platform for other trades.

7.5.2. *Disadvantages*

— The joints between members pose the greatest problem. Adequate design and detailing are required for a joint to be easily formed on-site while at the same time providing the required strength. As it is essential for precast components to be manufactured well in advance of the other building activities, flexibility for subsequent changes may be little. It is also essential for precast concrete components to be designed to function as part of a total structure but also to withstand stress conditions during handling, transport and erection.

— In order to obtain the greatest economy in precast concrete construction, a high degree of repetition is required. It is less suitable for structures with irregular or isolated features. In the case where precast units are gigantic, large capacity cranes and handling equipment are needed [3–4].

7.6. Prestressed Slab

Prestressing means the intentional creation of permanent internal forces and stresses in a structure or assembly, for the purpose of improving its behaviour and strength under service conditions.

Since concrete is strong in compression and weak in tension, prestressing the steel against the concrete would put the concrete under compressive stress that could be utilised to counterbalance tensile stresses produced by external loads.

The prestress or precompression may be induced by the use of steel or by means of external jacks. In the latter case, sufficiently solid abutments are essential. For greater efficiency, it is necessary to apply the prestress in the tensile zone only. By the selection of the appropriate pressure (which is kept to a minimum to economise on steel) and the point of application, the stress in the section may be apportioned in the desired manner [5–9].

7.6.1. *Advantages*

- It possesses the many advantages of reinforced concrete, such as free choice of shape and size, minimal maintenance and good insulation.
- It has better appearance and durability than reinforced concrete, because at working loads it is (or almost) crack-free.
- It utilises materials more efficiently than reinforced concrete in which only the concrete above the neutral axis is effective.
- It reduces the principle tension caused by shear, thus enabling use of webs similar in appearance to structural steel shapes.
- It allows for shrinkage and creep at the design stage itself.
- It employs high strength materials and therefore requires less of them for the same load, which results in lighter and more slender members (a saving in head room).
- Structures with longer spans are possible, reducing the need for columns.
- A prestressed concrete flexural member offers greater rigidity under working loads than a reinforced concrete member of the same depth.

— Prestressing raises the average stress in steel during repeated loads and this action tends to minimise effects of stress variations and so prolong the life of the material.
— Prestressing improves the ability for energy absorption under impact loads.

7.6.2. *Disadvantages*

— The initial cost is higher than that for reinforced concrete. This is due to the requirement of hardware (prestressing jacks, anchorages, bearing plates, more complicated formwork and higher labour cost).
— Higher strength materials necessary are more costly, harder to fabricate and require closer supervision.

7.6.3. *Pre-Tensioning*

— In this method, high-tensile steel wires are tensioned before the concrete is cast. When the concrete has attained sufficient strength, the wires are released inducing compressive stress in concrete.
— As strong abutments are required between the ends, pre-tensioning is invariably applied to precast units.
— It is usually carried out in a factory, although a prestressing bed may be set up on a site for a large contract so as to cut the factory overheads and the transport cost.
— Although pre-tensioning can be applied to individual members formed and stressed in their own moulds, the most usual method is the "long line" system in which the wires are stretched within continuous moulds between anchorages 120 m or more apart. After the concrete has hardened sufficiently the wires are released and are cut between each unit.
— At the extreme ends, the bond between steel and concrete is not fully developed. The wires contract considerably in their length with the consequence of loss of stress in the wires, the stress at the end

being zero. This contraction is accompanied by a lateral swelling which forms a cone-like anchor. The length over which this occurs (between 80 to 120 times the diameter) is termed the transfer length and this requires shear reinforcement.

— Small diameter wires are used as the larger surface area increases the bond (between 2 mm and 5 mm).
— It is essential that the wires be thoroughly degreased and allowed to rust slightly to increase the bond.
— Some forms of curing (e.g. steam curing) is normally applied to accelerate hardening. Rapid hardening admixtures are also commonly used.

Figure 7.18 shows the details of a typical pre-tensioning arrangement. Figure 7.19 shows the pre-tensioning bed and the jack for the stressing respectively.

7.6.4. *Post Tensioning*

— In this method the concrete is cast and permitted to harden before the steel is stressed.
— The steel, if placed in position before concreting, is prevented from bonding with the concrete either by being sheathed with thin sheet steel or tarred paper or being coated in bitumen.
— Alternatively, the prestressing steel can be introduced after the concrete has set by casting in duct-tubes at the appropriate positions which are extracted before the steel is inserted (Figure 7.20).
— The cable or bar is anchored at one end of the concrete unit and stressed by jacking against the other end to which it is then also anchored.
— The steel is subsequently grouted under pressure through holes at the ends of the unit to protect it from corrosion and to provide bond as an additional safeguard.

Figure 7.21 shows the post tensioning for a ground slab with blast links and ducts. Figure 7.22 shows an example of the grouting pipe.

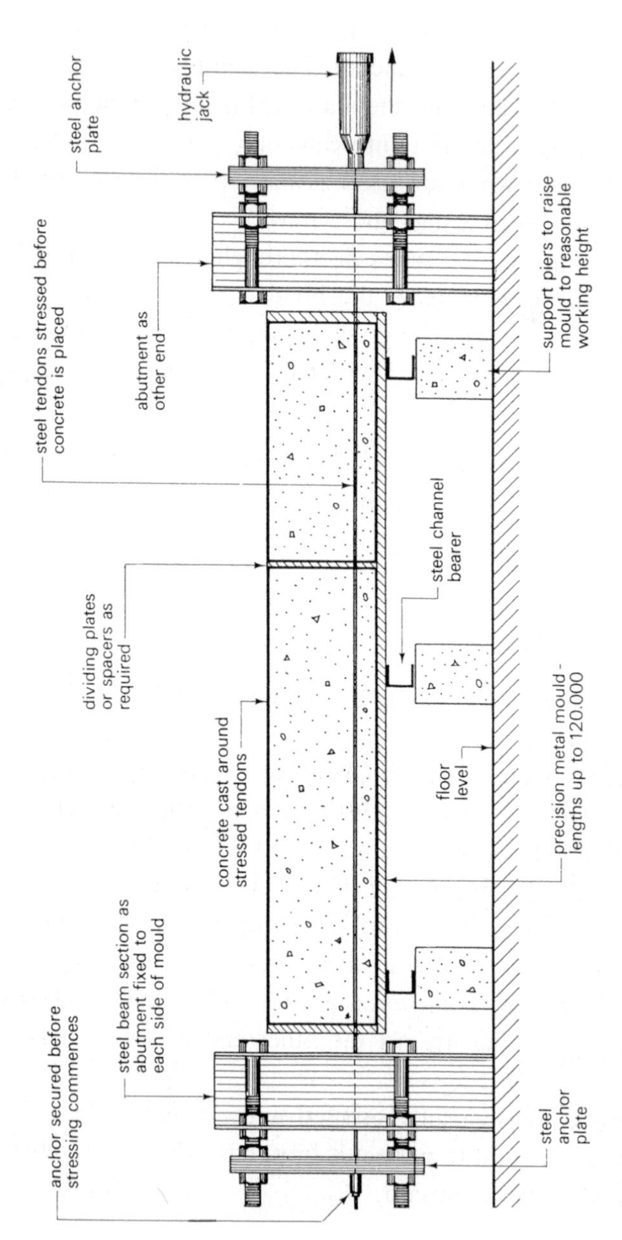

Figure 7.18. A typical pre-tensioning arrangement.

(a)

(b)

Figure 7.19. (a) A typical pre-tensioning bed. (b) Hydraulic jack for the stressing.

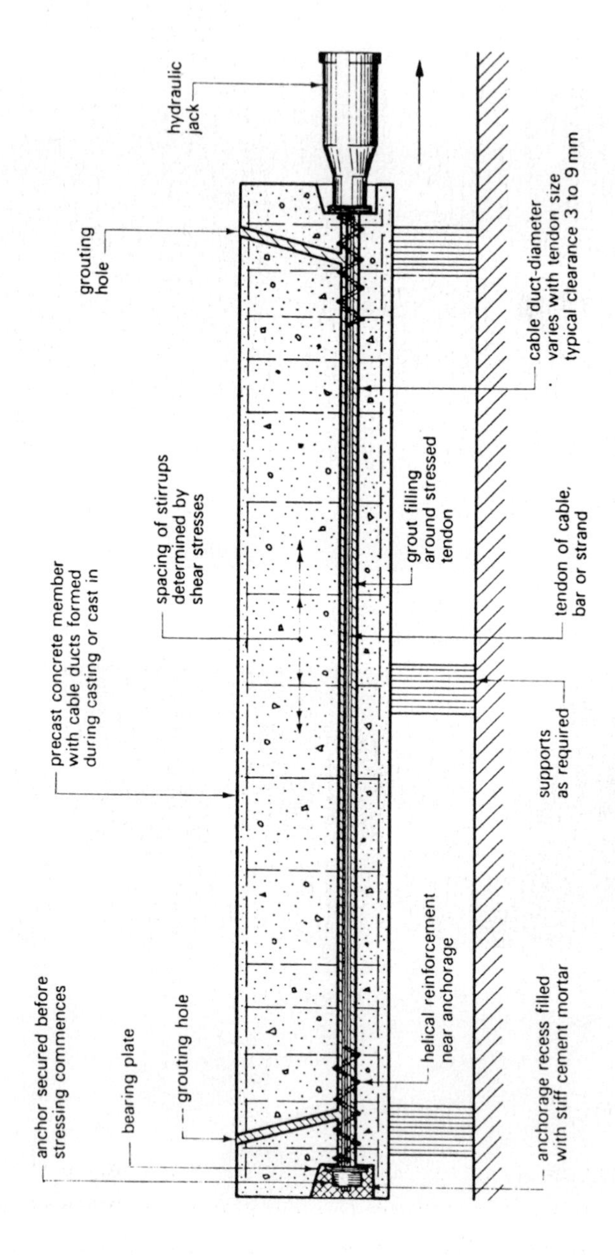

Figure 7.20. A typical post-tensioning arrangement.

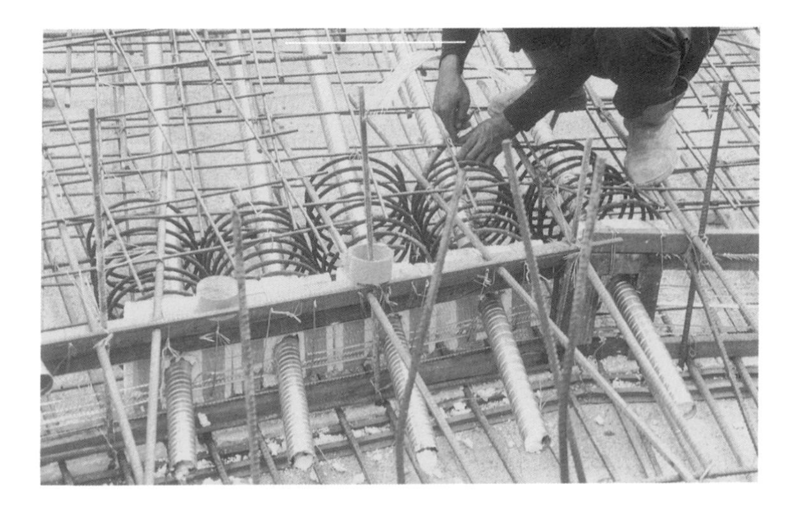

Figure 7.21. Post-tensioning with blast links at the edge.

Figure 7.22. A typical grouting pipe.

7.6.5. *Cable Profile*

— The maximum stresses due to external loads are at the point of maximum bending moment. At other points, the prestressing stresses may be excessive. The prestressing stresses can be reduced by varying the eccentricity.

— This can be achieved by:

(a) using straight cables and copying the section.
(b) by raising the centre of a beam of constant cross-section and using straight cables.
(c) by curving the wires upward from their lowest point.

— As the shear force tends to be higher at the supports, it is not advisable to reduce the section at the supports.

7.7. Steel Structures

7.7.1. *Structural Steel Frame*

The most common steel shapes used in building frame construction are S (standard) shapes, W (wide flange) shapes, channels, HSS (hollow structural) shapes, structural Ts, angles, HP shapes and plates. The two most common methods used for connecting steel members in structural steel frames are bolting and welding. In most structural steel buildings, both welds and bolts are used in combinations to produce the most practical and economical connections possible [10]. Details of the various connections are discussed by Smith [11].

Figure 7.23 shows a modular steel structure using a series of lattice structure across the whole building, which thus reduces the number of columns needed. Exposed structural steel is not allowed by the fire code and there is limited use of intumescent material due to the high relative humidity in the tropics. Figure 7.24 shows the use of sprayed vermiculite for fire protection.

The advantages of structural steel construction are obvious:

- *Speed of construction* — high level of prefabrication
- *Space constraints* — the finished products enable the application of the "just-in-time" management concept.
- *Construction cost* — reduce cost due to reduced construction time.
- *Increase load bearing capacity* — achieved through the use of castellated beams and lattice girders. The reduced size of beams and columns also increase the lettable space.

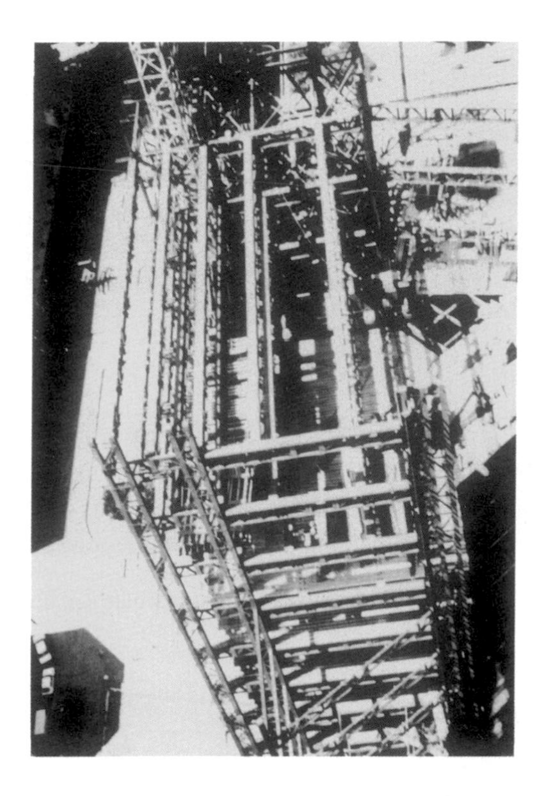

Figure 7.23. Modular steel structure.

Figure 7.24. Steel structures sprayed with vermiculite for fire protection.

7.7.2. *Steel Decking*

The use of steel decking as the permanent formwork for floors has also gained popularity, especially with structural steel frames. The steel decking is fastened to the steel frame by spot welding, plug welding or by self-tapping screws (holes must be predrilled). Connections to a concrete frame may be made by welding to cast-in connection strips or by the use of power-driven pins.

Figure 7.25 shows a 0.75 mm thick Bondek galvanised corrugated steel decking with standard size of 12 m × 2 m. Figure 7.26 shows the underside of the system supported by castellated beams. Generally, shear studs are first welded onto the beams and girders. The steel deckings are cut to the required shapes and holes are drilled to allow fixing onto the studs. The decking are then placed according to their positions on the girders and beams of the steel frame structures. Deckings and the steel studs are welded together to form an integral part of the girders and beams (Figure 7.27). Reinforcements are then placed on spacers and concrete is poured.

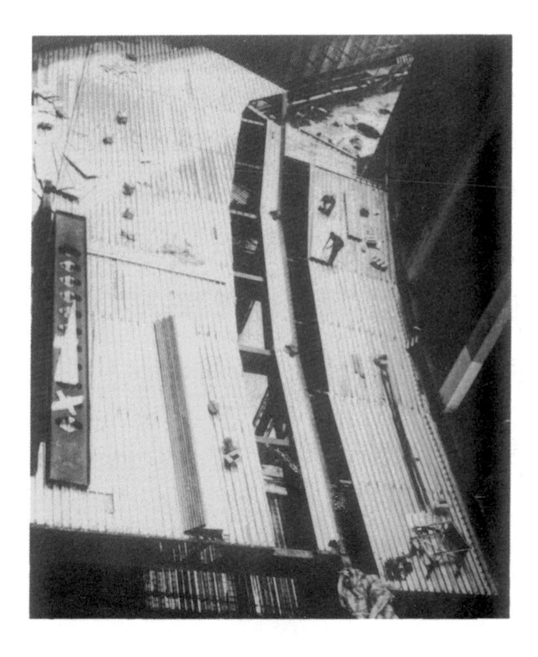

Figure 7.25. Galvanised corrugated steel decking as the permanent formwork.

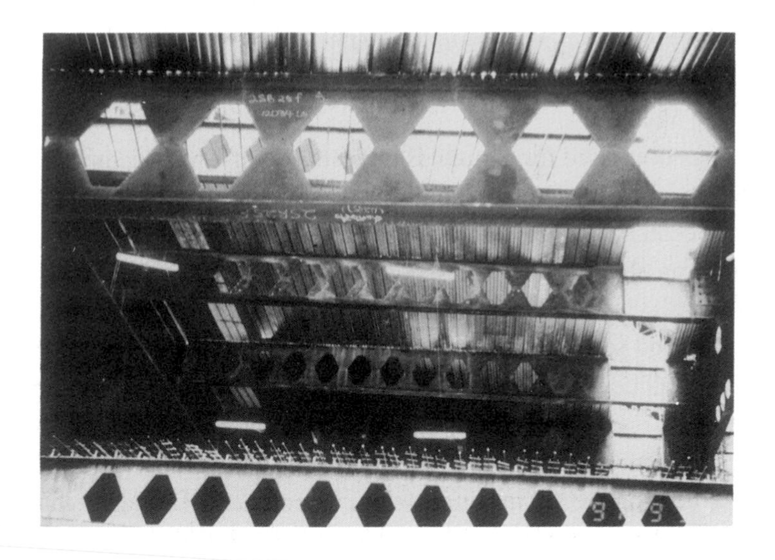

Figure 7.26. Castellated beams supporting the steel sheet.

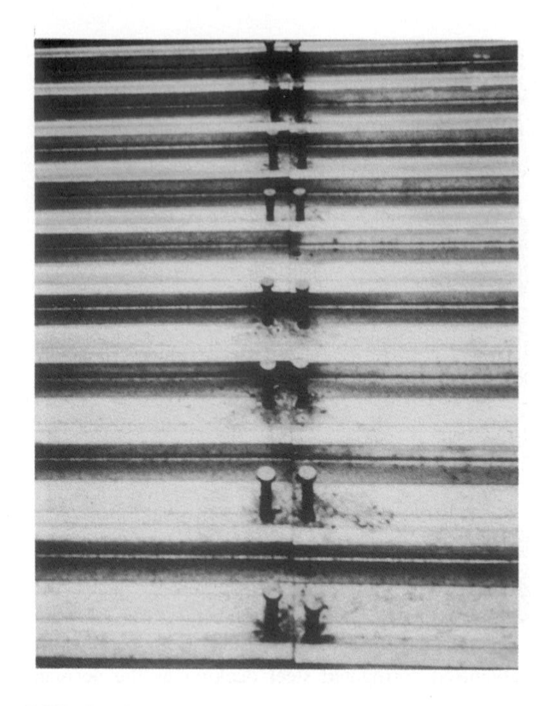

Figure 7.27. Steel studs welded to the decking, girders and beams.

7.8. Concrete

Concrete has been used for thousands of years. The use of structural reinforced concrete, made with Portland cement, dates back to the middle of the eighteenth century, gaining momentum at the turn of the nineteenth century and increasing sharply after the second world war, when steel was in short supply.

Concrete and steel are the two most commonly used structural materials in buildings. They sometimes complement one another, and sometimes compete with one another [12]. Steel is a homogeneous material and is manufactured under carefully controlled conditions. Concrete on the other hand, is a heterogeneous material, with cement, sand and aggregates interlocking with each other forming a matrix, of which the behaviour is harder to predict.

A good concrete structure must satisfy the requirements for strength, stiffness, stability, serviceability as well as durability. Adequate design must be accompanied with proper workmanship in particular the four "Cs", i.e. (a) mix constituents and proportion, (b) cover, (c) compaction and (d) curing [13].

7.8.1. *Concrete Admixtures*

Concrete admixtures can be described as materials other than cement, aggregates and water which are added to concrete just before or during mixing for the purpose of modifying selected properties of concrete in a beneficial manner. Modification can be in the form of workability, setting times, strength and durability [14]. Table 7.2 shows the applications of various types of admixture commonly used. Table 7.3 shows the acceptance test requirements as specified in SS 320:1987.

7.8.2. *Batching Plant*

Figure 7.28 shows the main components of a typical batching plant. The following factors will control the selection of plant:

(1) Topography of the site — boundaries, restrictions, noise, contours of land, soil conditions.
(2) The total volume of concrete required.
(3) The maximum amount of concrete required at any point at any one time.
(4) Availability of plant.
(5) Time of year in which concreting is to be carried out.
(6) Amount of space available for setting up plant.
(7) Quality of concrete required i.e. specification, varying mixes.
(8) Cost of producing concrete by various methods.

Table 7.2. Admixture/application/benefit chart [14].

Admixture type	Broad mechanism	Ultimate effect	Practical application and benefit
Accelerators	Speed the chemical reaction between the three main phases in cement and water. *'Interactive'*	Reduced setting times. High early strengths. Normal strength/time at low temperatures	Offsetting low temperatures and cold weather. Early removal of forms and moulds. Improved production schedules. Reduce energy needs.
Retarders	Controlled interference of Hydration reaction. *'Interactive'*	Retained workability. Extended setting times. Slower strength/time response. Higher ultimate strengths.	Offsetting the effects of high ambient temperature. Prevention of cold joints between pours. *Acceler'g/ water reducing*
Water reducing/ plasticising	Dispersion of cement. Increased rate of hydration. *'Adsorptive'*	High workability for given water content. Higher strengths for a reduced water content at a maintained workability.	Denser concrete. Stronger concrete. Improved standard deviation. Better compaction. Lower permeability. Cheaper concrete. *Retarding/ water reducing*
Air-entraining	Stabilised air volume throughout. Hydrating cement mass. *'Adsorptive'*	Improved workability. Improved consistence or 'fattiness'.	Good freeze thaw reponse. Offsetting poor sands and gap graded materials.
Superplasticisers	Extreme dispersion. *'Adsorptive'*	Very high workability for a given water content. High water reductions for a given workability.	Facile concrete placing in difficult situations. i.e. closely spaced reinforcement/ high early strength concrete/ time and energy savings/ non-shrink, non-bleed grouts.

Admixture		Slump with respect to control mix		Stiffening time from completion of mixing to reach a resistance to penetration of $0.5\ N\ mm^{-2}$		Stiffening time from completion of mixing to reach a resistance to penetration of $3.5\ N\ mm^{-2}$		Minimum compressive strength, % control mix: 1 day	7 days	28 days
Accelerating	(A)	Not more than 15 mm below	(A)	More than 1 hour	(A)	At least 1 hour less than control mix	(A)	125	—	95
Retarding	(A)	Not more than 15 mm below	(A)	At least 1 hour longer than control mix		—	(A)	—	90	95
Normal water-reducing	(A)	At least 15 mm above					(A)	—	90	90
	(B)	Not more than 15 mm below	(B)	Within ± 1 hour of control mix	(B)	Within ± 1 hour of control mix	(B)	—	110	110
Accelerating water-reducing	(A)	At least 15 mm above					(A)	125	—	90
	(B)	Not more than 15 mm below	(B)	More than 1 hour	(B)	At least 1 hour less than control mix	(B)	125	—	110
Retarding water-reducing	(A)	At least 15 mm above					(A)	—	90	90
	(B)	Not more than 15 mm below	(B)	At least 1 hour longer than control mix		—	(B)	—	110	110
High-range water-reducing⁺							(A)	—	90	90
	(B)	Slump not more than 15 mm below	(B)	Within ± 1 hour of control mix	(B)	Within ± 1 hour of control mix	(B)	140	125	115
Retarding high-range water-reducing⁺⁺							(A)	—	90	90
	(B)	Slump not more than 15 mm below	(B)	1 to 4 hours longer		—	(B)	—	125	115

*(A) Mix water content as control mix concrete. (B) Mix water content 92% of control mix concrete for water-reducing admixtures, (B) Mix water content 84% of control mix concrete for high-range water-reducing admixtures.

⁺Additional requirements: Mix A – (i) flow 510 to 620 mm, (ii) slump at 45 minutes not less than, and at 3 hours not more than, that of the control mix at 10 to 15 minutes.

⁺⁺Additional requirements: Mix A – (i) flow 510 to 620 mm, (ii) slump at 3 hours not less than that of the control mix at 10 to 15 minutes.

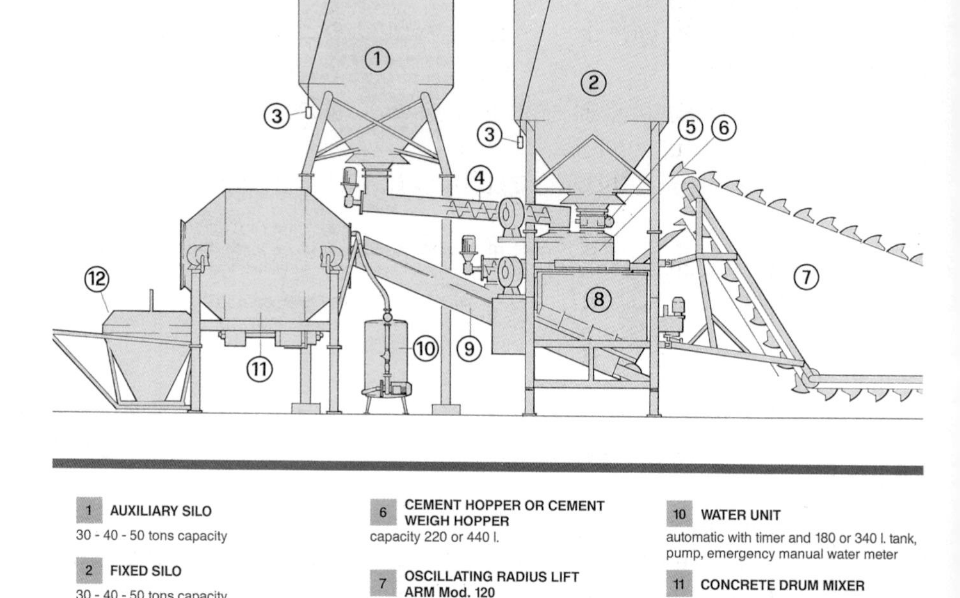

Figure 7.28. The main components of a typical batching plant (Acma).

7.8.3. *Ready-Mixed Concrete*

— It may be economical to use ready-mixed concrete if setting up a plant on site is not justified.

— Concrete purchased from a ready-mixed plant can be provided in any one of the following ways:

(a) <u>Central-Mixed Concrete</u>

Concrete is fully mixed in a stationery mixer and agitated during transit.

(b) Shrink-Mixed Concrete
Concrete is partially mixed in a stationery mixer and then mixed completely in a truck mixer (usually en route to the site).

(c) Truck-Mixed Concrete (Transit-Mixed Concrete)
Concrete is completely mixed in a truck mixer, with 70 to 100 revolutions to be at a speed sufficient to completely mix the concrete.

(d) Dry-Batched Concrete
Water is added at the site and mixed.

— Concrete may be ordered in several ways:

(1) Recipe Batch:
- The mix design is done by the purchaser and the specifications are given to the supplier.
- Under this approach, the purchaser takes the responsibility for the resulting strength and durability, providing the stipulated amounts are furnished as specified.

(2) Performance Batch:
- The purchaser specifies the requirements and the supplier assumes full responsibility for the proportions of the various ingredients that go into the batch.

(3) Part Performance and Part Recipe:
- The purchaser generally specifies some requirements such as minimum cement content, the admixtures to be used, allowing the supplier to proportion the concrete mix within the constraints imposed.
- This allows the supplier some flexibility to supply the most economical mix.

7.9. Formwork

The various common system formworks have been discussed in earlier sections. This section describes the general methods and accessories used to overcome the hydrostatic pressure exerted by wet concrete on formwork.

7.9.1. *Lateral Pressure of Fresh Concrete*

Loads imposed by fresh concrete against wall or column forms differ from the gravity load on a horizontal slab form. The freshly placed concrete behaves temporarily like a fluid, producing a hydrostatic pressure that acts laterally on the vertical forms. This lateral pressure is comparable to a full liquid head when concrete is placed full height within the period required for its initial set. The effective pressure is influenced by the weight, rate of placement, temperature of the concrete mix, setting time, shape and size of aggregates, and method of vibration.

The weight of concrete has a direct influence since hydrostatic pressure at any point in a fluid is created by the weight of superimposed fluid. The hydrostatic pressure is the same in all directions at a given depth in a fluid, and it acts at right angles to any surface which confines the fluid. However, lower hydrostatic pressure is expected in concrete as it is a mixture of solid and liquid. The pressure is further reduced due to aggregate interlogging, setting of the matrix etc.

The rate of placing is referred to as the rate of rise of the poured concrete. As the concrete is placed, lateral pressure at a given point increases as concrete depth above this point increases. The rate of placing has a primary effect on lateral pressure, and the maximum lateral pressure is proportional to the rate of placing, up to a limit equal to the full fluid pressure.

Internal vibration is a common method of consolidating/compacting concrete. It results in temporary lateral pressure locally which can be 10–20% greater than those from simple spading because it causes concrete to behave as a fluid for the full depth of vibration. The setting time i.e. duration when the concrete remains to behave as fluid, is influenced by the temperature of the concrete during placing, the inclusion of admixtures such as superplasticisers (water reducing agent) and retarders.

7.9.2. *Column Form*

The method of building form panels depends on the materials being used as well as the means of clamping or yoking the columns. Figure 7.29 shows

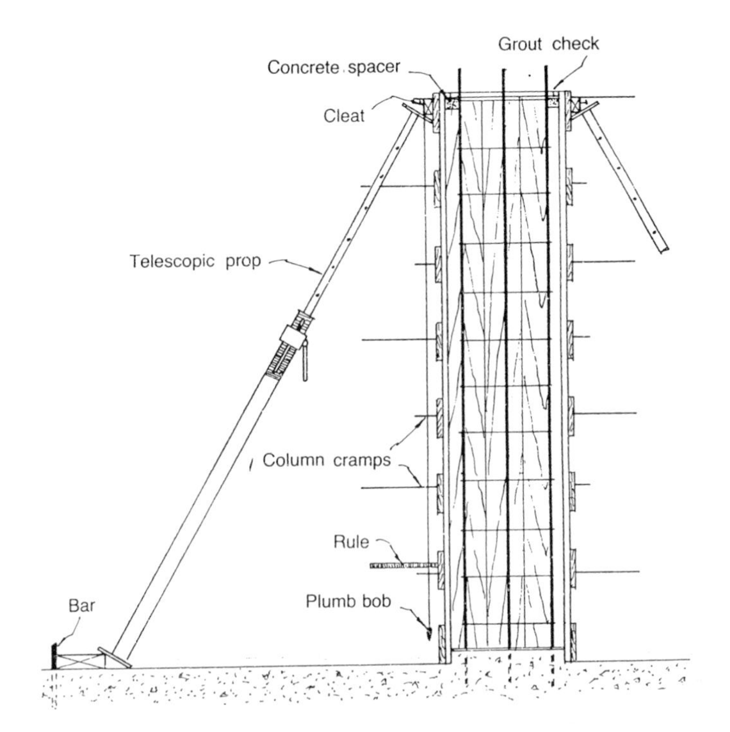

Figure 7.29. Formwork arrangements for a light column.

formwork construction suitable for light columns held together with a combination of wood and bolt yoke. Battens attached to the plywood side panels are a part of the yoke, and ties or bolt with washers form the other two sides of the yoke. Rakers are used to ensure verticality and also form additional lateral supports to the formwork.

Heavier column forms are commonly tied with adjustable ready-made column clamps (Figure 7.30). The column form can be placed directly into the preformed reinforcement cage in which case the spacers have to be rigidly tied and adequately spaced to ensure accurate placement and to achieve sufficient concrete cover. Telescopic or adjustable props are used to adjust verticality as indicated by plumlines and the like (Figure 7.31).

7.9.3. *Wall Form*

In the case of a wall, due to its length, clamping is not efficient. The use of ties through the width spaced along the length is needed (Figure 7.32). Many forms of ties are commercially available, some are retrievable while others are not and end up forming part of the concrete structure after the concrete is set. Details on the placement of reinforcement, formwork and concreting are reported elsewhere [15–21].

7.10. Scaffoldings

Scaffoldings are temporary erections, constructed to support a number of platforms at different heights, to enable workmen to reach their work and to permit the raising of materials [22–33].

The definitions of scaffold members, fittings and other terms in common used according CP14 [26] are as follows:

Baseplate/soleplate: A plate for distributing the load from a standard or raker.

Bay: The space between the centre lines of two adjacent standards along the face of the scaffold.

Brace: A diagonal member incorporated in a scaffold to prevent distortion and collapse of the structure.

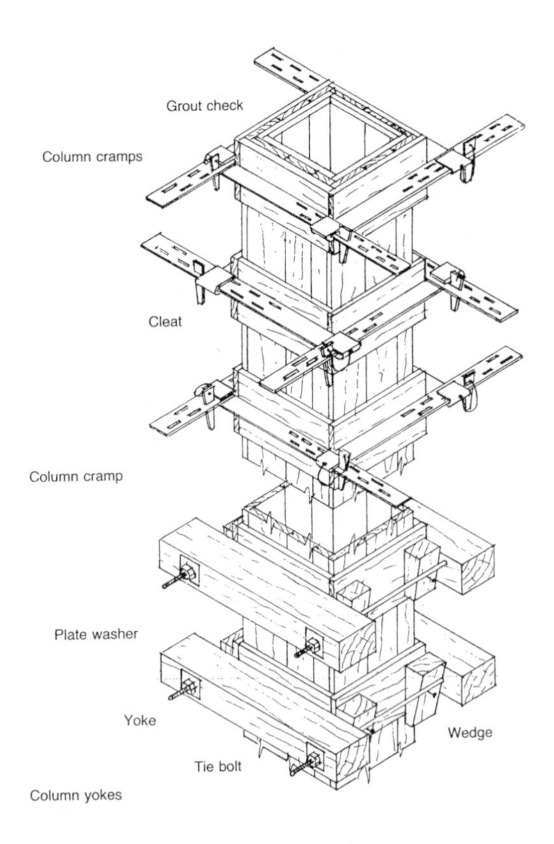

Figure 7.30. Formwork arrangement for a heavy column.

(a)

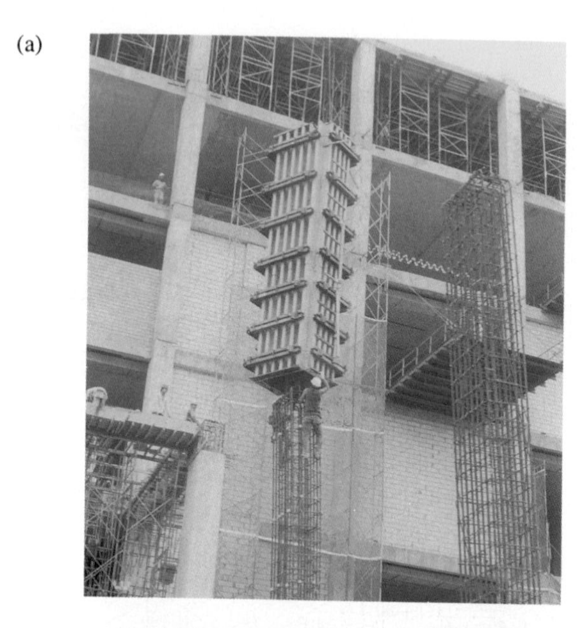

(b)

Figure 7.31. (a) Installation of a column formwork on a preformed reinforcement cage. (b) The use of adjustable props to ensure verticality.

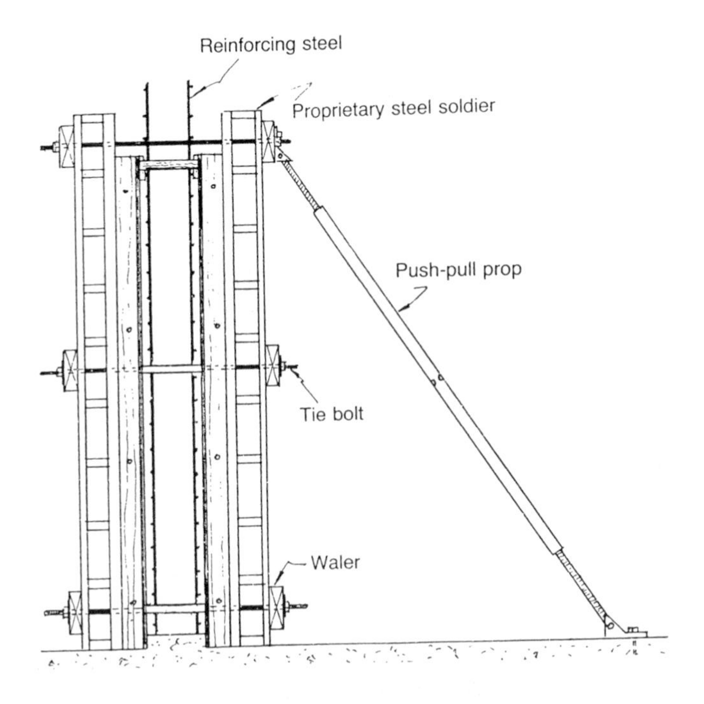

Figure 7.32. Section showing ties through the width of a wall.

Couplers: See Section 7.10.3.

Guard rail/handrail: A safety rail to prevent persons from falling.

Independent scaffold: A scaffold supported by a double row of standards and independent of wall support.

Joint pin: An internal fitting for joining two tubes end to end.

Ledger: A longitudinal horizontal supporting member.

Lift: The height from the ground or floor to the lowest ledger, or the vertical distance between adjacent ledgers.

Reveal pin: The fitting used for tightening a tube between two opposing surfaces.

Reveal tie: A member wedged between two opposing rigid surfaces, e.g. window opening to form a friction anchor to which the scaffold may be tied.

Standard/column: A vertical supporting member for transmitting load to the ground.

Toeboard: A board set on edge used to prevent tools, materials or feet from slipping off the platform.

Transom: A horizontal cross-member spanning from outer to inner ledgers.

Considerations must be given to the following loads in the design of scaffoldings:

(a) The self weight of the scaffold structure.
(b) The dead weight and live loads imposed on the scaffold structure during the course of the work.
(c) Lateral loads (wind, inaccuracy of construction).

The design and construction criteria, quantitative recommendations, checklists for safety inspection etc. are clearly detailed in the various codes of practice [26–29].

Industrial safety on the use of scaffoldings has been a big issue as repeated whole or partial collapses have led to high fatality rate (Chapter 2). The three main causes for such collapses are (a) lack of or ineffective ties,

(b) lack of or inefficient bracing, and (c) poor foundations or the undermining of them as work proceeds. It is difficult to pinpoint on one particular cause for a collapse for it may be the result of one or a combination of many defects. It is assured, however, that good design, materials and workmanship backed by thorough checking at all times is the only safe policy for all scaffolds.

7.10.1. *Bintangor, bamboo*

This is the most common type of wooden scaffolding used in the region. It is readily available and its low initial cost has made it popular for small projects. It possesses high strength capacity in relation to the weight and is easily attached and removed using lashings (Figure 7.33).

Figure 7.33. Bintangor scaffolds.

7.10.2. *Tubular Scaffolding*.

It comprises pairs of *standards* typically 1.25 m apart and spaced along the building between 1.2 and 2.4 m apart. The pairs of standards are connected horizontally parallel to the building with horizontal tubes called the ledgers. Ledgers are normally spaced vertically at the working height of 2 m or slightly less. The inside and outside grids so formed are connected with short tubes known as *transoms*. A layer of ledgers, transoms and board bearers is referred to as a *lift*. The scaffold is completed with ledger bracing or diagonal bracing and facade bracing. Facade bracing runs diagonally up the facade to provide stability along the structure.

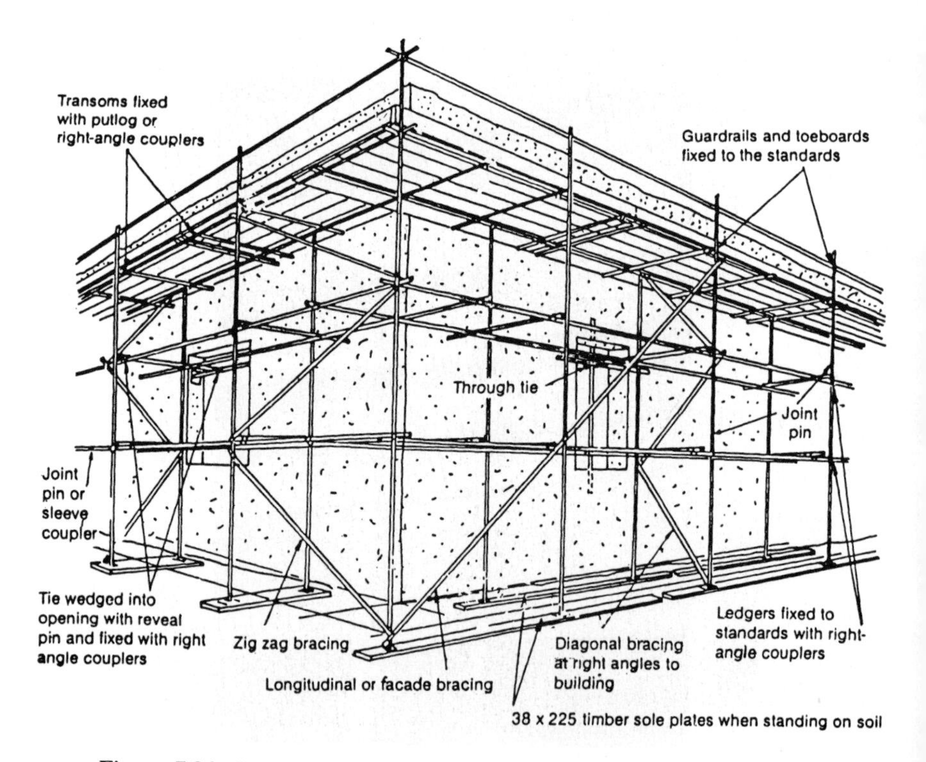

Figure 7.34. Schematic diagram of a typical tubular system scaffold [4].

Figure 7.34 shows a schematic view of a typical system tubular scaffolding. Figure 7.35 shows the use of tubular scaffolds supported on angles bolted into the concrete frame. The arrangement is to provide access into the building at the ground level. Figure 7.36 shows the elevation and section with details of the various elements of a metal frame scaffold system. Peripheral overhead shelter shall be incorporated with the scaffold (Figure 7.37).

Figure 7.35. Tubular scaffolds supported by angles bolted into the concrete slab.

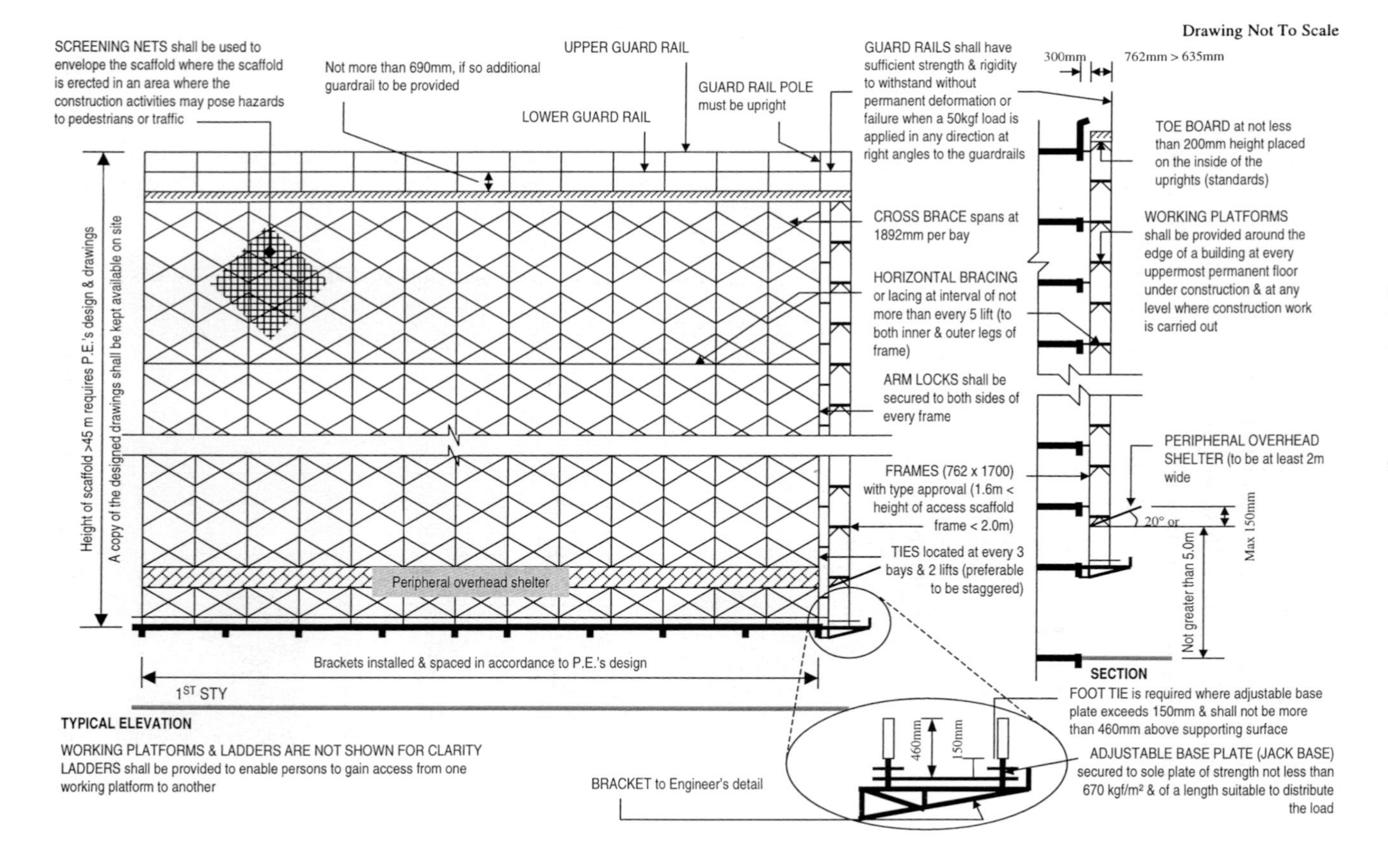

Figure 7.36(a). Details of a metal frame scaffold system.

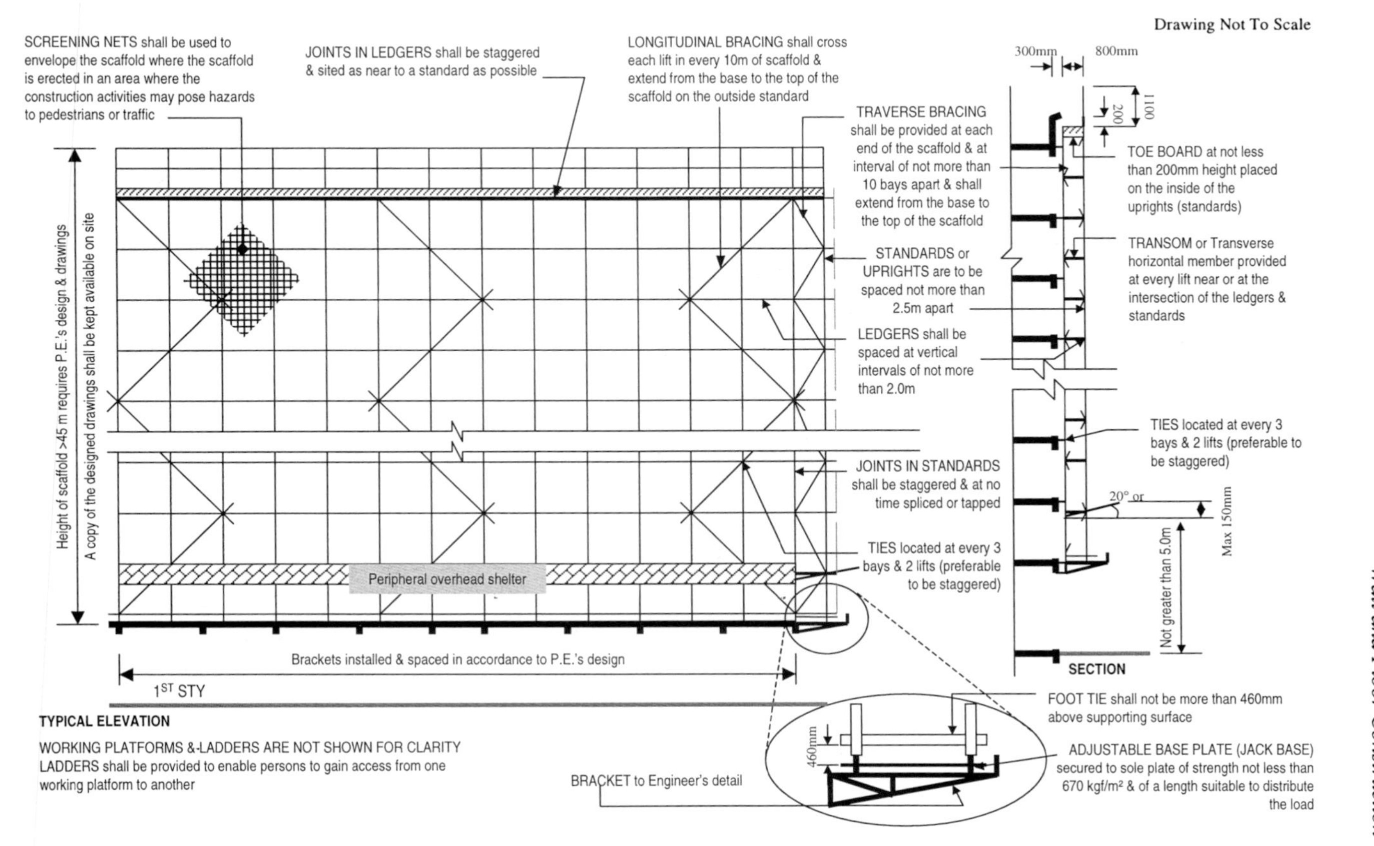

Figure 7.36(b). Details of a tubes and fittings system.

Figure 7.37. An example of a peripheral overhead shelter (catch platform).

7.10.3. *Couplers*

Couplers are used to attach one tube to another at different angles. *Right angle couplers* have a failure load in excess of 1270 kg in slip. Normally a safety factor of 2 is applied, giving a safe working load of 635 kg. *Putlog couplers* also provide right angle coupling. They enable a short tube or board bearer to be fixed above a ledger. It does not project upwards, or it would prevent the boards laying flat. *Brace couplers* provide higher strength than putlock couplers, approaching that of right angle couplers. *Swivel couplers* are capable of rotating through 360°. It has the same slip capacity as the right angle couplers but not the same rigidity. Figure 7.38 shows examples of right angle and swivel couplers.

(a)

(b)

Figure 7.38. Examples of (a) right angle and (b) swivel couplers.

7.10.4. *Ties*

- *Ties* should be placed sufficiently close together so that the scaffold structure is strong enough to span horizontally and vertically between them.
- *Reveal ties* — when scaffolding existing structures it may be impractical to go through windows. An alternative anchorage is to use a short length of tube and a reveal screw which is tightened between the sides of a window opening (Figure 7.34). The scaffold is tied to this with another tube. It is normal to place timber packers at each side, to reduce the risk of damaging the window openings.
- *Through ties* — in this case the ties goes horizontally through a window or other opening. Tubes are placed at the outside and the inside of the opening and fixed with right angle couplers (Figure 7.39).

Figure 7.39. Example of a through tie.

- *Box ties* — In this case two ties go either side of a column and two lateral tubes prevent inward or outward movements.
- *Drilled anchorage* — It is possible to drill and fix a screwed eye bolt into the facade. If the facade is of concrete, ample strength will almost always be obtained.

References

[1] W. Schueller, *The Vertical Building Structure*, Van Nostrand Reinhold, 1990.
[2] R. E. Fischer, *Engineering for Architecture*, Halliday Lithograph, 1980.
[3] "Guidelines for Waterproofing of Precast Concrete Structures", Prestressed Precast Concrete Society, Singapore, 1987.
[4] J. Feld, *Construction Failure*, 2nd Edition, John Wiley, New York, 1996.
[5] E. G. Nawy, *Prestressed Concrete: A Fundamental Approach*, 2nd Edition, Prentice Hall, New York, 1996.
[6] L. Spiegel, *Reinforced Concrete Design*, 4th Edition, Prentice Hall, New York, 1998.
[7] B. C. Gerwick, *Construction of Prestressed Concrete Structures*, 2nd Edition, Wiley, New York, 1993.
[8] FIP, "Prestressed Concrete: Safety Precautions in Post-Tensioning", FIP Guide to Good Practice, Telford, 1989.
[9] A. Hillerborg, *Strip Method Design Handbook*, E & FN Spon, New York, 1996.
[10] Nippon Steel, "Quality Controlled Steel Structure In Super-Highrise Building", Video on the Construction of the Treasury Building by Nippon Steel, 1990.
[11] R. C. Smith and C. K. Andres, *Principles and Practices of Heavy Construction*, Prentice-Hall, 1986.
[12] A. M. Neville, *Properties of Concrete*, Sir Isaac Pitman & Sons, 1972.
[13] G. Somerville, "The design life of concrete structures", *The Structural Engineer*, Vol. 64A, No. 2, February 1986, pp. 60–71.
[14] CIDB, SCI, "Concrete Admixtures — Properties & Applications", CIDB, SCI, 1992.
[15] M. Marosszeky and M. Y. L. Chew, "Site investigation of reinforcement placement on building and bridges", *Concrete International*, Vol. 12, No. 4, April 1990, pp. 59–70.

[16] M. Y. L. Chew, M. Marosszeky, M. Arioka and P. Peck, "Textile form to improve concrete durability", *Concrete International*, Vol. 15, No. 11, November 1993, pp. 37–42.

[17] M. Marosszeky and M. Y. L. Chew, "Importance of Workmanship on Concrete Durability", Concrete Durability Seminar, National Building Technology Centre, Sydney, September 1987.

[18] M. Y. L. Chew, "High Strength Concrete in Singapore", "Utilisation of High Strength Concrete", Norwegian Concrete Association, 1993.

[19] M. Y. L. Chew, "Repair of Reinforcement Corrosion Induced Failure", XIXth IAHS World Congress, Ales, France, September 23–27, 1991.

[20] M. Y. L. Chew, "Laboratory Evaluation of Concrete Repair Systems", 16th Conference on Our World Concrete & Structures, 26–27 August 1991.

[21] M. Y. L. Chew, "Efficient Maintenance: Overcoming Building Defects and Ensuring Durability", Conference on Building Safety, The Asia Business Forum, Kuala Lumpur, 4 & 5 April 1994.

[22] M. Grant, *Scaffold Falsework Design to BS5975*, Viewpoint Publication, 1982.

[23] J. R. Illingworth, *Temporary Works: Their Role in Construction*, T. Telford, London, 1987.

[24] MOL, "Code of Practice for Examination and Test of Suspended Scaffolds for Approved Persons, Singapore, Ministry of Labour, 1989.

[25] S. Champion, *Access Scaffolding*, Longman, Chartered Institute of Building, Harlow, 1996.

[26] Singapore Institute of Standards & Industrial Research, "Singapore Standards Code of Practice 14 — 1980: Scaffolding", Singapore Institute of Standards & Industrial Research, 1980.

[27] Singapore Institute of Standards & Industrial Research, "Singapore Standards Specifications 280 — 1984: Frame scaffoldings", Singapore Institute of Standards & Industrial Research, Singapore, 1984.

[28] Singapore Institute of Standards & Industrial Research, "Singapore Standards Specifications 311 — 1986: Steel Tubes & Fittings Used in Tubular Scaffolding Detail", Singapore Institute of Standards & Industrial Research, 1986.

[29] Singapore Institute of Standards & Industrial Research, "Singapore Standards Code of Practice 20 — 1981: Suspended Scaffolding Detail", Singapore Institute of Standards & Industrial Research, 1981.

[30] R. Doughty, *Scaffolding*, Longman, London, 1986.

[31] Premier Conference, "Metal Scaffolding Conference", 24–25 August 1981 Singapore, Premier Conference, 1981.
[32] ICE, "Access Scaffolding: ICE Works Construction Guide Detail", Thomas Telford, London, 1981.
[33] R. E. Brand, *Falsework and Access Scaffolds in Tubular Steel*, McGraw-Hill, UK, 1975.

CHAPTER 8

EXTERNAL WALL CONSTRUCTION

8.1. General

A building envelope serves the functions of weather and pollution exclusion, thermal and sound insulation. It also provides adequate strength, stability, durability, fire resistance, aesthetics appeal, etc. (Figure 8.1).

The external walls of traditional buildings are mainly made of masonry and/or reinforced concrete. They are usually finished with cement render and painted or finished with various cladding materials.

Cladding materials ranged from small tiles/mosaics fixed to the substrate by adhesion, to large metal/stone panels by anchorage methods. They form the covering or dressing for the substrate/wall behind them. Tiles of various modular sizes are either hand-laid or incorporated into the surface during the prefabricating process. Details on the design and construction of tiling works are given in [1]. The use of tiles based on adhesives, e.g. ceramic tiles and mosaics on external walls of tall buildings has lost its popularity due to cases of adhesive failures. The main causes for the falling of tiles are as follows [2]:

- Deformation of the adhesive (or mortar) onto which the tiles have been laid due to shrinkage etc.
- Differential movements between the tile, adhesive and the immediate substrates, due to heat, moisture or other effects.
- The strength of the original rendering (substrate) is inadequate.
- Structural movements, vibrations or settlement problems.

- Improper surface preparation such as inadequate cleaning, no provision of proper keys.
- Improper design and selection of materials.
- Improper sequence of work.

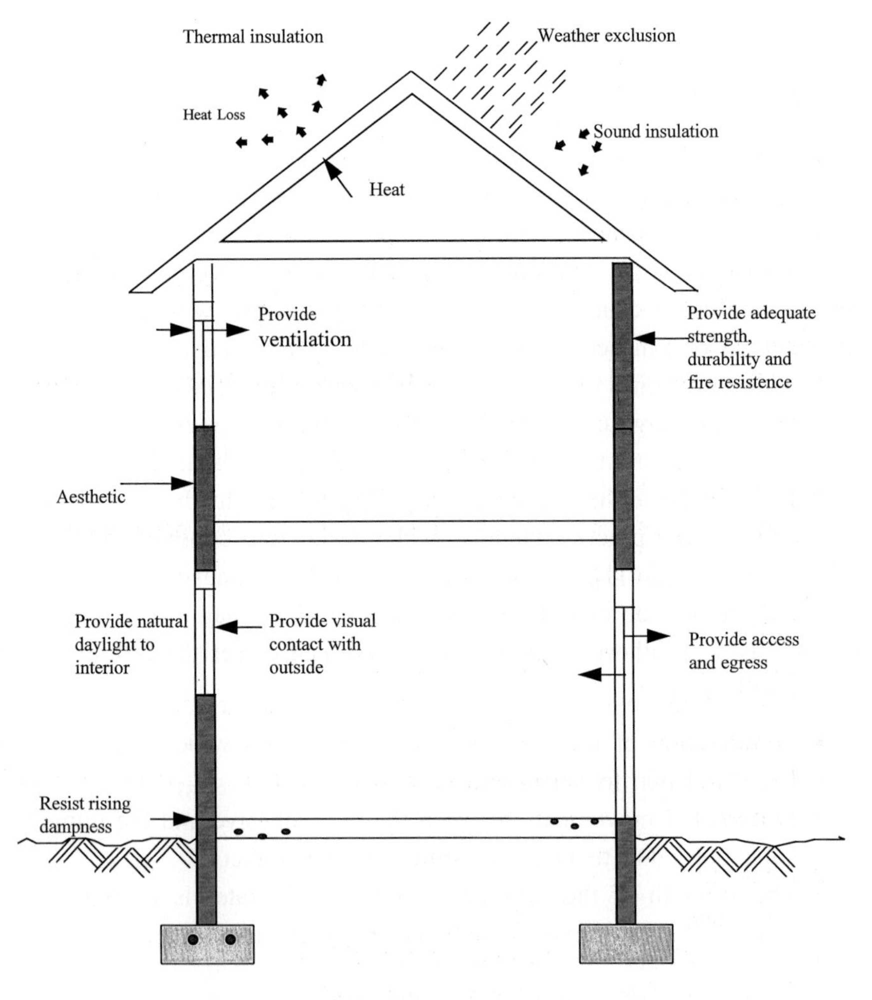

Figure 8.1. Functions of the building envelope.

With the advancement of prefabrication, large panel curtain walls have become popular especially for tall commercial buildings. A curtain wall may be defined as a thin wall, supported by the structural steel or concrete frame of the building, independent of the wall below, and is designed to envelop the building and act as a barrier to elements of nature.

Considerations to be given to the design, construction and maintenance of external walls include:

Visual

- Panel shape and size.
- Joint locations, joint sizes.
- Daylighting, nightlighting.
- Blinds, shades.
- Materials, colours, finishes.
- Integration with interior design, e.g. cabling behind partitions.

Integrity

- Air and water tightness — sealing, drainage, indoor air quality.
- Loading — static, dynamic, fatigue.
- Movements — load, thermal, moisture.
- Exceptional loads — blast, intrusion, impact.
- Fire — resistance, reaction, spread vertically and horizontally.

Physics/Environment/Comfort

- Heat transfer.
- Lighting.
- Sound transmission — noise from street, next room.
- Ventilation.
- Moisture — rainwater, humidity, condensation, degradation, mould growth.

Buildability

- Tolerance.
- Pre-assembly — stick, unitised, panelised.
- Quality — QA, factory work, site work.

<u>*Maintenance*</u>

- Access — cleaning, inspection, repair, replacement.
- Life cycle — component life, inspection cycle, repair cycle.
- Serviceability — cleaning, repairability, replaceability.

8.2. Precast Concrete Panel

This is a common form of precast cladding and is used for both loadbearing and non-loadbearing purposes. The main advantages of using precast as opposed to *in situ* cladding are (a) speed of erection, (b) freedom from shuttering on site and (c) better quality. Cranes are used to hoist the concrete panels into position and bolted and/or welded onto anchorages (Figure 8.2). The joints between precast concrete panels are sealed with either one-stage or two-stage joints. One-stage joints are those which are simply sealed against water penetration by the use of a sealant near the outer surface. Two-stage joints are those which have a sealant near the outer surface and an air-seal usually close to the inner face of the panels. Between the two is a chamber which must be vented and drained to the outside (Figure 8.3). Figure 8.4 shows two panels butt jointed together with a neoprene gasket in between.

Although dimensional inaccuracies in manufacturing are small (±3 mm), those arising during assembly due to the deviations in the frame and erection deviations on the panels (possibly ±25 mm) and movement deviations (usually in the order of 1 mm per 3 m) necessitate the allowance of tolerance in the method of fixing and in the clearance between the panels and the structure.

8.3. Glass Reinforced Polyester (GRP)

This is a composite material with resin and glass fibre treatments. It is characterised by its high strength, lightweight, and low modulus of elasticy. GRP exhibits good corrosion and weather resistance making it suitable for external use. Generally, a panel is fixed rigidly to one length while the other attachment points allow movements to occur. There must also be tolerance and provisions for the fixings to move to accommodate any inaccuracies during the assembly.

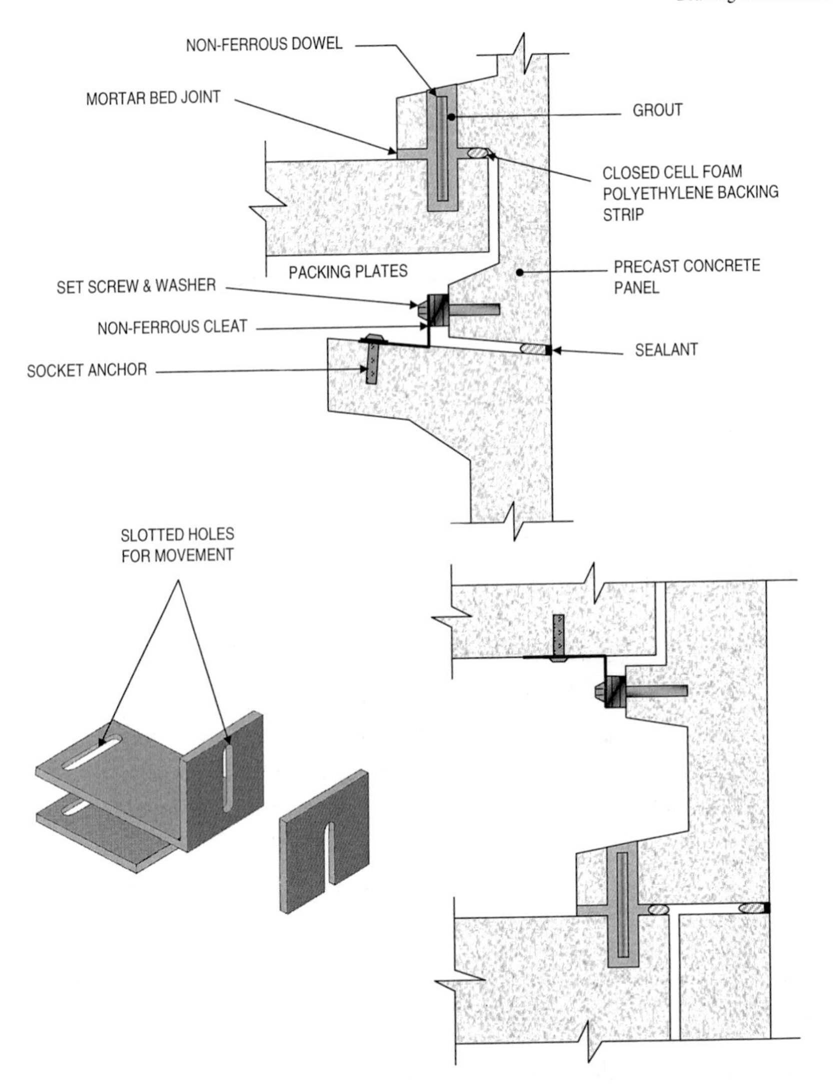

Figure 8.2. Fixing details of precast reinforced concrete panel.

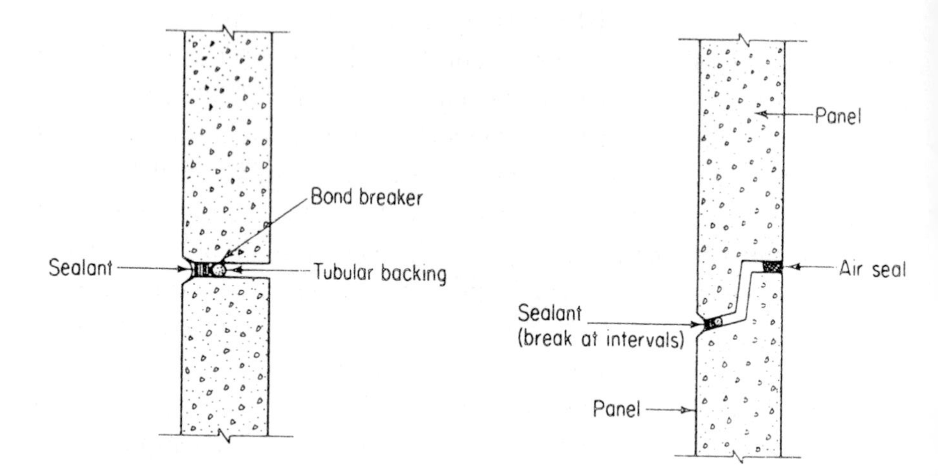

Figure 8.3. One-stage joint and two-stage joint [17].

Figure 8.4. Butt joint with neoprene gasket between two panels.

8.4. Glass Fibre Reinforced Concrete (GRC)

This is a composite material consisting of ordinary Portland cement, silica sand and water mixed with alkali-resistant glass fibres. The inclusion of glass fibres in the mix increases the compressive and tensile strengths as well as durability. The composite is non-combustible, with low thermal and moisture movements and has a higher strength to weight ratio. Thinner panels are hence possible with smaller concrete covers compared with the normal reinforced concrete. Fixing of GRC panels using angle cleats and dowels are similar to those used for precast concrete panels. Allowance for movement is imperative as GRC panels exhibit movements approximately double those of precast concrete. It is therefore usual to find that panels are supported at the base and restrained at the top by slotting them into oversized holes with frictionless washers or neoprene bearing pads.

8.5. Curtain Walling

Curtain walls are non-loadbearing external walls of buildings composed of repetitive factory assembled elements, with its dead weight and wind loading transferred to the structural frame through anchorage points. The most common materials for curtain walling of tall commercial buildings are natural stone, metal sheeting and glass.

Colomban [3] reviewed the history of the development of curtain walling by classifying it into three generations.

8.5.1. *First Generation (1800–1960)*

Generally based on the fixing of vertical mullions to which horizontal rails or transoms, frames and insulated panels are attached as shown in Figure 8.5. This system with mullions and transoms or more commonly called the stick system has been found to have the following limitations:

— It involves mainly *in situ* works which requires high standard of workmanship and quality control.

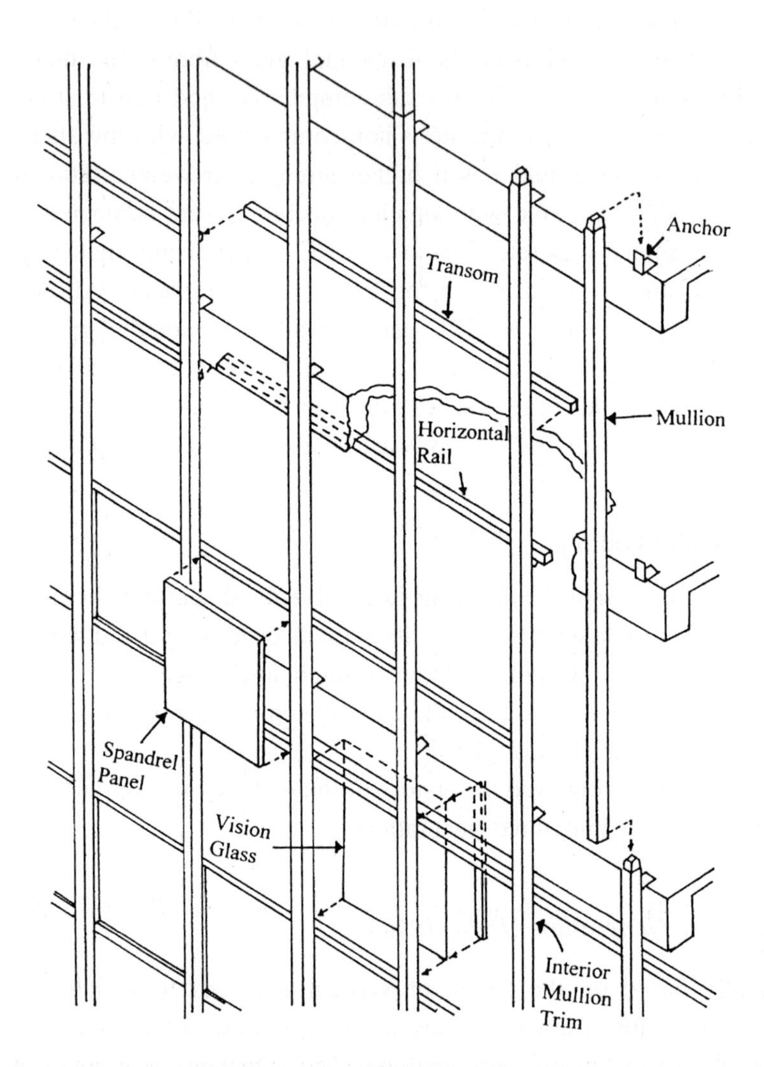

Figure 8.5. A typical "stick system".

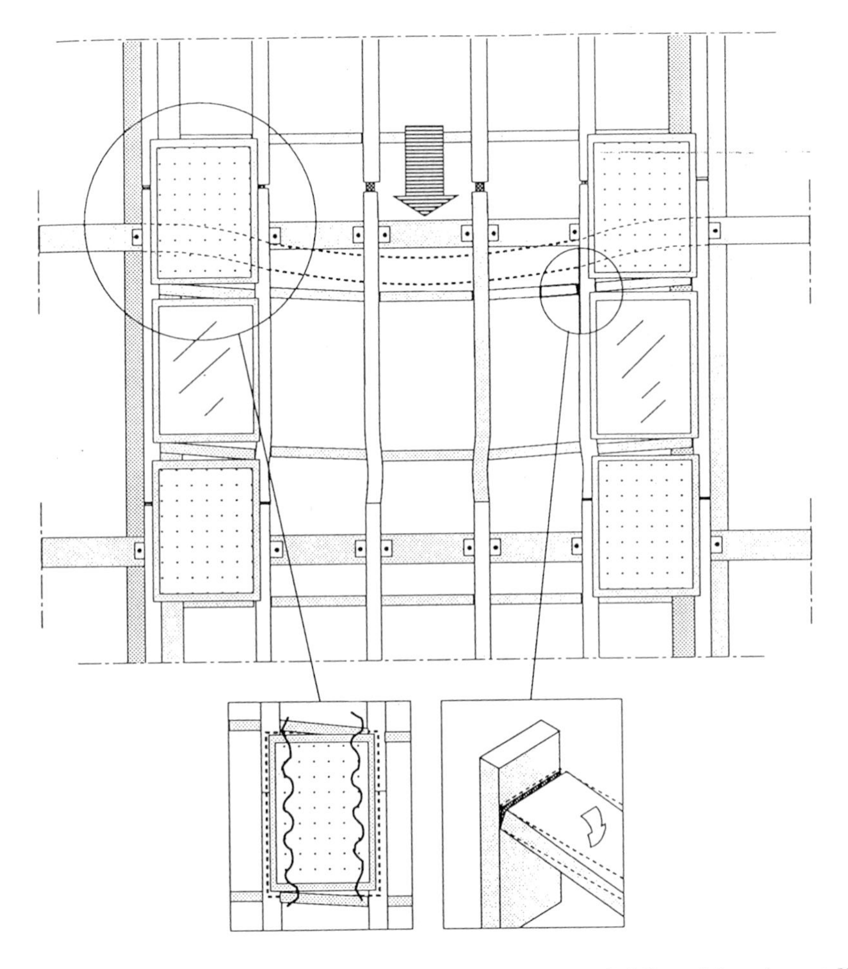

Figure 8.6. Movements due to moisture, temperature, creep and differential settlement [3].

— It provides insufficient allowance for movements due to temperature, moisture, creep and differential settlement (see Figure 8.6).
— Water resistance relies solely on gaskets and sealants.
— Lack of floor-to-floor flashing which makes it almost impossible to locate a leak caused by water entry at mullion sleeves (see Figure 8.7).

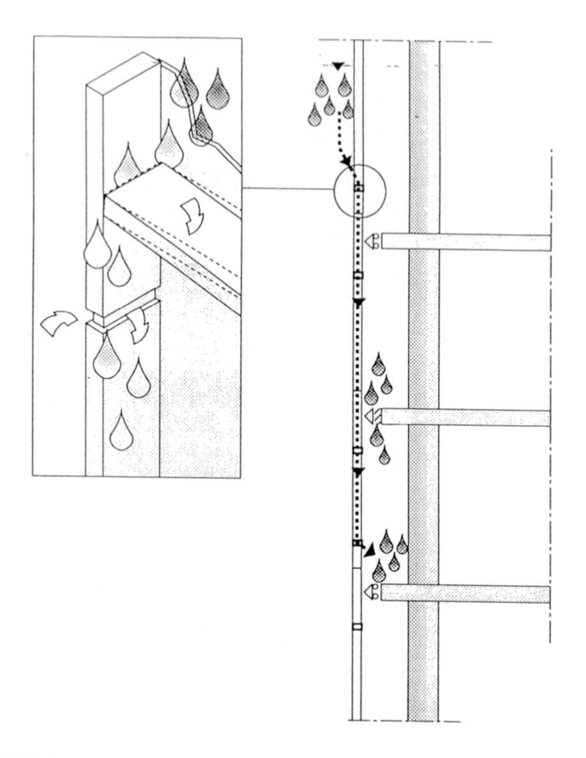

Figure 8.7. Moisture movements from floor to floor without flashing [3].

8.5.2. *Second Generation (1960–1980)*

The second generation is characterised by:

(a) *pressure equalisation system* which no longer relies totally on the closure of all holes but rather on the equalising pressure in the cavity between external and internal skin. The principle of pressure equalisation system is through eliminating the pressure difference at the level of the external joint as shown in Figure 8.8. The external wall is not sealed to create an inner chamber that has equal pressure as the outside. The inner wall is sealed to prevent both air and water penetration. Thus the concept employs:

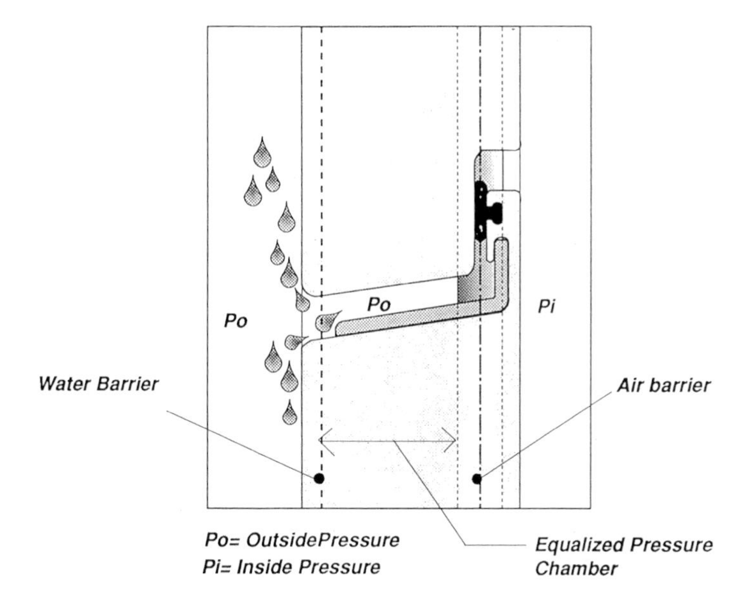

Figure 8.8. Pressure equalisation system [3].

- A rainscreen* to stop the initial flow of water.
- An inner chamber where the pressure is similar or equal to the outside.
- An inner seal.

Water that enters the chamber being heavier than air will fall and exit the chamber through weep holes. Positive drainage should be designed to control the water within the confines of each horizontal area.

---

*A rainscreen is the outer skin or surface of a wall or wall element — the part that is exposed to the weather. The rainscreen principle prescribes how penetration of this screen by rainwater may be prevented. Rainscreens contrast directly with face-sealed systems in that they are constructed with open joints instead. The rainscreen design assumes that the joints admit water.

(b) *panel or unitised system* (see Figure 8.9) which is completely finished at the factory with consistent quality control.
(c) water barrier between floors (see Figure 8.10).

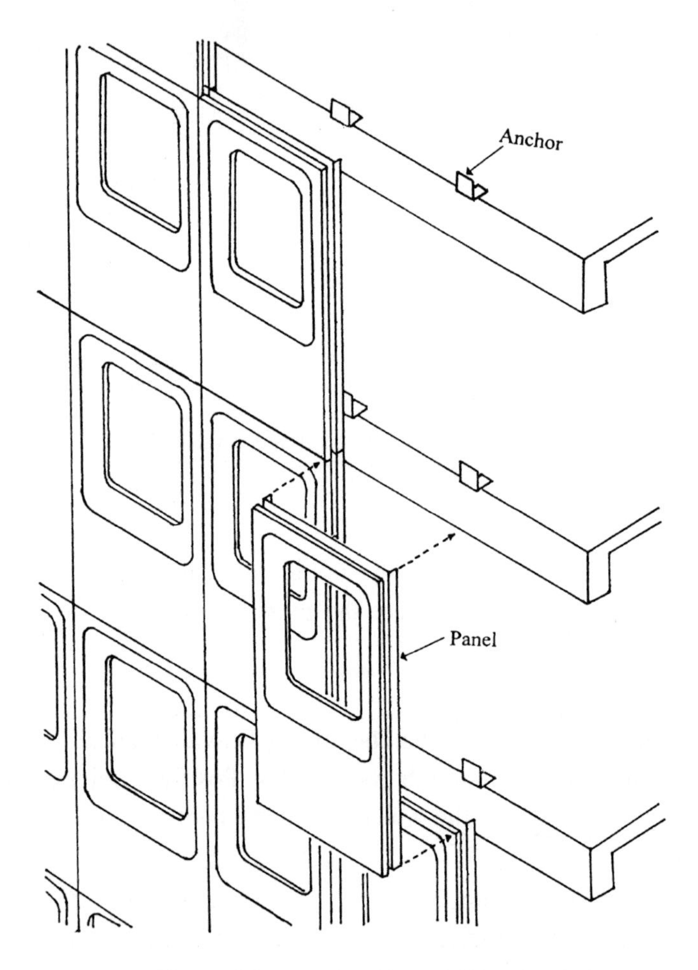

Figure 8.9. A typical "panel system".

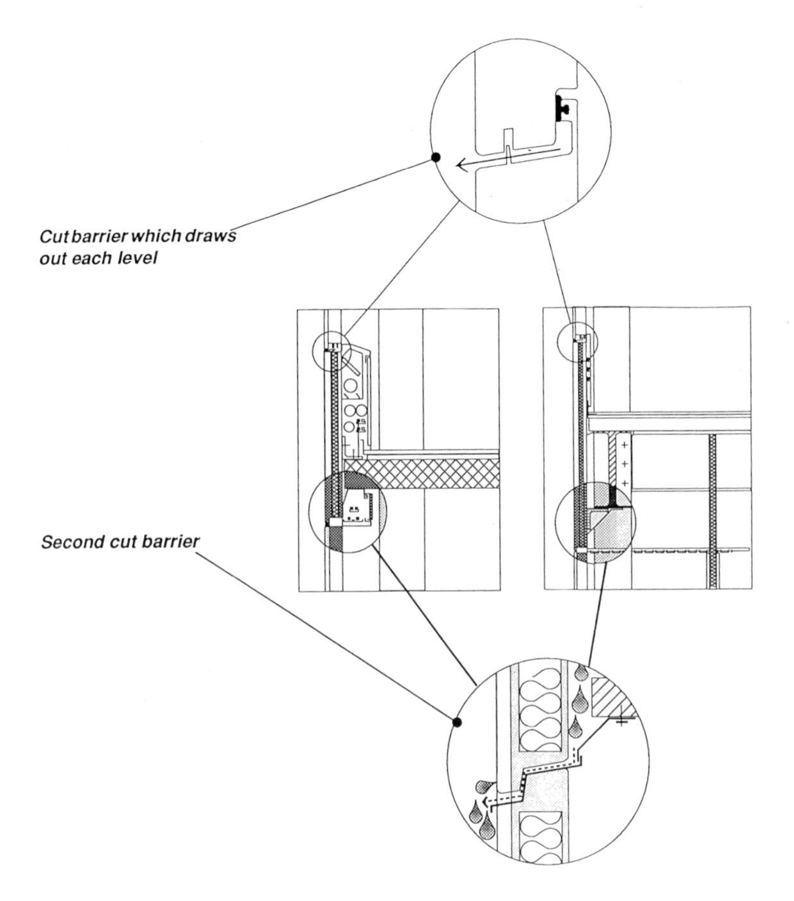

Figure 8.10. Water barrier between floors [3].

8.5.3. *Third Generation (1980–1990+)*

The third generation is characterised by an improvement of the techniques and by diversification of their use. The wide use of structural sealants and adhesives was observed especially on glazed curtain walls where glass is bonded to the frame.

Colomban's envisages the fourth generation or the curtain walling of the future to be of active walls designed and fitted with devices such as photocells, fluids, fans, exhausts, etc. to accommodate changes in thermal, air quality, lighting, etc.

Figure 8.11 shows the construction of facade for UOB Towers, currently the highest building in Singapore (280 m). It is clad with 45000 m^2 of curtain walling in 30 mm thick natural granite, 23000 m^2 of 26 mm thick double glass and 700 tons of aluminium extrusions as a pressure equalising system which allow for the downflow of rainwater through a channel.

8.5.4. *Typical Stick System*

The installation procedure of a typical stick system is as follows:

1. Set out main marking.
2. Cast anchorage angle plate to floor.
3. Fix fasteners.
4. Install vertical mullion.
5. Install horizontal transom.
6. Install spandrel panel.
7. Install window unit.
8. Install fire stop and insulation.
9. Insert cover plate.
10. Install gypsum board or partition.
11. Cleaning.

The anchorage angle plates are cast-in with the slab construction. A hook shape metal rod is welded to the anchorage angle plate to provide bonding with the concrete floor slab. Figure 8.12 shows the fastening detail at the end of a vertical mullion. A "T" piece is slid into the tubular mullion. Figure 8.13 shows the fastening detail at the mid-span of a vertical mullion welded to the "T" piece. Figure 8.14 shows the cross sectional view of a vertical mullion. After the installation of vertical mullions, horizontal transoms are fixed between them. Figure 8.15 shows the junction where vertical mullion and horizontal transom meet.

Spandrel panels and window units are loaded to position and screwed to the mullions and transoms. Figure 8.16 shows the handling and positioning

Figure 8.11. Facade of UOB Towers, Singapore, under construction.

Figure 8.12. Fastening of a vertical mullion to the concrete slab.

Figure 8.13. A vertical mullion welded to a "T" plate.

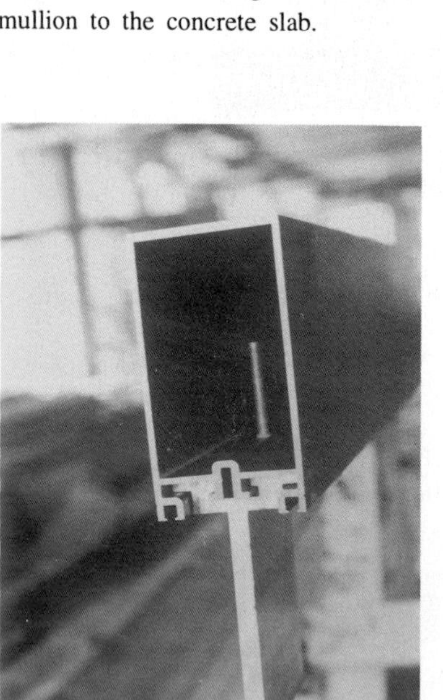

Figure 8.14. Cross-section of a mullion.

Figure 8.15. Jointing of a vertical mullion with a transom.

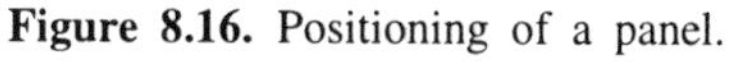

Figure 8.16. Positioning of a panel.

Figure 8.17. Gap between panels and slab filled with insulators.

of a panel for fixing using suction pads. Rockwool or similar insulation materials are installed behind the panel and the gap between the panel and the slab for thermal insulation and fire protection (Figure 8.17). The gap between the panel and the slab must be sealed as part of a fire protection to prevent fire and smoke from spreading from one floor to another.

Various fasteners can be used to cater for movements in various directions caused by wind (x-axis), weight (z-axis), thermal/moisture (y-axis), etc. Figure 8.18 shows a fastener allowing movement along the x-axis.

Figure 8.18. Allowance for movements using fasteners.

8.5.5. *Typical Panel/Unitised System*

The typical installation procedure of a typical unitised system is as follows:

1. A monorail system which is a small crane mounted on rails is fixed along the beam or slab edge of a building (Figure 8.19). A typical monorail crane can usually cover up to 10 floors of installation before it has to be re-sited.
2. A crane is used to hoist up the curtain walling panels which are usually packaged in crates, to the respective loading dock.
3. Other common methods of handling panels are the use of a gear and cog track pulley system, manual crane, etc. (Figure 8.20).
4. Panels are then transferred to their respective locations by monorails or other methods.
5. The alignment is checked before installation.
6. A panel is slided into its neighbour to ensure weather-tightness (Figure 8.21).

Figure 8.19. A typical monorail system.

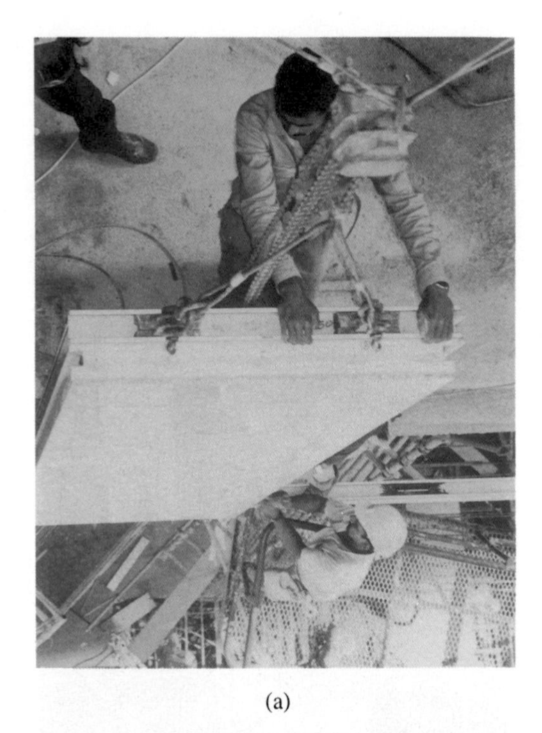

(a)

(b)

Figure 8.20. (a) The use of gear and cog track. (b) The use of a manual crane.

Figure 8.21. Sliding of a panel into position.

8.6. Water Penetration

The action of rainwater on a building facade depends on the rate of rainfall, the effects of wind, the building's shape and size, the surface configuration, and the moisture storage capacity of the surface materials. The action of rainwater on buildings with traditional facades such as cement rendering, masonry, etc. would be quite different from claddings made up of highly impervious materials, e.g. glass, metal etc. In the former case, rainwater will be absorbed, forming a film of saturation on the wall surface. After the rain stops striking the wall, this film of water starts to dissipate and the absorbed water evaporates. In the latter case, as there is little or no absorption, a substantial film of water flows across the outer face. Depending on the rate of rainfall, the effects of wind, etc., this film of water under the combined effects of gravity and wind action will concentrate at places of irregularities such as joints (Figure 8.22). Wind causes a lateral migration of the water film, which concentrates the downward flow along the lines of vertical protrusions and depressions [4–6]. This explains the extensive staining problem found even on relatively new facades.

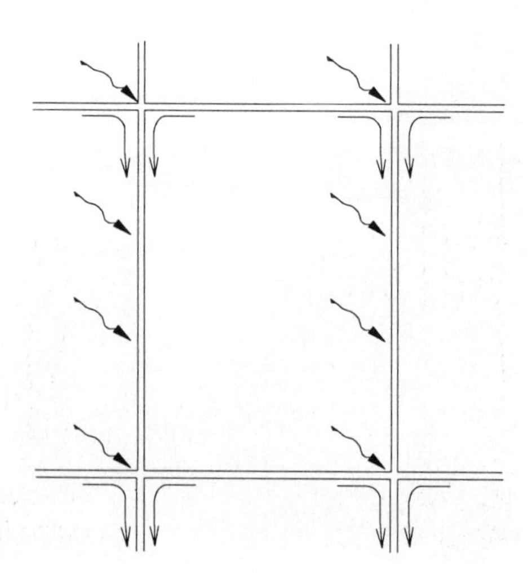

Figure 8.22. Concentration of rainwater flow on joints.

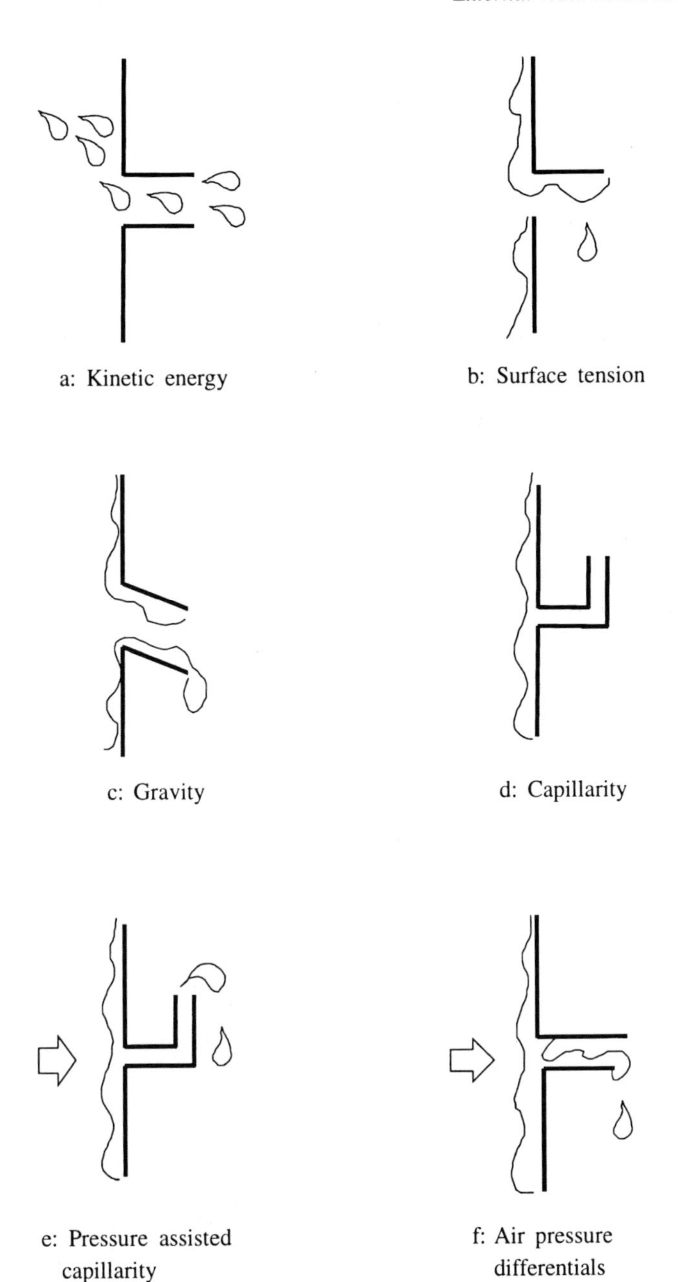

Figure 8.23. Mechanism of water penetration.

The mechanism of water penetration through joints involves the effects of kinetic energy, surface tension, gravity, pressure-assisted capillarity and wind-induced air-pressure differentials. Anderson *et al.* [7] describes the mechanism as shown in Figure 8.23.

It has been observed long ago that to achieve weather-tightness by attempting to maintain a completely unbroken impervious membrane at the outer wall surface was difficult due to continual movements. Other methods developed which have been shown to be successful include:

- Internal drainage system/secondary defence system — provides within the wall itself a system of flashing and collection devices, with ample drainage outlets to the outdoor face of the wall.
- Pressure equalisation/rainscreen principle — provides a ventilated outer wall surface, backed by drained air spaces in which pressures are maintained equal to those outside the wall, with the indoor face of the wall being sealed against the passage of air (Section 8.5.2).

The successful use of these methods depends on a clear understanding of the action of wind driven rain, careful detailing and proper installation. Ample weep holes or drainage slots, strategically located and properly baffled are important (Figure 8.24).

Figure 8.24. Weep holes strategically located for internal drainage.

8.7. Joints

There are three jointing methods for cladding panels: (a) filled joint, (b) gasket joint and (c) drained joint. The first two methods are classified as single-stage joints where the joint material functions both as a rain and air barrier. Drained joints, on the other hand, are classified as multiple-stage joints where the rain and air barriers are separated. Figure 8.25 shows the details of a filled joint, gasket joint and drained joint.

In a single-stage joint, in particular a filled joint, the use of sealant to completely seal off the joint is required. It is the simplest way of jointing and the sealant is usually applied flushed with the surface of the element. To reduce the effect of direct exposure to ultraviolet radiation, such joints may be set back from the face of the element, providing shade.

Gaskets, unlike sealants, are not adhesive and rely on pressure compressing each other to provide a weathertight seal. It is necessary for the material to be compressible, impermeable to moisture, with high resistance to environmental agents, and high elastic recovery. It is important to ensure dimensional accuracy during manufacture and to obtain the required pressure in-service to ensure effectiveness. Slight deformation of the panel or unevenness in the joint surface can result in complete failure of the joint.

The drained joint uses the neoprene baffle to divide it into two sections, with the outer section draining away most of the water which enters, and the air barrier in the inner section providing a second line of defence to any water by-passing the baffle in the first section. The incorporation of a wind and air barrier behind the baffle acts to reduce wind pressure difference across a cladding. The wind pressure differential, which provides the force required to drive water into the panel, is the primary mechanism causing water penetration. Hence, the incorporation of an air chamber which acts as a pressure equalisation compartment in drained joints is successful in ensuring weather tightness. In addition, the position of the air barrier away from the face of the joint ensures that it is shielded from the effects of sunlight. Another advantage of multiple-stage joints is that it does not require a close dependence on tolerance limits, and the risk of loss of adhesion of the jointing material is less critical.

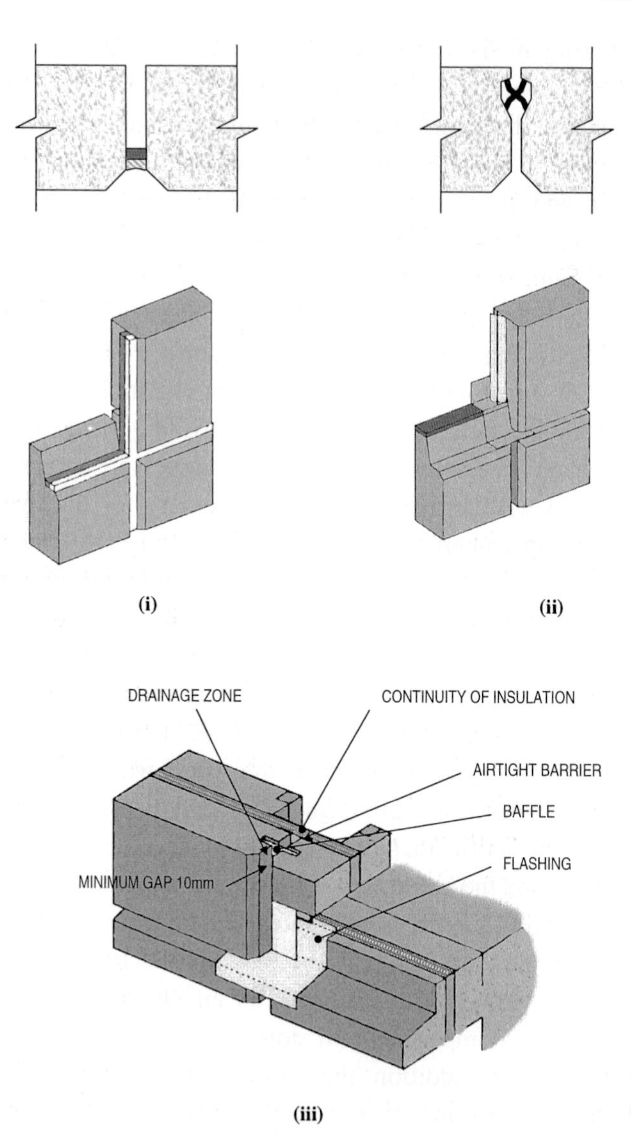

Figure 8.25. Details of a (i) filled joint, (ii) gasket joint and (iii) drained joint.

8.8. Wind Pressures

A significant characteristic of modern building design is lighter cladding, more flexible structural framing and unusual shape of the building in order to fulfil the architectural and aesthetic requirements. These features produce an increased vulnerability to large deflection and stresses in the building frame and glass and cladding become prone to wind damage.

When adjusted for building shape and height, the wind pressure values may result in local pressures or suctions on the building two to three times greater. Wind gusts lasting 3 to 5 seconds can exert forces on a building in excess of 2500 Pa. Wind loading may be modified by nearby structures which can produce beneficial shielding or adverse increase in loading. Overestimating lead in uneconomical design; underestimating may result in cladding or window failures.

In the case of a vented cavity, the variation in wind pressure along the four walls would cause air to flow inside the cavity as it moves from a region of positive pressure to one of negative pressure. This in turn influences how the wall performs under adverse weather conditions.

Figure 8.26(a) shows the plan view of the top floor of an office building. The dashed line represents the vented cladding, the solid one represents the wall's air barrier. It is assumed that the fan pressurisation account for the 100 Pa internal pressure, while the 160 km/h wind is responsible for the 1000 Pa positive pressure on the windward side. With continuous cavity all around the building and openings uniformly distributed in all four walls, −300 Pa negative pressure is assumed. The figure shows that where the wind is relatively free to circulate within a wall cavity, much of the wind pressure is resisted by the cladding and not by the air barrier. The flow of air can cause rainwater to circulate and can also damage insulation etc. [8].

This flow of air can be controlled by dividing the cavity into compartments both horizontally and vertically as shown in Figure 8.26(b). The air pressure acting on the cladding is the same as that acting on the air barrier, thus eliminating the pressure difference across the cladding. Rain cannot enter the cavity transported by the air as in non-compartmentalised cavities.

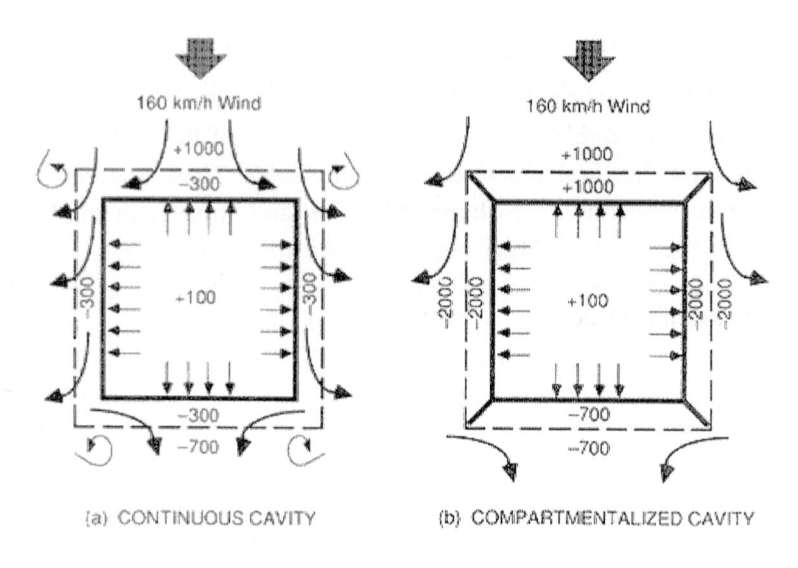

Figure 8.26. Cavity air flow, pressures and compartmentation [8].

Techniques have been developed for wind tunnel modelling of proposed structures which allow prediction of wind pressures on cladding and windows, as well as overall structural loading. Accurate knowledge of the intensity and distribution of the pressures on the structure permits adequate but economical selection of cladding strength to meet selected maximum design winds and overall wind loads for the design of the frame for flexural control.

Modelling of the wind loading on a structure requires special consideration of flow conditions in order to achieve similarity between model and prototype. A detail discussions on the similarity requirements and wind tunnel implementation can be found in [9–12]. In general, the requirements are that the model and prototype are geometrically similar, that the approach mean velocity at the building site has a vertical profile shape similar to the full scale flow and that the Reynold's number for the model and prototype is equal.

These criteria are satisfied by constructing a scale model of the structure and its surroundings and performing the wind tests in a wind tunnel specifically designed to model atmospheric boundary layer flows. Reynold's number similarity requires that the $(ulr)/h$ is similar for the model prototype, where: u = wind velocity; l = linear dimension fixing the scale; r = air density and h = viscosity of air.

Since h, kinematic viscosity of air, is identical for both, Reynold's numbers cannot be made precisely equal with reasonable wind velocities. To accomplish this the air velocity in the wind tunnel would have to be as large as the model scale factors times the prototype wind velocity, a velocity which would introduce unacceptable compressibility effects. However, for sufficiently high Reynold's numbers ($>2 \times 10^4$) the pressure coefficient at any location of the structure will be essentially constant for a large range of Reynold's numbers. Typical values encountered are 10^7–10^8 for full scale and 10^5–10^6 for wind tunnel model. In this range acceptable flow similarity is achieved without precise Reynold's number equality.

Wind pressure p_w, on the surface of a wind tunnel model will be given by an expression of the form:

$$p_w/q = f(ulr)/h) = C_p$$

where,

$$p_w = \text{wind pressure}$$

$$q = \text{velocity pressure}$$

$$C_p = \text{pressure coefficient}$$

When an airstream blows against an object, the pressure, p, at any point on its surface may be regarded as consisting of two parts — the static pressure, p_s which in a natural wind is the barometric pressure; and the excess $p - p_s$, caused by the presence of the object. This excess $p - p_s$ arises solely from the motion of the air with reference to the model. It will be called simply the wind pressure and will be denoted by p_w. If there is no wind or no object present then $p = p_s$ everywhere and the wind pressure is everywhere zero. The wind pressure may be either positive or zero; that is p, which by definition of p_w is equal to $ps + pw$ may be either greater or less than p_s or equal to it. In comparison with wind pressure carrying the negative sign, the lower numerical value corresponds to the higher absolute pressure. Thus -0.2 is a higher pressure than -0.4.

The expression applies only to geometrically similar bodies. The wind pressure, p_w could be measured in any convenient units, but there are advantages in using the velocity pressure as the unit. For bodies with sharp corners C_p is practically independent of the wind speed and the size of the model (that is, $f((ulr)/h)$ is a constant for any station) so that for a single value of it for any given shape of body the pressure at a corresponding point on a similar body of any size at any wind speed can be computed with the aid of a table of velocity pressures. The ratio C_p is a pure number independent of the units used so long as the pressures are all measured in the same units.

In the presentation of data for the distribution of wind pressure over the exterior surface of the structure it is commonly assumed that the interior is at a constant pressure equal to the static pressure, p_s that is p_w is zero for the interior.

The wind tunnel investigation is generally performed on a building or building group modelled at scales ranging from 1:150 to 1:400. The building model is constructed of plywood or clear plastic fastened together, glued or screwed. The structure is modelled in detail to provide accurate flow patterns in wind passing over the building surfaces. The building under test is often located in a surrounding whereby terrain may provide beneficial shielding or adverse wind loading.

The test model, equipped with pressure tap, is exposed to an approximately modelled atmospheric wind in the wind tunnel and the pressure at each tap measured. The model is rotated at different angles and another set of data recorded for each pressure tap.

Data are recorded, analysed and processed by a computer. Pressure coefficients are calculated by the computer for each reading for each wind direction on each pressure tap, and are printed in tabular form as computer readout. Using wind data applicable to the building site, representative wind velocities are selected for combination with measured pressures on the building model. Integration of test data with wind data results in prediction of peak local wind pressures for design of glass or cladding and include overall forces and moments on the structure for design of the structural frame. Pressure contours are drawn on the developed building surfaces showing the intensity and distribution of peak wind loads on the building. These results may be used to divide the building into zones where lighter or heavier cladding or glass may be desirable.

Figure 8.27 shows an example of a laboratory wind tunnel, modelling a 137 m high commercial building, undergoing a wind tunnel test.

Figure 8.27. A typical wind tunnel with a model under test.

8.9. Common Defects on Building Facades

An increasing number of defects on building facades have been reported in the last decade [12–19]. These include:

— cracking
— staining/discolouration
— sealant failures
— efflorescence
— rising damp/water penetration

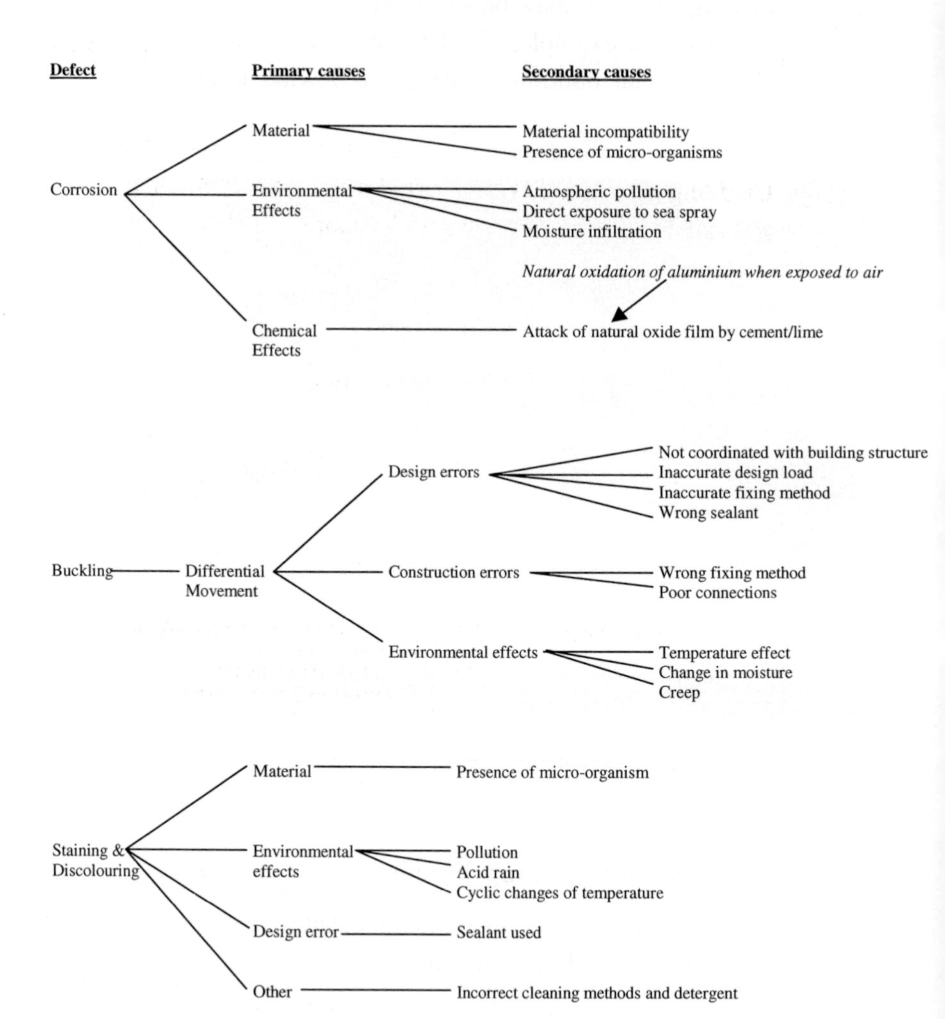

Figure 8.28. Defects and common causes of aluminium façade.

— abrasion
— corrosion
— tile delamination

Among the common causes (Figures 8.28–8.30) responsible for these defects are:

a. Action of wind pressures underestimated at the design and construction stage. Cladding elements can be affected by short duration high speed gusting, local high wind pressure and suction effects.

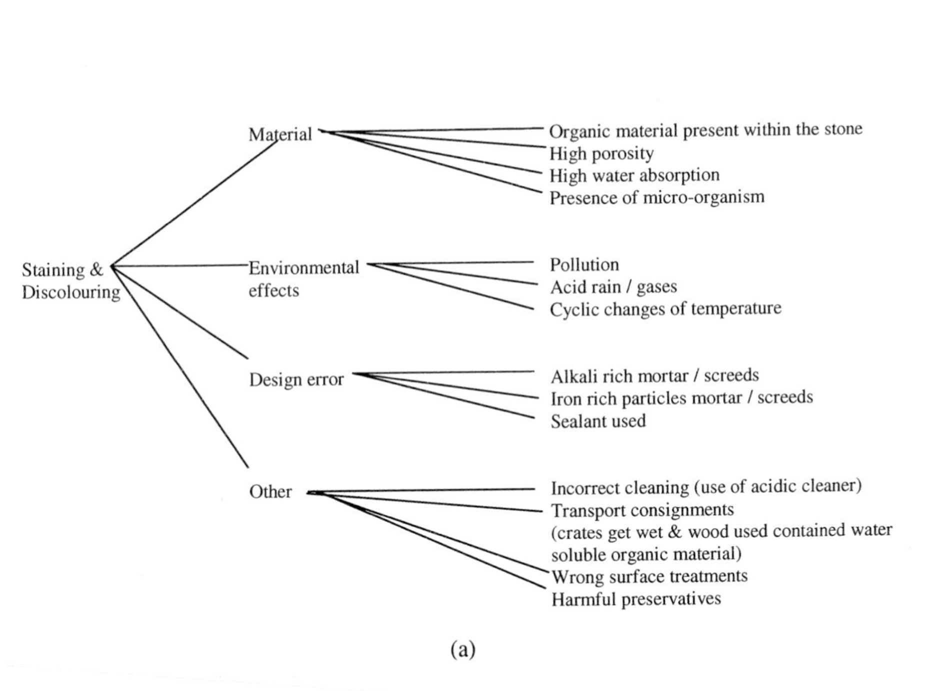

Figure 8.29. Defects and common causes of natural stone façade.

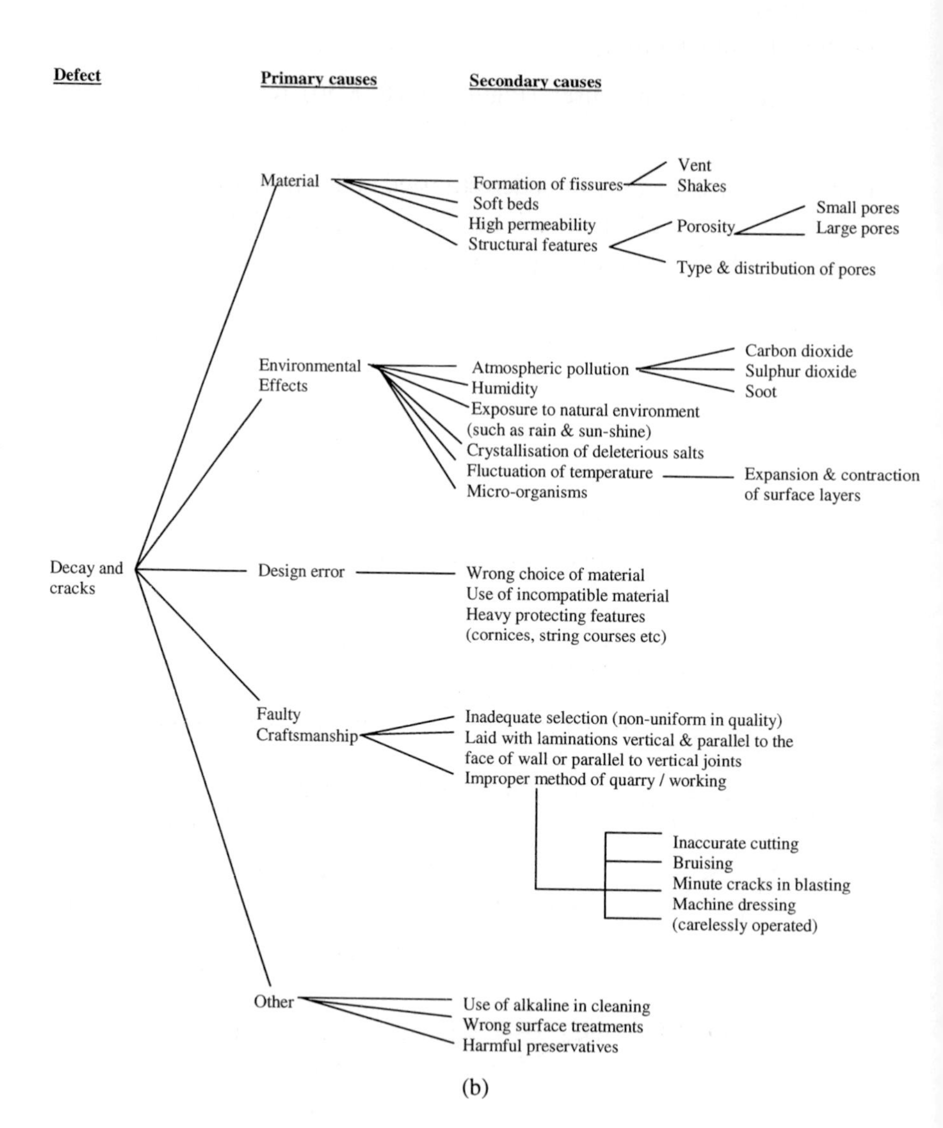

Figure 8.29. (*Continued*).

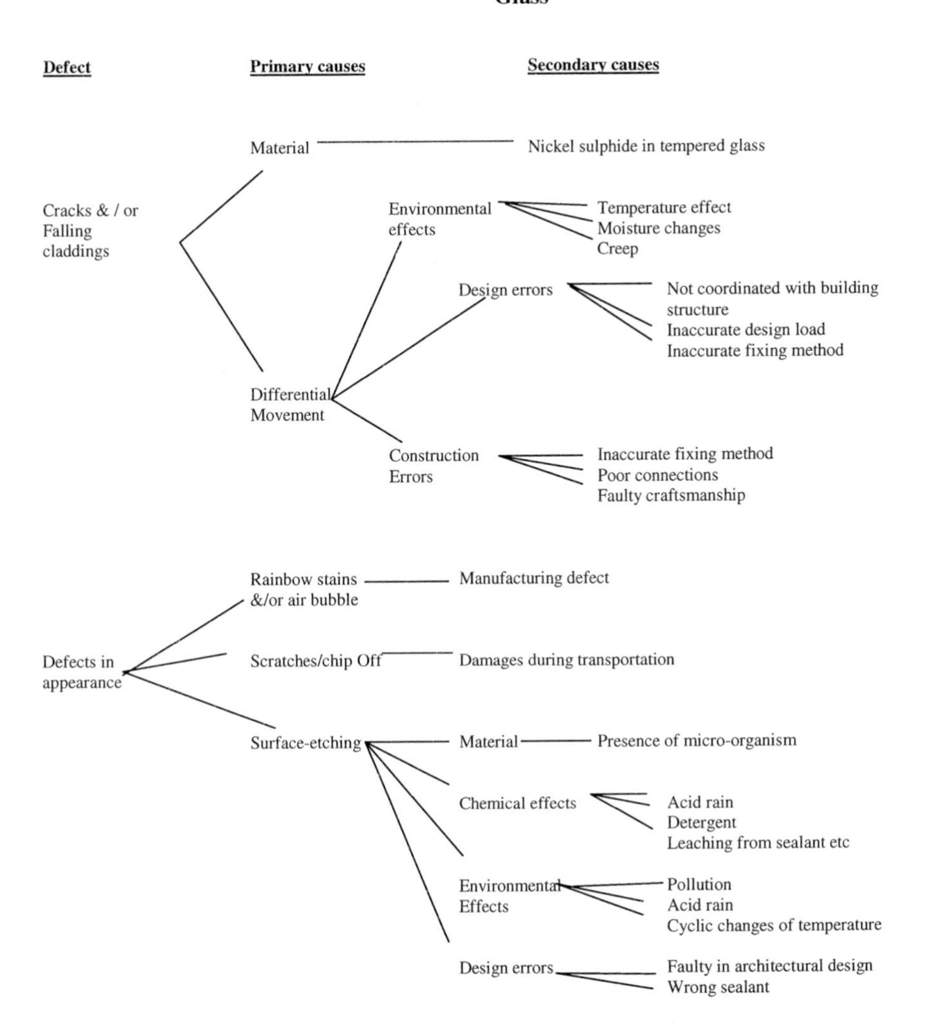

Figure 8.30. Defects and common causes of glass façade.

b. Insufficient knowledge on construction/fixing of the cladding elements. Even if suppliers of these fixing elements have already tested the elements to a certain standard, the actual use of such fixings may still need to be verified by appropriate tests depending on the situations.
c. Insufficient knowledge on the durability of fixings. The importance of design in affecting durability is often overlooked.
d. Failure to carry out necessary maintenance or incorrect identification of the cause of defect resulting in more damage done.
e. Failure of joint sealants due to: (1) poor joint design; (2) inappropriate choice of sealant; (3) poor surface preparation and workmanship in installation of the sealant; (4) rapid thermal/moisture changes resulting in excessive movement; (5) ultraviolet radiation and high temperature and humidity.
f. Inaccurate information from manufacturers resulting in inappropriate choice and use of materials.
g. Designer lacks the knowledge of the physical properties, performance and potential deterioration of the materials.

The manifestation of defects on high-rise building have necessitated even more regular inspection and maintenance to be carried out on these buildings.

References

[1] "Code of Practice for Wall and Floor Tiling", Singapore Productivity and Standard Board, 1998.

[2] M. Y. L. Chew, "The study of adhesion failure of wall tiles", *Building & Environment*, Vol. 27, No. 4, pp. 793–499, 1992.

[3] M. Colomban, "History and technical development of curtain walling", Proceedings on Building Envelope Systems and Technology, NTU, 1994.

[4] O. Birkeland, "Curtain walls", Norwegian Building Research Institute, Oslo, 1962.

[5] G. K. Garden, "Rain penetration and its control", CBD40, National Research Council, Canada, Ottawa, 1963.

[6] M. R. M. Herbert, "New ways with weatherproof joints", CP90/74, Building Research Establishment, Garston, Watford, 1974.

[7] J. M. Anderson and J. R. Gill, "Rainscreen cladding", CIRIA, Butterworths, 1988.

[8] U. Ganguli, "Wind and Air Pressures on the Building Envelope", BSI 86, National Research Council, Canada, 1986.

[9] A. J. Dutt, "Wind loading on a multiple hyperbolic paraboloid shell roof structure", Proc. International Conference on Space Structures, University of Surrey, Elsevier, 1984.

[10] P. Sachs, *Wind Forces in Engineering*, Pergamon Press, 1972.

[11] A. J. MacDonald, *Wind Loading on Buildings*, Applied Science Publishers, 1975.

[12] British Standard Code of Practice, CP3: Chapter V, Part 2: 1972.

[13] M. Y. L. Chew, C. W. Wong and L. H. Kang, "Durability of building facades under tropical conditions", Final Report, National University of Singapore, April 1996.

[14] M. Y. L. Chew, "Efficient maintenance: Overcoming building defects and ensuring durability", Conference on Building Safety, The Asia Business Forum, Kuala Lumpur, 4 & 5 April 1994.

[15] M. Y. L. Chew, "The study of adhesion failure of wall tiles", *Building & Environment*, Vol. 27, No. 4, pp. 493–499, 1992.

[16] M. Y. L. Chew, C. W. Wong and L. H. Kang, *Building Facades: A Guide to Common Defects in Tropical Climates*, World Scientific, 1998.

[17] R. C. Smith and C. K. Andres, *Principles & Practices of Heavy Construction*, Prentice-Hall, 1986.

[18] A. J. Brookes, *Cladding of Buildings*, Longman 1990.

[19] P. Ng, "External cladding" in *Building Maintenance Technology*, ed. C. Briffett, Singapore University Press, NUS, 1995.

CHAPTER 9

ROOF CONSTRUCTION

The roof forms the top part of the facade system to protect a building's interior. Its functions are thus similar to that of the facade. In addition, it has to be designed to take account of the following loads:

- Dead load
- Imposed load
- Wind load
- Loads incidental to construction
- Access and roof traffic
- Deflection and differential movement
- Lateral restraint and shear diaphragm

9.1. Structural Forms

There are essentially three main types of roof: (a) flat, (b) pitched and (c) shell. In general, a flat roof is chosen if the roof is to be used for storage, maintenance, etc. A pitched roof is chosen for simplicity in design and construction, for the relatively less maintenance and the good thermal insulation it provides. A shell roof is chosen mainly for aesthetic and architectural reasons.

The various structural forms of roof are illustrated in Figure 9.1 [1].

9.2. Beams and Slabs

This is a simple form of roof structure similar to that of a typical floor structure. It is the common structure forming flat roofs for many buildings.

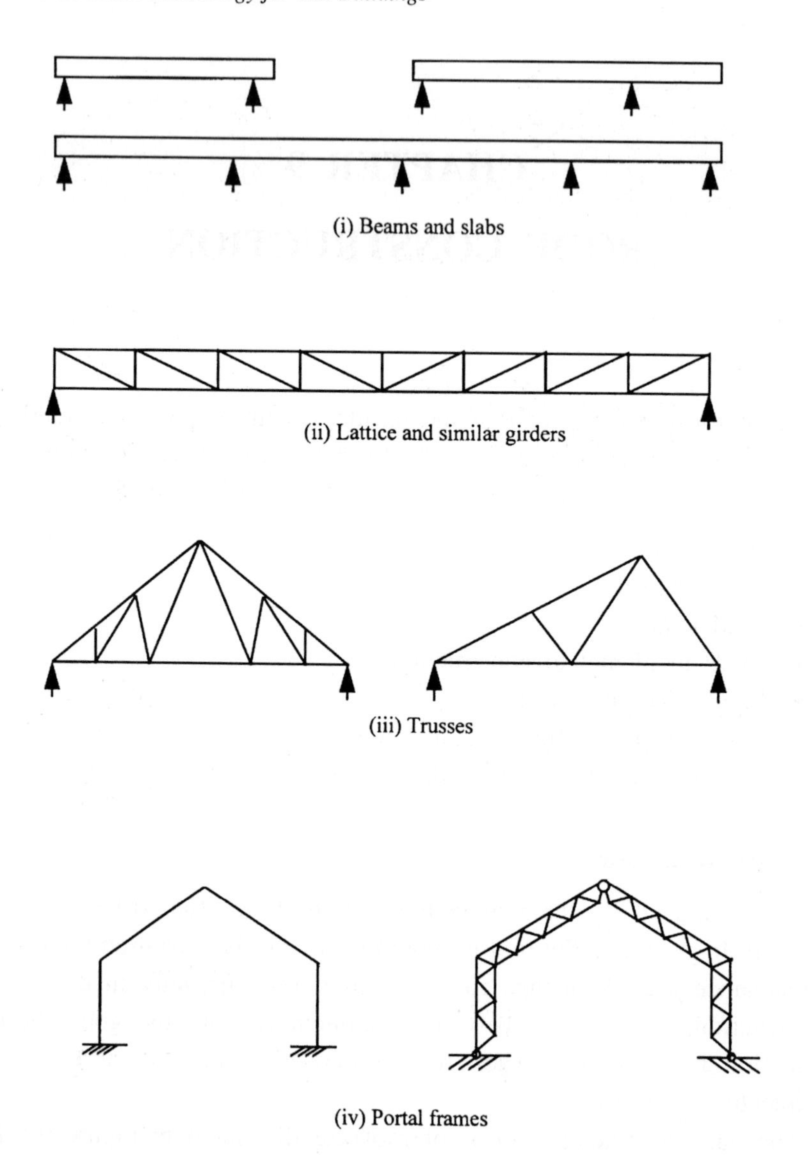

Figure 9.1(a). Various structural forms of roofs (i) beams and slabs, (ii) lattice and similar girders, (iii) trusses, (iv) portal frames.

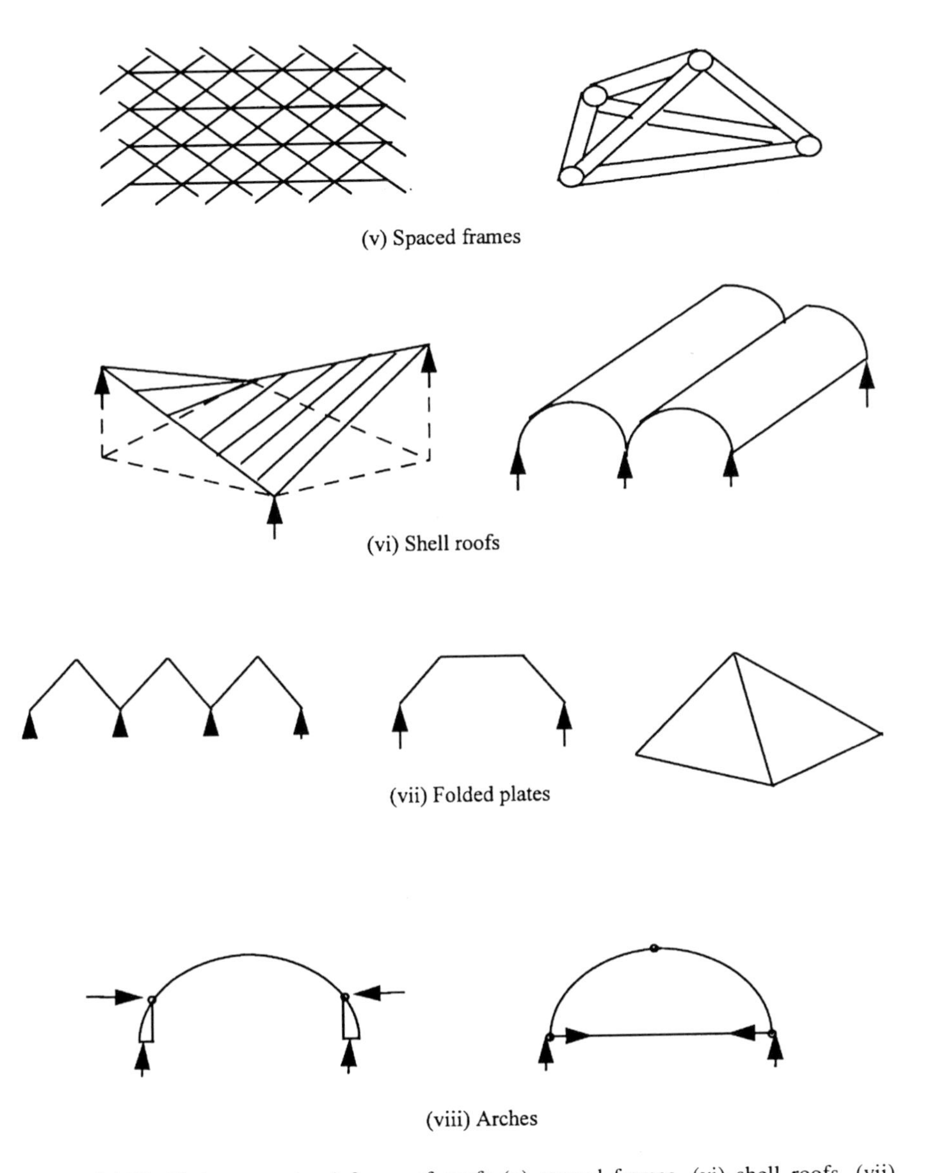

Figure 9.1(b). Various structural forms of roofs (v) spaced frames, (vi) shell roofs, (vii) folded plates, (viii) arches.

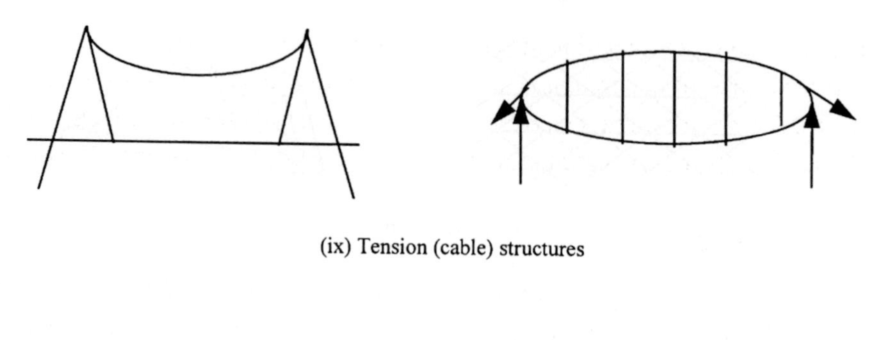

(ix) Tension (cable) structures

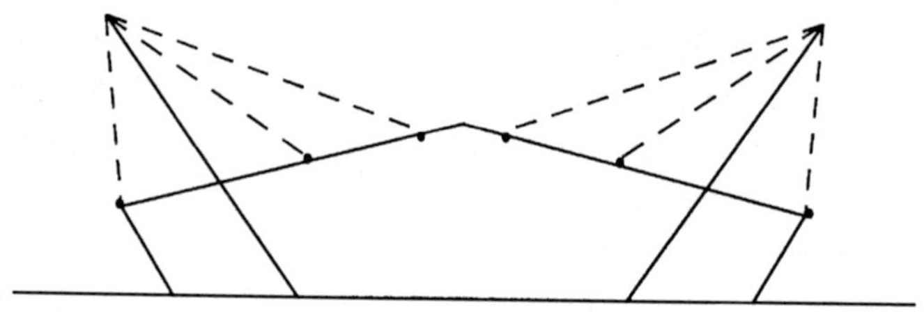

(x) (Cable-stayed (suspended) roofs

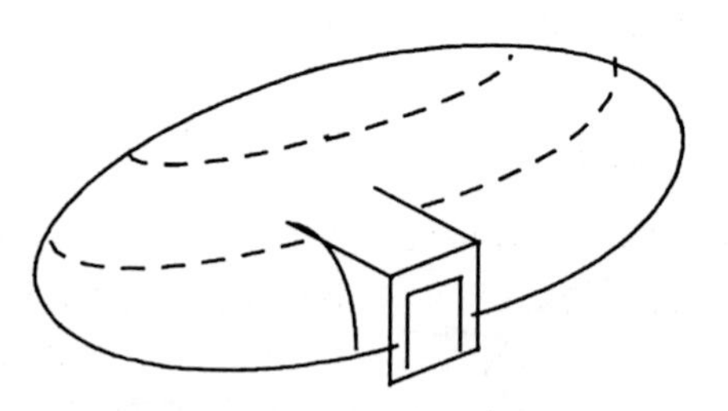

(xi) Inflated membrane structures

Figure 9.1(c). Various structural forms of roofs (ix) tension structures, (x) cable-stayed roofs, (xi) inflated membrane structures.

The most common materials used for this form of roof is concrete and steel or a composite of concrete and steel. The various section profiles are shown in Figure 9.2. Various types of steel and concrete sections have been used. Figure 9.3 shows an example of a prestressed concrete joist roof system.

9.3. Flat Roofs

A flat roof is normally described as a roof having a pitch of less than 10°. It is by far the most popular type of roof for tall buildings in the region for practical reasons. It is commonly used to house services such as water storage tanks, cooling towers, lift motor rooms, monorail for gondola, etc. (see Figure 9.4). It is thus necessary that it is designed to withstand all the superimposed loads, self load and any additional stresses resulting from the uses to which the roof is put.

It is characterised by a continuous waterproof layer over the whole roof area. The roof frame supports a roof deck which is covered with a waterproof roofing system. The roof frame can be a structural steel roof frame or a concrete roof system depending on the structural system employed for the superstructure. In some cases, for architectural reasons, a flat roof may be covered with other roof coverings/architectural features but is essentially still considered a flat roof.

9.4. Lattice and Girder

Square and hollow steel sections are common for this form of roof although precast prestressed concrete sections have also been used. Similar to a typical floor construction of a structural steel frame, the lattice and girder support the roof slab on the top and ceiling on the bottom, leaving a space between for services. Figure 9.5 shows the various profiles for a lattice and girder acting as the support for a roof slab.

9.5. Steel Sheet Roofs

It uses steel channels to support metal roofing sheets. A layer of reflective insulation sheet is laid to the entire underside of the covering. This system has been used extensively at one stage for residential flats.

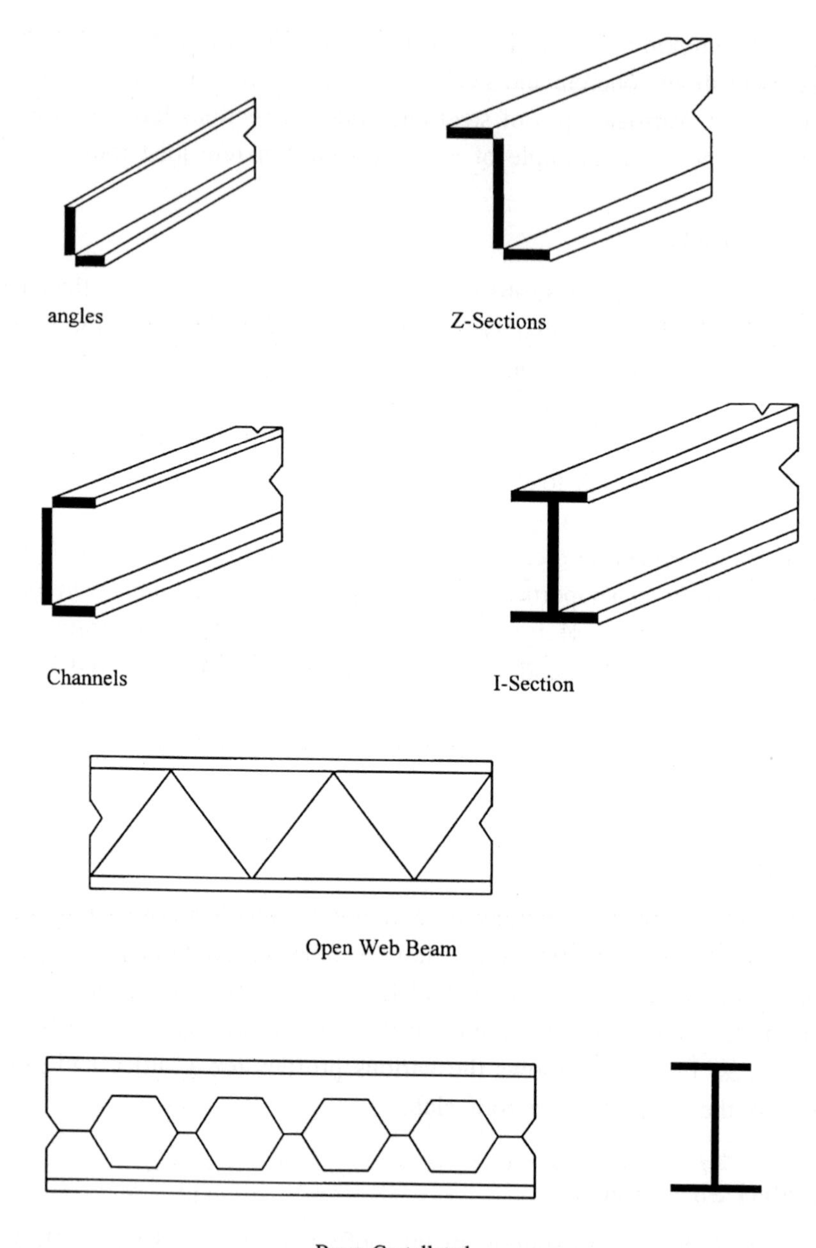

Figure 9.2(a). Section profiles of steel beams and slabs.

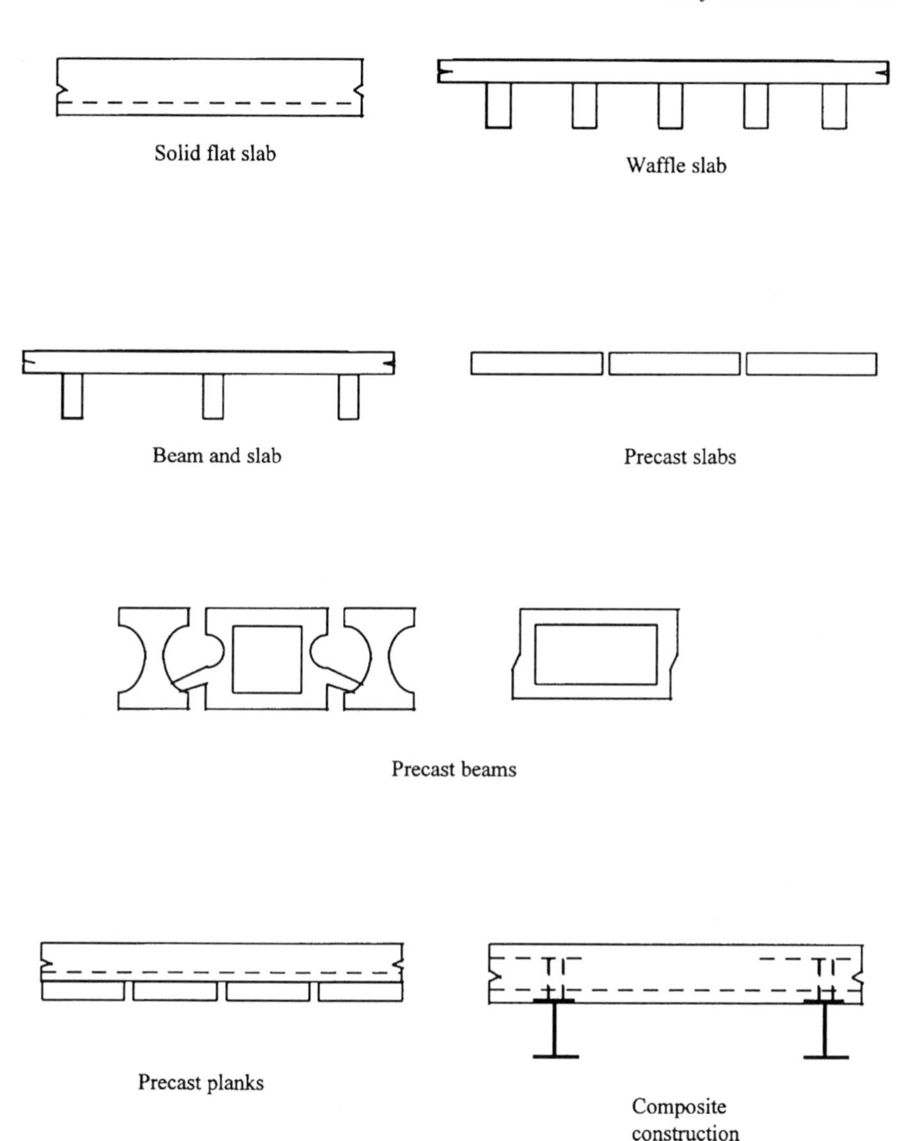

Figure 9.2(b). Section profiles of concrete beams and slabs.

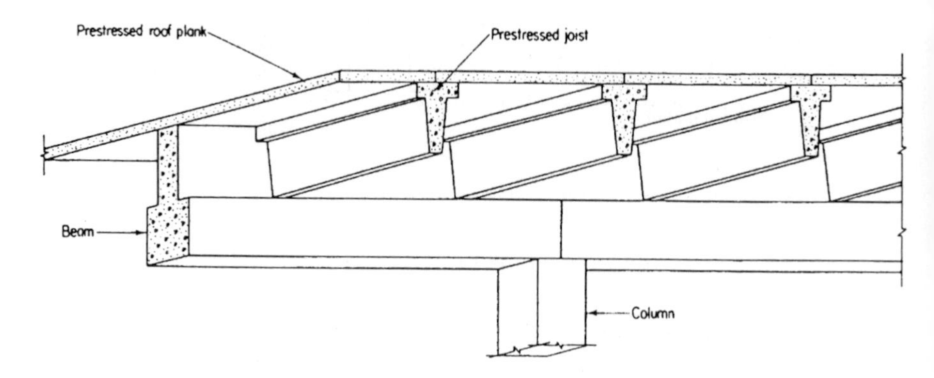

Figure 9.3. A typical prestressed concrete joist roof system [2].

Figure 9.4. A flat roof with cooling towers and other devices.

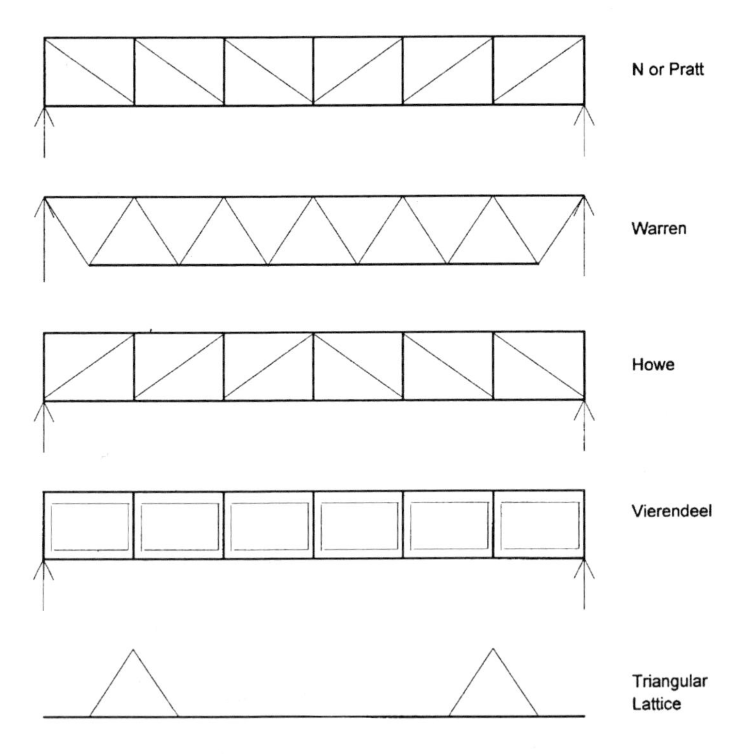

Figure 9.5. Section profiles of lattice and girders.

9.6. Truss Roofs

Truss roofs are generally made of straight members arranged and fastened together in triangular form, so that the stresses in the members caused by loads at the panel points are either compressive or tensile. The basis of the triangulation is the fact that if three members are joined together at their ends to form a triangle, no matter how loosely, that shape will remain unaltered unless one or more of the members is stretched or buckled. Figure 9.6 shows the various types of truss roofs [2–10].

Depending upon the materials and the structural systems used, trusses can span up to 30 m or more. It is therefore popular in situations where a large, clear, unobstructed floor areas are needed.

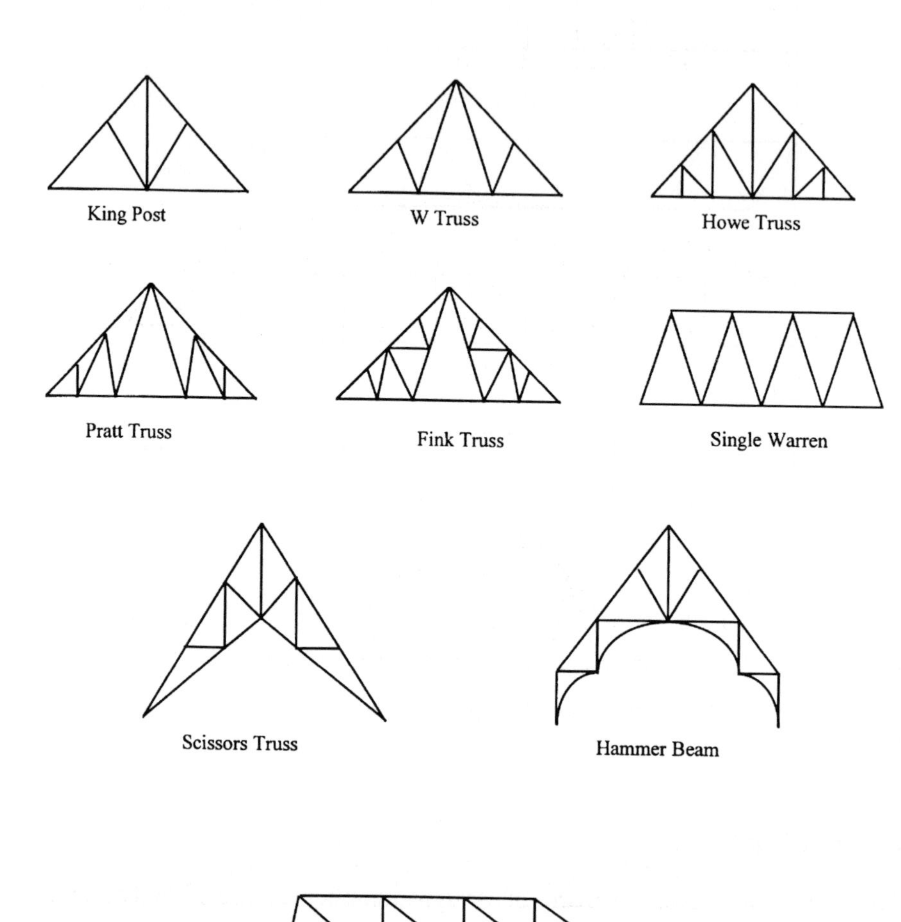

Figure 9.6(a). Section profiles of trussed roofs.

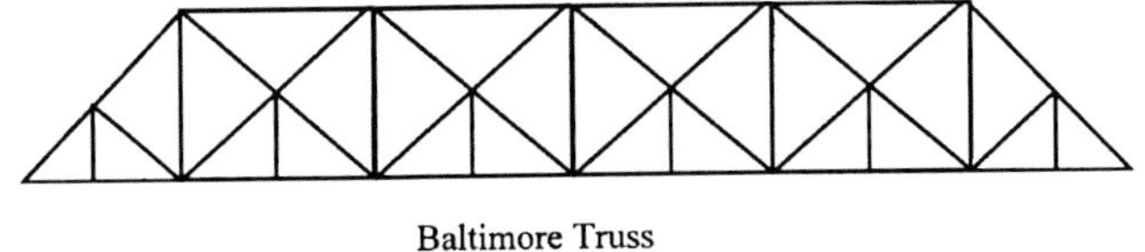

Baltimore Truss

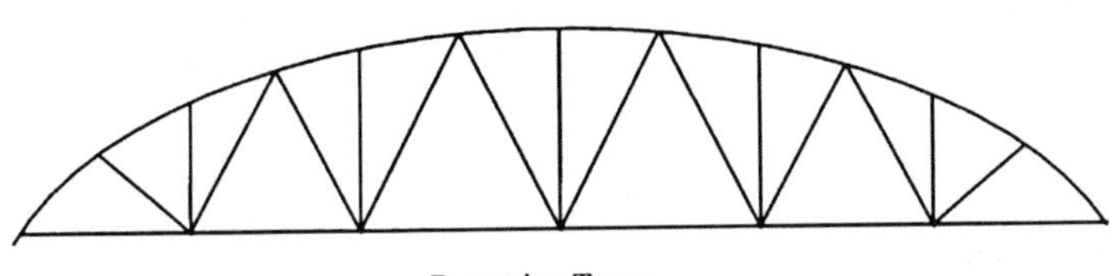

Bowstring Truss

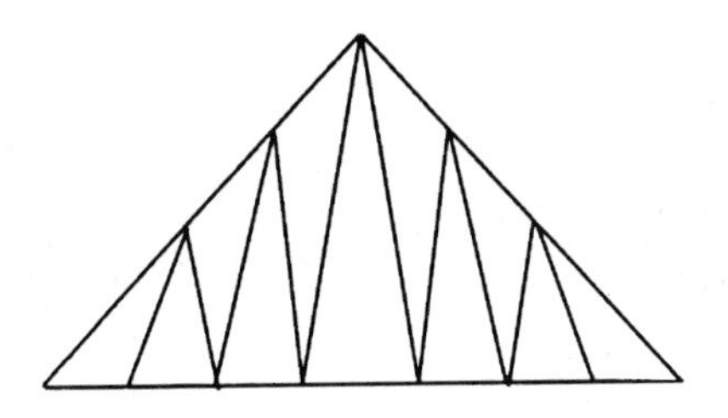

Belgian Truss

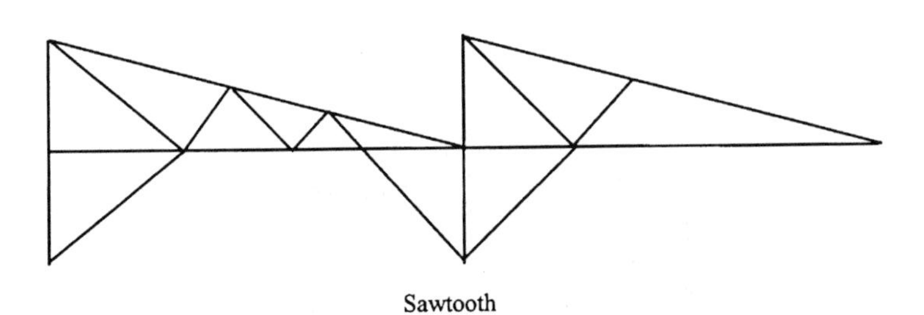

Sawtooth

Figure 9.6(b). Section profiles of trussed roofs.

The construction sequence of a roof using steel trusses is shown below. The roof structure weighs 167 tonnes and include 15 steel trusses braced by purlins on which a metal roof finishes was placed. Each truss unit has a clear span of 36 m, with no intermediate column.

Figure 9.7. Temporary storage area for steel trusses.

Figure 9.8. Hoisting of steel trusses.

- Trusses were fabricated, tested, painted in the factory and brought to the site (Figure 9.7).
- The steel trusses were hoisted to their required position by the crane (Figure 9.8).

Figure 9.9. The steel truss sits temporarily on base plates.

Figure 10. Connection of purlins between the main trusses.

- The steel trusses sat temporarily on the base plates secured by hold down bolts preformed during the casting of roof beams (Figure 9.9).
- Purlins were connected between the main trusses to act as lateral bracing between the trusses (Figure 9.10).
- Installation of the fascia trusses (Figure 9.11).
- Tack welding of the main trusses to the base plates (Figure 9.12).

Figure 9.11. Installation of fascia truss.

Figure 9.12. Tack welding of the main trusses to the base plates.

9.7. Space Frames

Space frames are three-dimensional grid structures or "space structures" constructed with lattice or grid frameworks. These structures usually provide a simple and economical method of covering very large areas without internal intermediate support. Grids are stiff and is useful for situations where deflection and not load is the criterion, e.g. in the case where the span is great and the load is small. Space frames possess a positive aesthetic quality by having a molecular-like structure. Extra loads may be applied

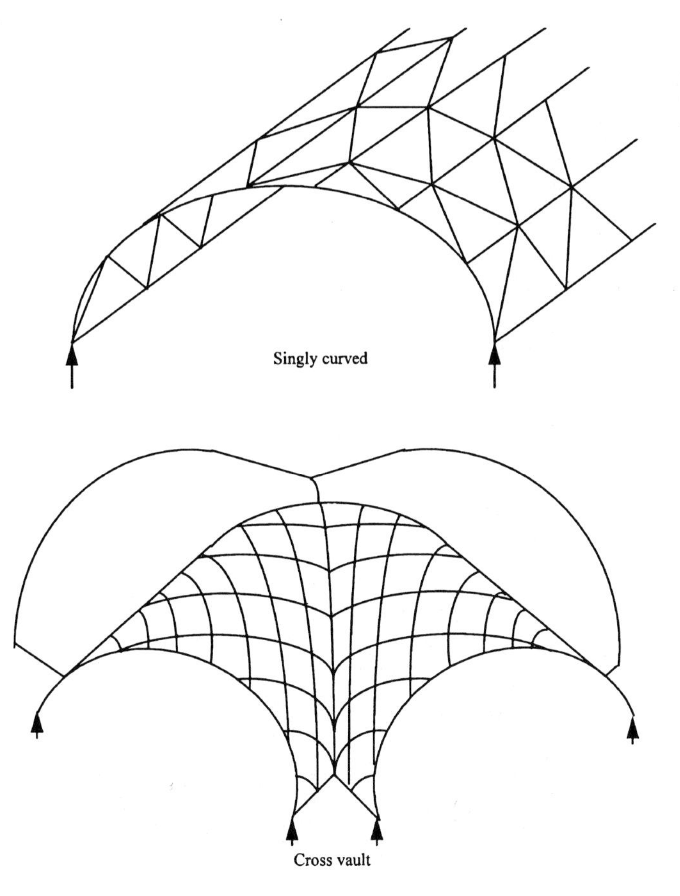

Figure 9.13(a). Section profiles of a single-layer grid spaced frames.

without inducing very severe concentrated loads in individual structural member. Provided the system is not overloaded, excess load is generally transmitted to adjacent members, which results in establishing and achieving a new condition of equilibrium. As a system, space frames comprise of great rigidity, inherent redundancy, and relatively higher factor of safety [10]. Other than being aesthetically pleasing, structurally efficient, great versatility in design and economically efficient especially for long span, space frames allow more considerable design freedom for alteration and extension as compared to the conventional steel trusses due to its unimpeded future extensions in any direction. The various profiles of a space frames in shown in Figure 9.13. The frame can take the form of a cylinder, intersecting cylinder, hyperbolic, dome, prism, etc.

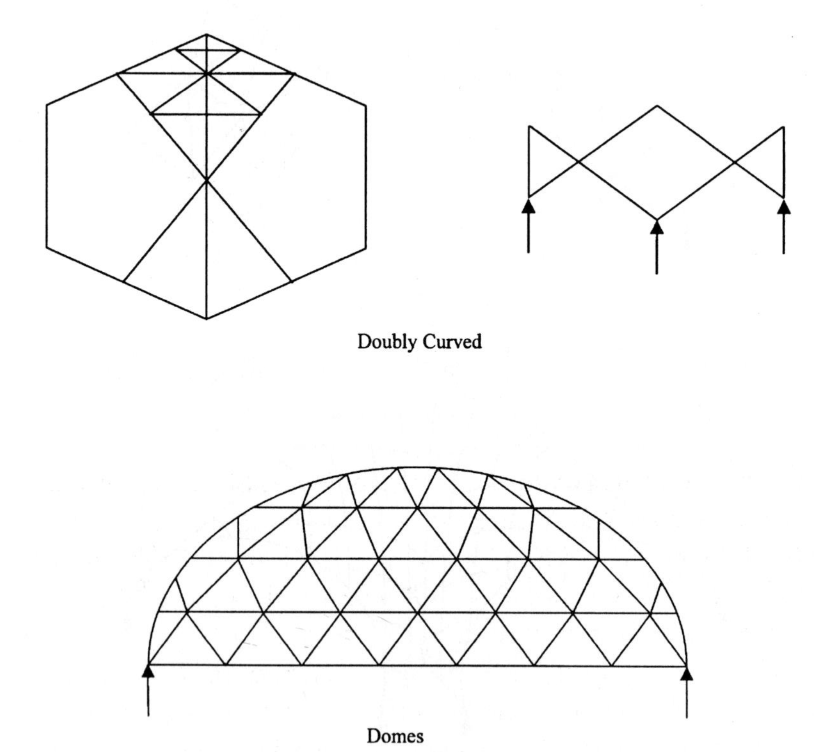

Figure 9.13(b). Section profiles of single-layer grid spaced frames.

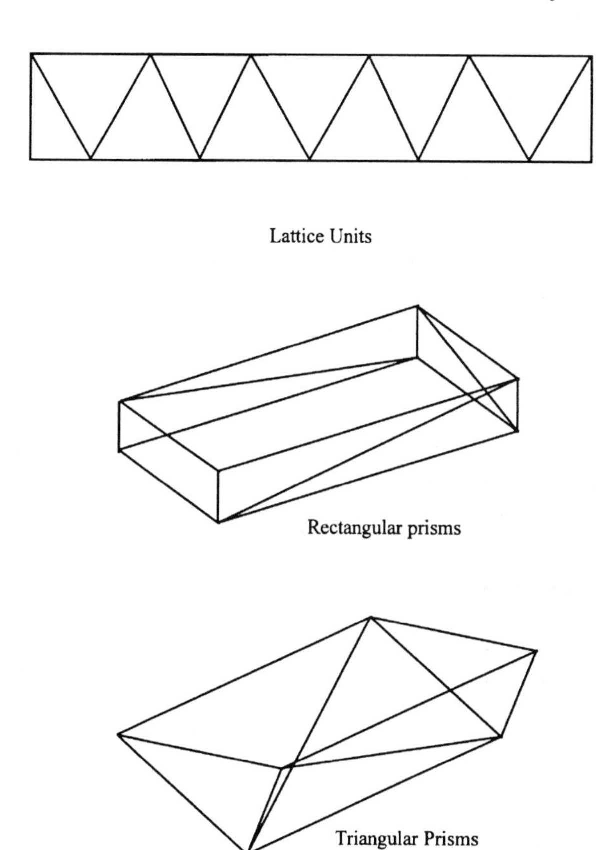

Figure 9.13(c). Section profiles of double-layer grid spaced frames.

An example of the use of space frame roof is the Singapore International Convention and Exhibition Centre (SICEC). It has an overall dimension of 172.8 m by 144 m and weighs 2400 tonnes. It has one of the world's largest clear spans of 86.4 m (Figure 9.14). The roof structure consists of an external fully exposed exoskeleton space frame and a series of secondary roof structure suspended from the exoskeleton to which the roof cladding systems are attached. The roof exoskeleton is supported at a total of 28 supports, 18 along the perimeter of the building and 10 internally (Figure 9.15). At each of these locations, the frame is supported on a single disc or confined elastomeric bearing with load capacities up to a maximum of 1340 tonnes. The bearings are required to accommodate rotations and translations due to vertical loading and thermal changes of the exoskeleton. Secondary roof structures are supported from hangers projecting below the exoskeleton nodes.

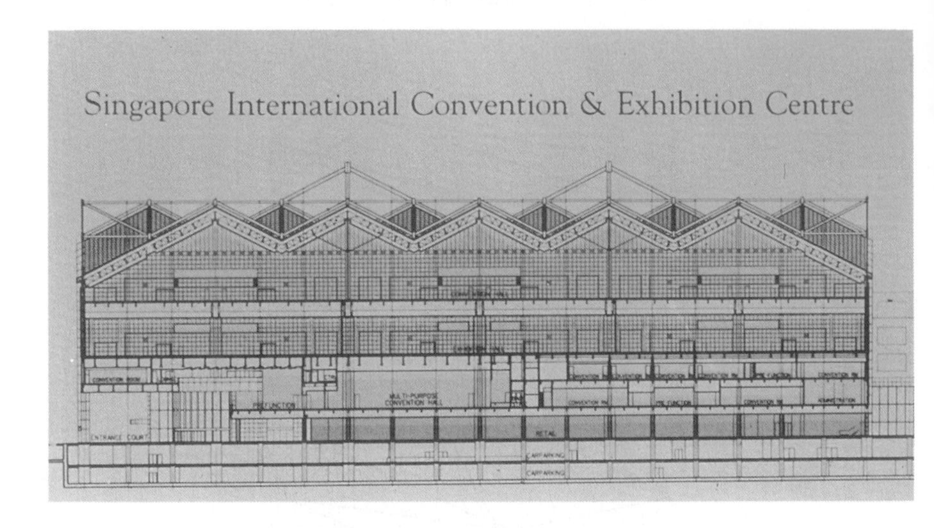

Figure 9.14(a). Architectural view of SICEC.

Figure 9.14(b). Space frame roof after the installation of roof cladding

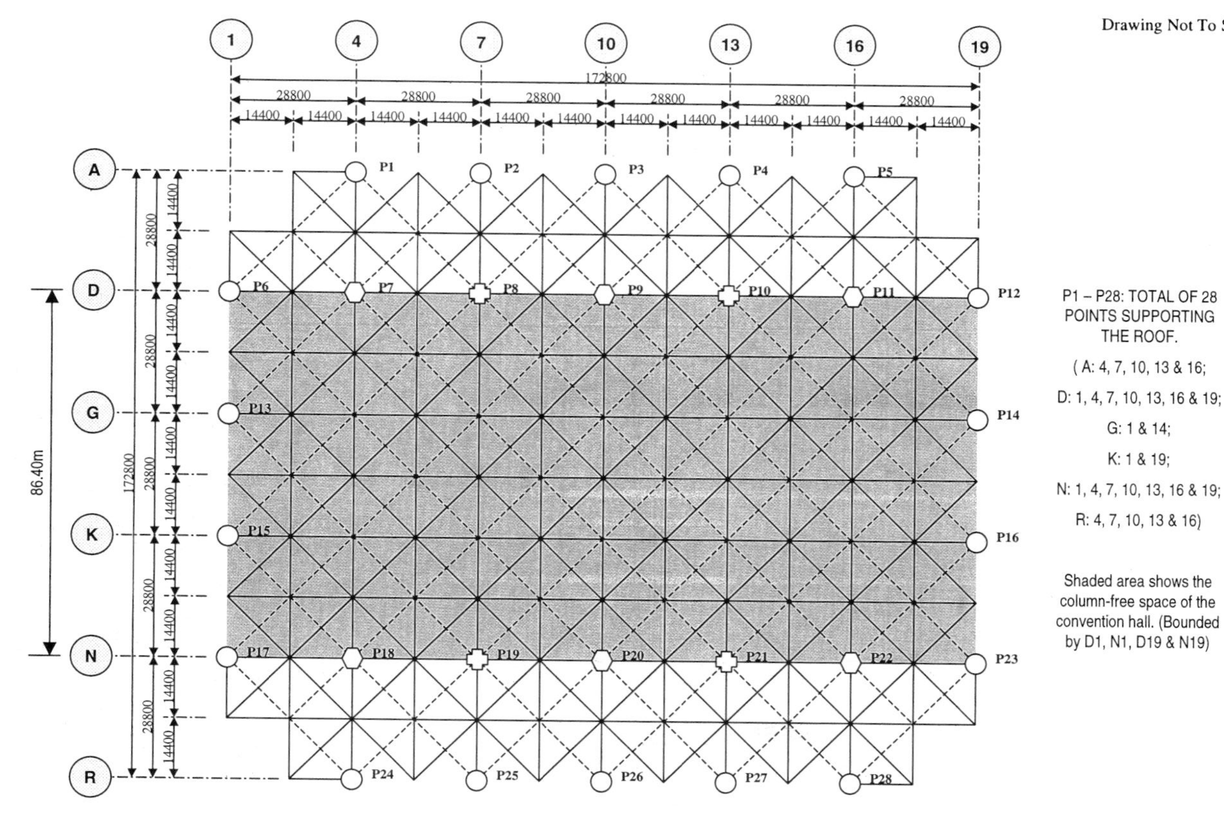

Figure 9.15. Layout of support positions.

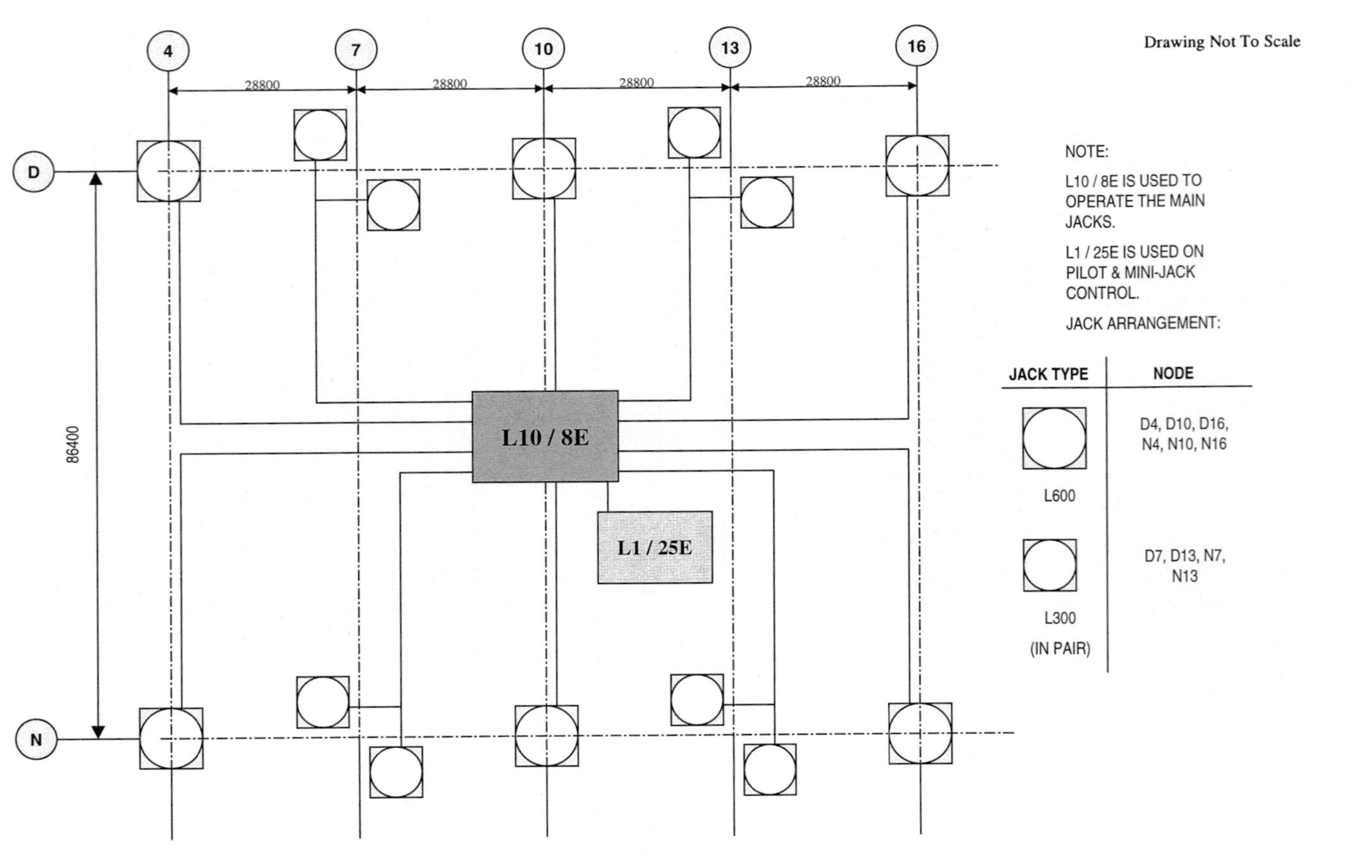

Figure 9.16. Plan of lifting jacks.

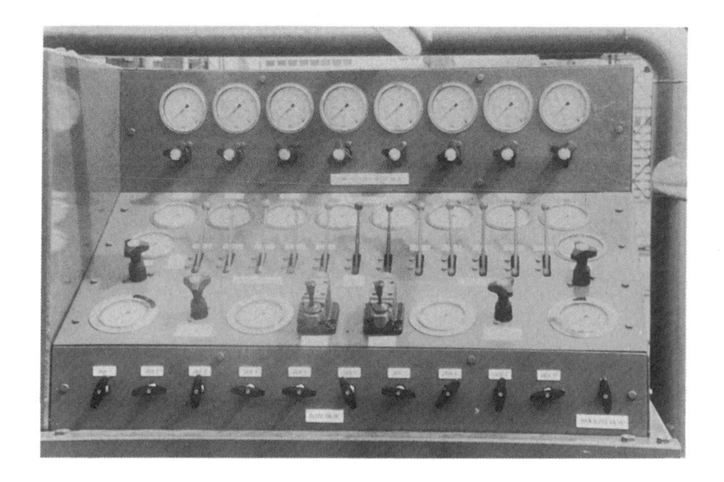

Figure 9.17. Electro-hydraulic power packs.

The hangers and associated branch elements are arranged to suit the many geometric requirements of the different secondary roof structural types and their location within the roof. The installation involved:

- *Marking and setting* — For placement of temporary node support towers, jacking towers and lifting jacks.
- *Erection of temporary node support towers, jacking towers and lifting jacks* — Ten sets of jacks of type L600 and L300 were used (Figure 9.16). These jacks were controlled by two electro-hydraulic power packs (Figure 9.17) which automatically synchronise operation of all the jacks. These lifting jacks were supported on temporary steel jacking towers (Figure 9.18). Each of these 14.2 m high jacking towers comprises four legs which are braced together at the top and include a series of temporary bracing to be removed and re-inserted according to the progress of the lifting operation. Figures 9.19 and 9.20 show the jacking towers for L600 and L300 jacks respectively. Temporary node support towers were erected to act as scaffolding for the welding of nodes to pipes, etc. (Figure 9.21).

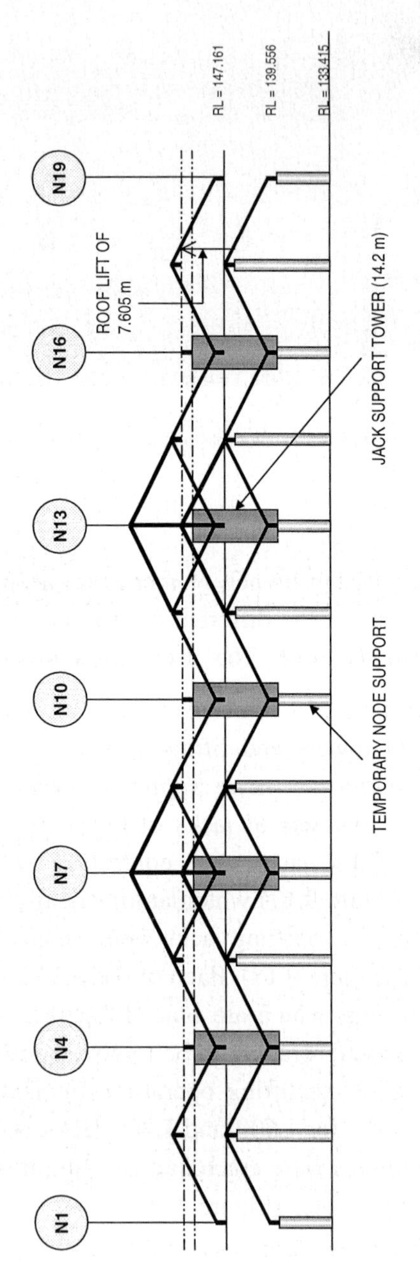

Figure 9.18. Support section at grid line N.

Figure 9.19. The L600 centre hole jacks.

Figure 9.20. Two L300 lifting jacks fixed to temporary steel jacking tower.

Figure 9.21(a). Layout of temporary node support towers.

Figure 9.21(b). Node assemblies sitting on temporary node support towers.

Figure 9.22. Completion of the lifting operation.

- *Assembly of space frame* — The shop fabricated steel components for the space frame were assembled on the sixth floor slab.
- *Lifting* — The roof was lifted over a three-day period. On the first day, it was lifted to 900 mm and the behaviour of the structure under load was monitored. Similarly, on the second day, it was lifted to a height of 4 m. On the third day, the frame was lifted to approximately 8 m from its temporary supports, slightly higher than its final level and stay for the next two weeks to allow for the erection of 28 permanent steel columns from the sixth floor to above eighth floor level before the space frame wás lowered to its permanent position (Figure 9.22).
- *Assembly of secondary roofs* — The secondary roof structure and roof cladding systems and M&E were installed.

Another example is shown in Section 9.10.

9.8. Shell Roofs

Shells are three-dimensional structures constructed with a curved solid slab or membrane acting as a stressed skin to transfer loading to a point of support. The main characteristic of a shell construction is the very curved membrane made structurally possible by providing restraint at the edges such that bending stresses in it are so small as to be negligible or are completely eliminated. The various profiles of a shell roof are shown in Figure 9.23.

An example of a shell roof in the form of barrel vaults is the Expo Gateway, World Trade Centre (Figure 9.24). The roof structure is of steel truss construction covered with a fabric membrane.

- *Assembly of trusses and bracing members* — the trusses were fabricated off-site and assembled using a 300 tonne crane (see Section 9.6).
- *Preparation of fabric membrane* — The membrane for the roof covering were manufactured in two pieces 23.8 m × 81 m weigh 2.45 tonne each. The roof covering were installed after the erection of the steel trusses and before installation of end wall glazing, aluminium and capping, smoke vents, lighting protection and sprinkler system. Temporary tensioning hardware were bolted into place around the perimeter to receive the fabric (Figure 9.25).
- *Lifting of fabric membrane* — The membrane was hoisted into position and lifted using a pulley system (Figure 9.26). The fabric was evenly spread, secured and tensioned (Figure 9.27).

9.9. Folded Plate Roofs

A folded plate roof is one in which the roof slab has been formed into a thin, self-supporting structure, usually made of concrete. Two of the most common shapes are "W" and "V" shapes. This type of roof is suitable when large areas of column-free space is needed. It is capable of long spans, and the cantilever projections can be used to advantage in exterior design, as well as to counterbalance the span. Figure 9.28 shows the various profiles.

9.10. Cable Supported Roofs

A cable supported roof derives its structural support from the tension capacity of steel cables. The steel cables are usually anchored to a compression ring

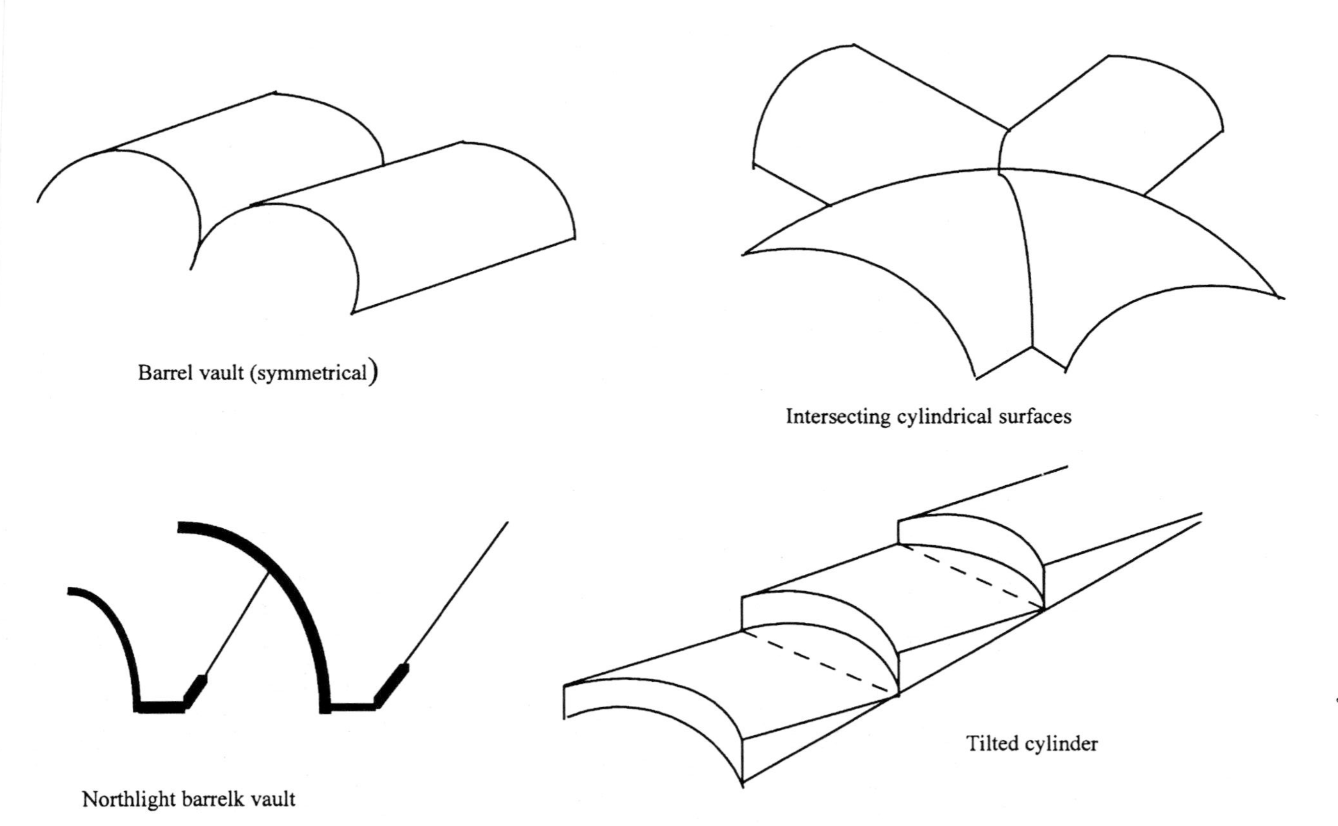

Figure 9.23(a). Section profiles of shell roofs — singly-curved shells.

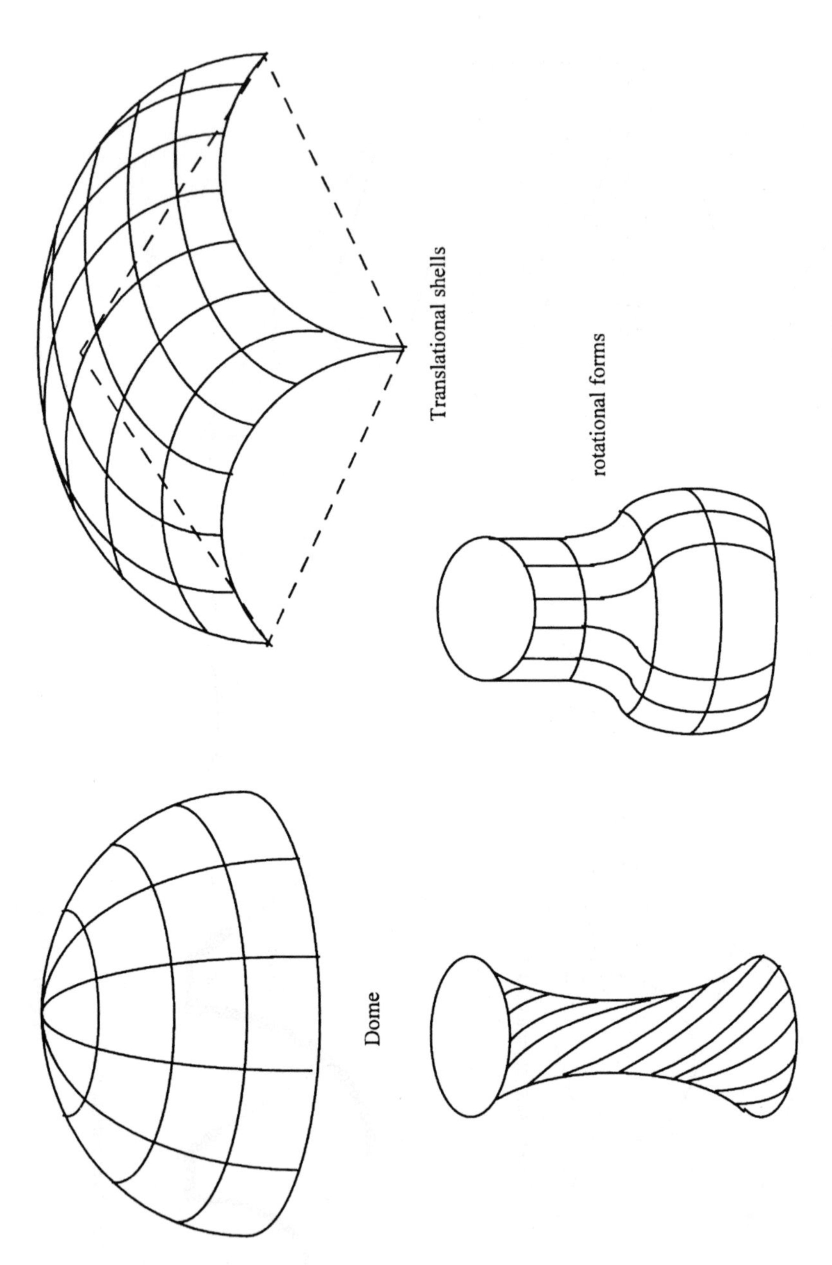

Figure 9.23(b). Section profiles of shell roofs — rotational shells.

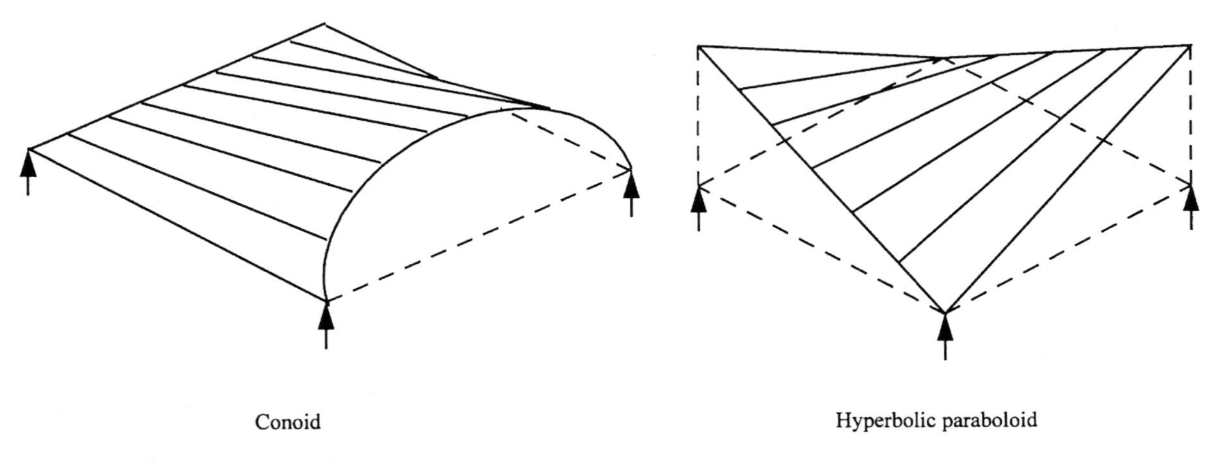

Figure 9.23(c). Section profiles of shell roofs — anticlastic shells.

Figure 9.24(a). Expo Gateway utilising a shell roof.

Figure 9.24(b). The supporting trusses, smoke vents etc viewing upwards.

Figure 9.25. Around the perimeter of the roof with tensioning devices for the fabric membrane.

Figure 9.26. The lifting of the fabric membrane.

Figure 9.27. The tensioning of the fabric membrane.

built into the exterior walls of a building. This arrangement produces a roof support system that has minimum depth, unlike other roof systems such as trusses and space frames which create a large volume of space that is of minimum use but still requires heating and ventilating. Figure 9.29 shows the various profiles of a cable supported roof.

An example of using the concept of cable stayed roof which utilises stainless steel rods to suspend the space frame roof to create a large column-free space is Keppel Distripark. The project required the assembly and erection of six blocks of space frame roofs stayed by stainless steel rods. The construction sequence employed is summarised as follows:

- *Marking and setting for placement of the pedestals.*
- *Erection of steel columns* — six steel columns were erected followed by the erection of perimeter columns and cladding (Figure 9.30).
- *Assembly of space frame* — The members of the space frame were assembled on the top floor slab. The adjustable pedestals were placed according to the location of the bottom nodes (Figure 9.31).

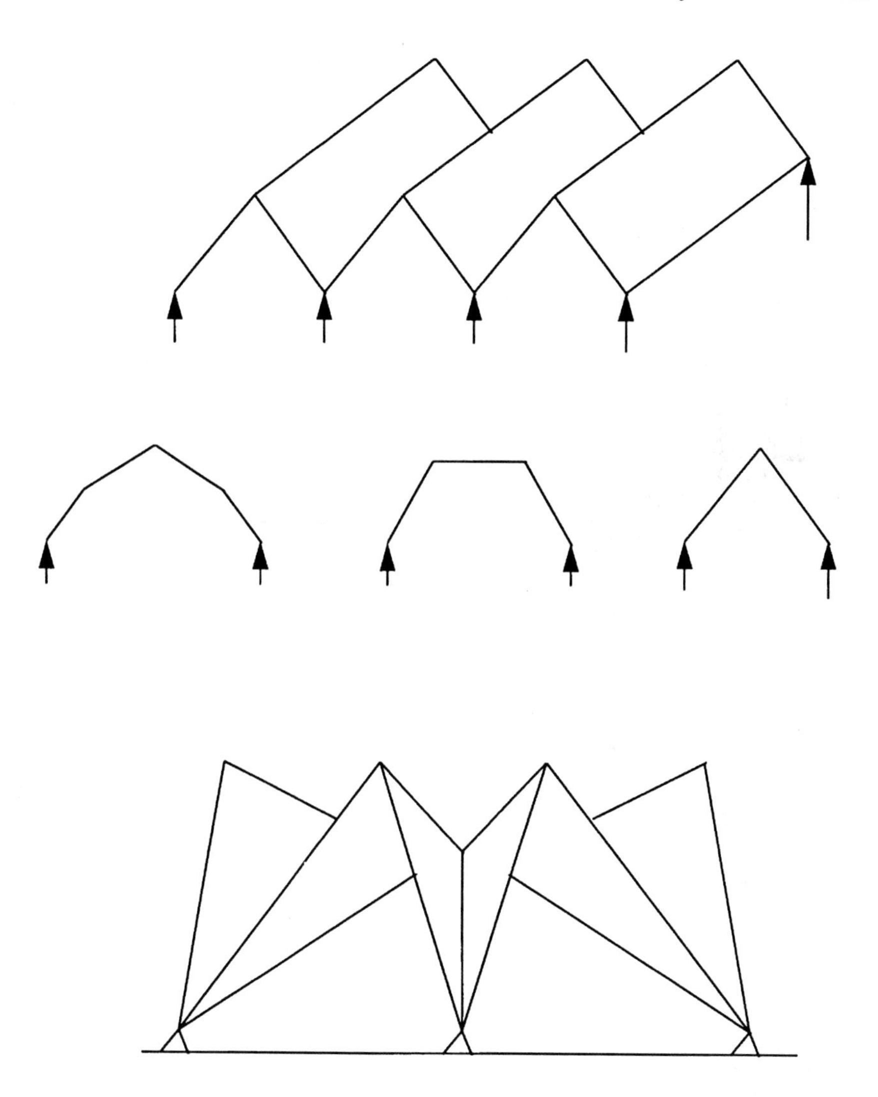

Figure 9.28(a). Section profiles of folded plate roofs — prismatic folded.

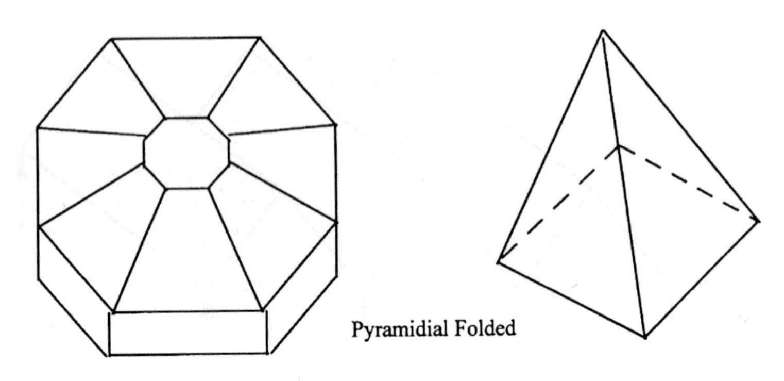

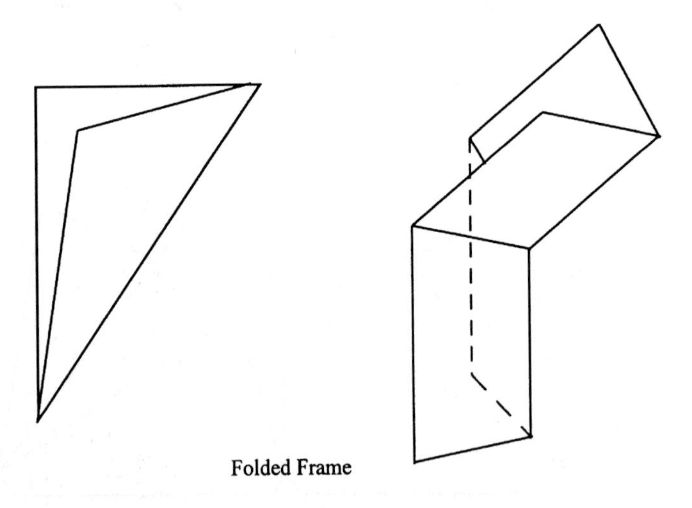

Figure 9.28(b). Section profiles of folded plate roofs — pyramidal folded and folded frame.

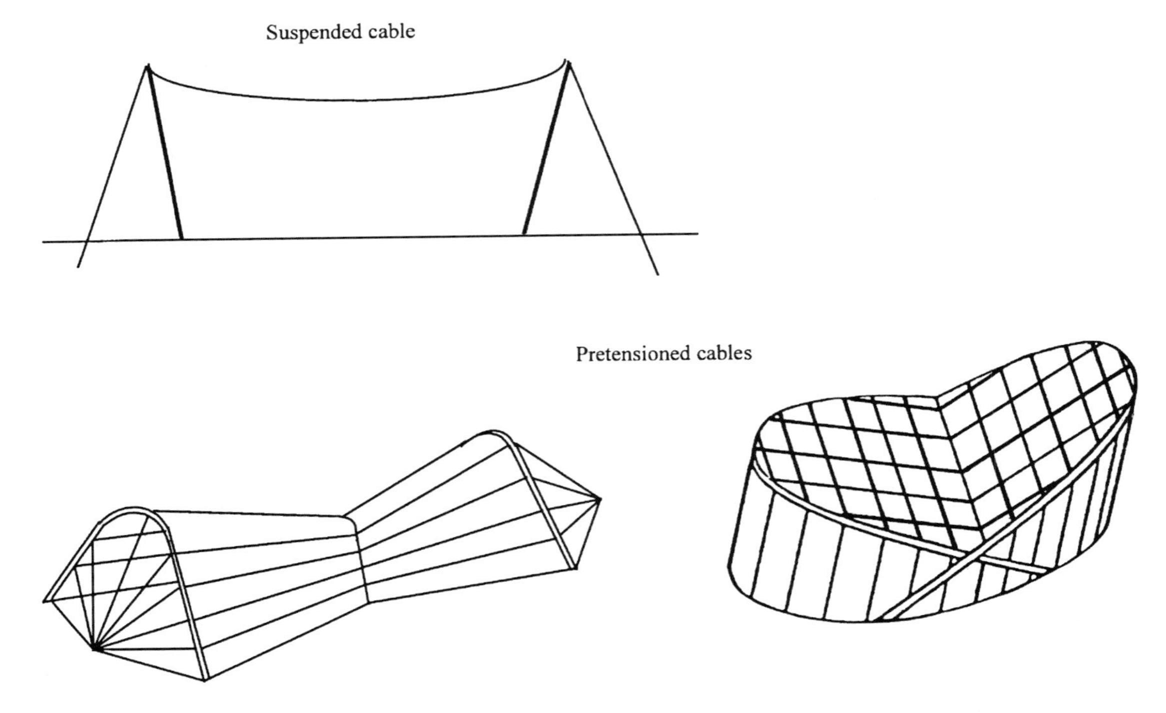

Figure 9.29(a). Section profiles of tension (cable) supported roofs.

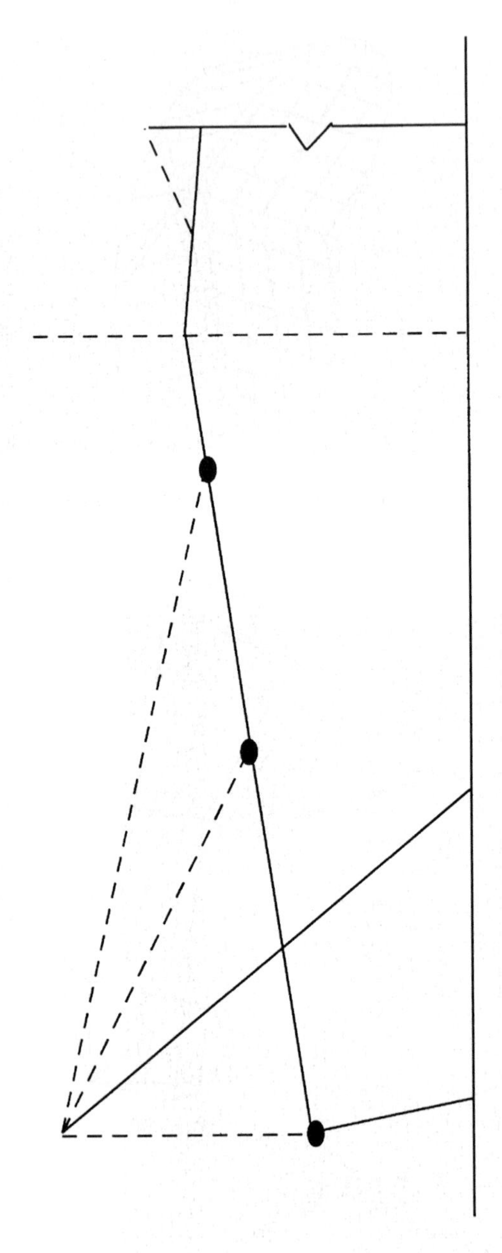

Figure 9.29(b). Section profiles of cable-stayed (suspended) roofs.

Figure 9.30. The erection of steel column, cladding and perimeter columns.

Figure 9.31. Space frames on adjustable pedestals.

- *Assembly of lifting tools* — The main lifting point specified for this roof was located at the connecting nodal point of the stainless steel rods and the nodal point of the space frame (Figure 9.32). The bottom pulley was erected and connected to the steel column, followed by connecting the lifting ropes and lashing of the lifting point (Figures 9.33 and 9.34). After that, the initial tensioning of the pulling and lifting ropes were carried out (Figure 9.35).
- *Lifting* — The space frame was lifted up at a constant speed of approximately 0.5 m/min, held temporarily at every one metre height for adjustment so that the synchro-discrepancy is controlled within a range of 50 mm (Figure 9.36).
- *Installation of cantilever frame* — Figure 9.37.
- *Erection of stainless steel rods* — There are four sets of stainless steel rods fixed to each steel column. These stainless steel rods were fixed to and lifted with the space frame. After initial positioning of the space frame, two 5T pulley blocks were used to fasten one end to the lower side of the space frame and the other to the lower lifting ear of the rod. The lower lifting ear was pulled close to the steel ball at lifting point. During the pull, the position of the space frame of the lower lifting ear was adjusted to align with the same axial line of upper lifting ear. The lifting ear was point welded to the steel ball of lifting point followed by the welding of each group of the four lifting ears to the lifting point. The tension of the rods were adjusted to comply with the design requirements.
- *Positioning* — Final positioning was carried out and lifting ropes were removed. The supports around the periphery and on the steel columns were all welded fast to bearing plates after positioning (Figure 9.38).

9.11. Roof Decks

The construction of a flat roof is similar to that of a floor. It is made up of layers of materials to satisfy the thermal and waterproofing requirements. It is usually covered with a screed of lightweight concrete to provide the necessary fall. The screed can also serve as a protective layer to the

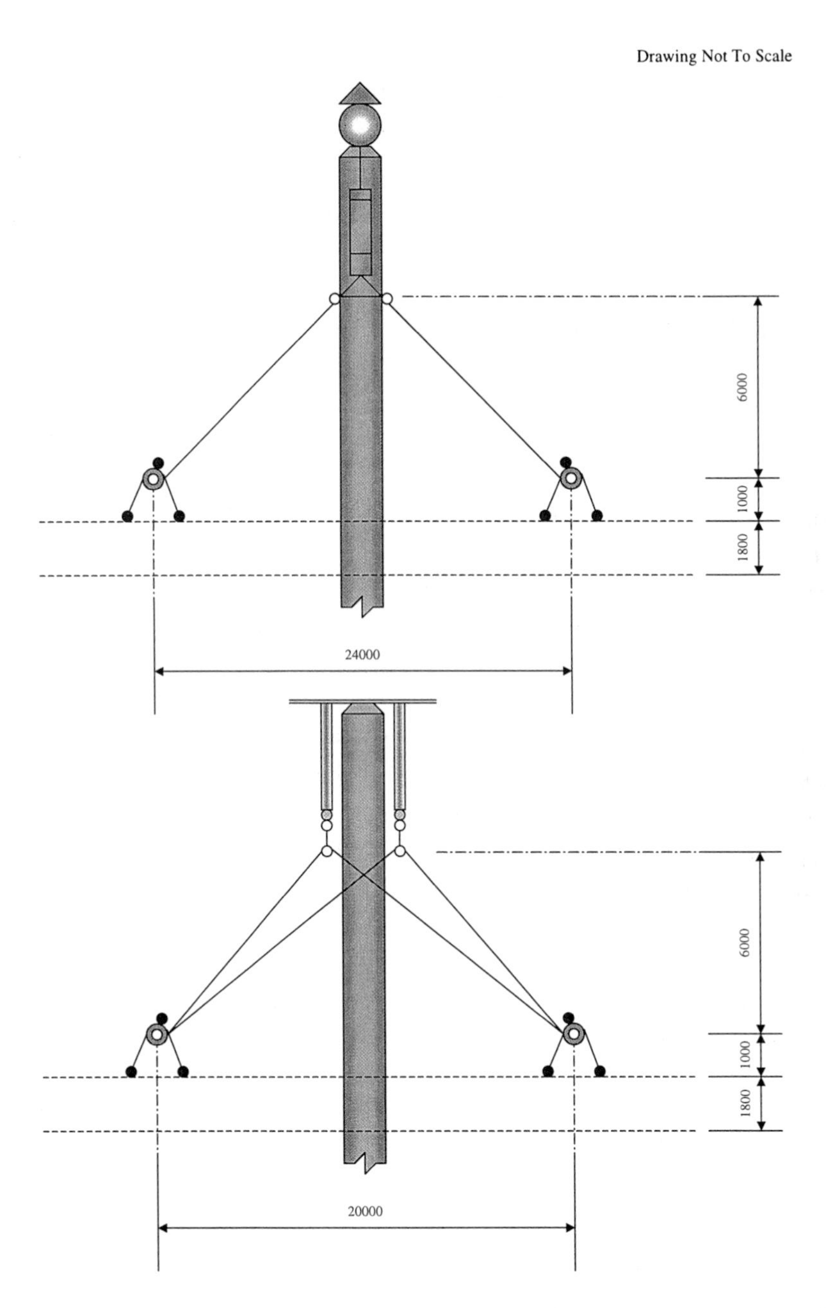

Figure 9.32. Elevations of lifting ropes.

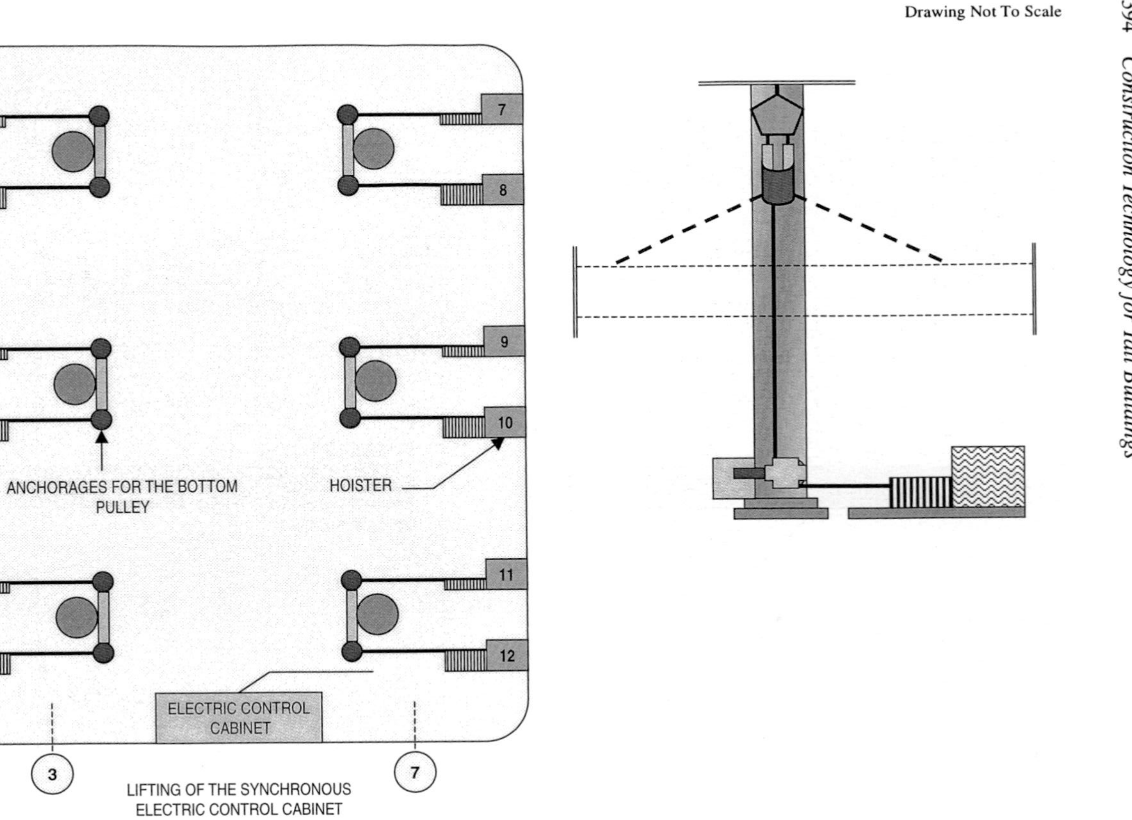

Figure 9.33. Arrangement of motor hoisters and electric control cabinet.

Figure 9.34. Anchorage detail of bottom pulley.

Figure 9.35. Initial tensioning of the pulling and lifting ropes.

Figure 9.36(a). Lifting of space frame.

Figure 9.36(b). Completion of the lifting process.

Figure 9.37(a). Cantilever members at the perimeter.

Figure 9.37(b). The frame section to be connected to cantilever members.

Figure 9.38. The completed space frame roof structure.

waterproofing membrane and thermal insulation. Figure 9.39 shows the construction details of a traditional Housing Development Board (HDB) roof system. The waterproofing layer comprises a bituminous waterproofing treatment over the roof. It is made up of six coats of material, one layer of fibre glass and one coat of bituminous aluminium paint. The system also includes a secondary roof which consists of lightweight precast concrete panels (1500 × 600 × 50) placed on hollow concrete stools (150 × 150 × 225) laid 225 mm above the main roof structure to serve as a heat insulation barrier (Figure 9.40). The amount of heat and the speed with which it penetrates into the roof slab is greatly reduced by the panel as well as the air gap. The heat penetration through joints in the precast secondary roof slabs are sealed with an elastic form of hot bitumen which is levelled and cleaned, and bent to the shape of a v-groove 12 mm thick. The system works as long as the concrete stools and slabs remain intact. Deterioration (breaking off) of the concrete stools have been reported in many cases and if the debris (mainly sand) is not cleaned off, serious drainage blockage may occur.

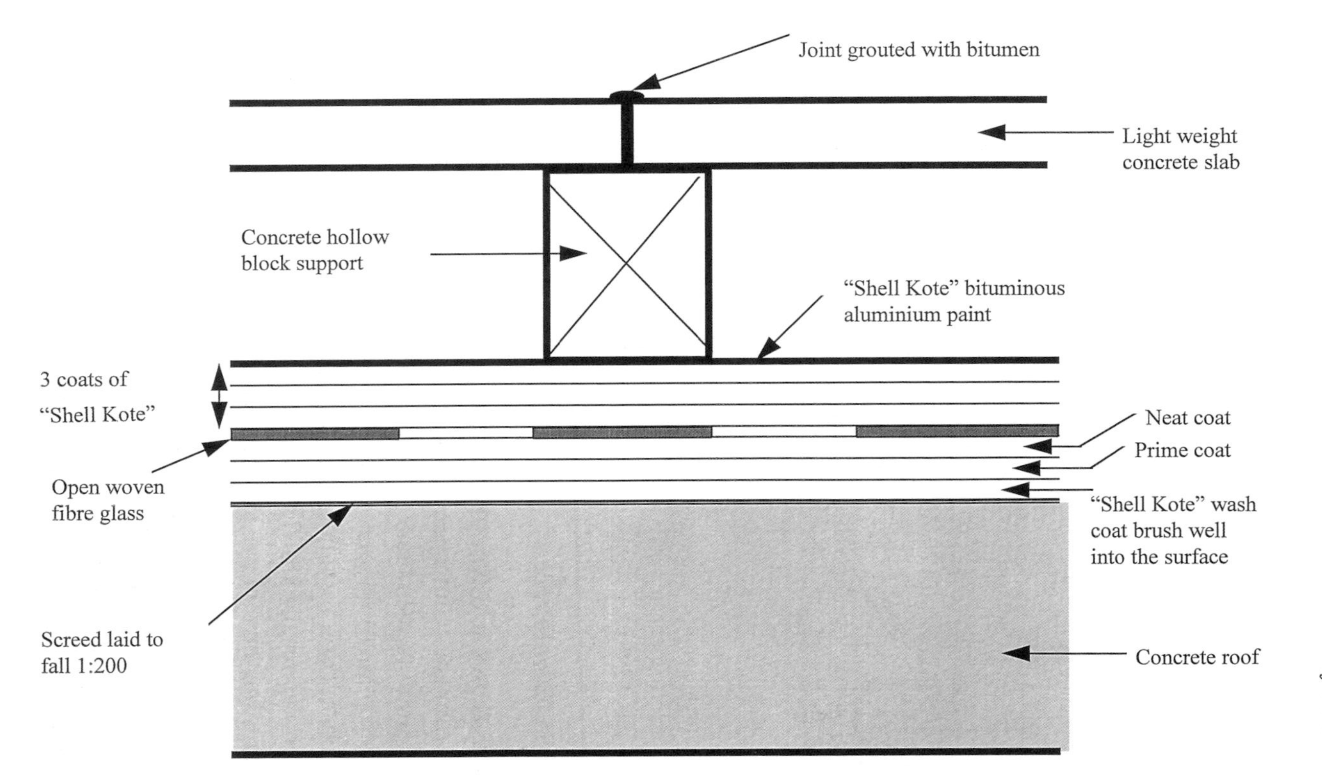

Figure 9.39. Sectional detail of waterproof treatment over a typical HDB flat roof.

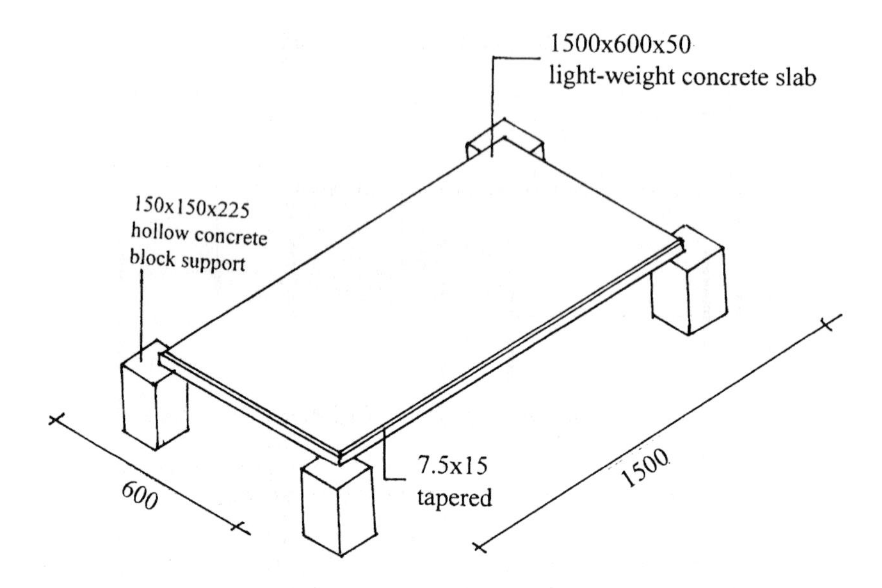

Figure 9.40. Lightweight precast slabs supported by hollow concrete blocks for thermal insulation.

9.11.1. *Installation Process*

Figure 9.41 shows two typical decking systems for commercial buildings. The systems mainly comprise: (a) the structural slab, (b) screeding to falls, (c) a waterproofing membrane, (d) insulation, (e) protective screeding, and (f) finishes (optional). Figures 9.42 to 9.46 show the operations of a typical flat roof construction for a commercial tall building. Figure 9.42 shows the priming process on the structural concrete slab prior to the application of the waterproofing membrane. Figure 9.43 shows the application of the rolled waterproofing membrane. Figure 9.44 shows the ponding process to test the watertightness. Insulation is then installed. Popular insulation materials used include polystyrene, polyurethane, etc. Figure 9.45 shows that the screeding process protects the insulation and also acts as the adhesive for the finishes. In this example, ceramic tiles are used (Figure 9.46). Note the installation of the monorail system for handling the gondola for the long term maintenance and cleaning of the building facade.

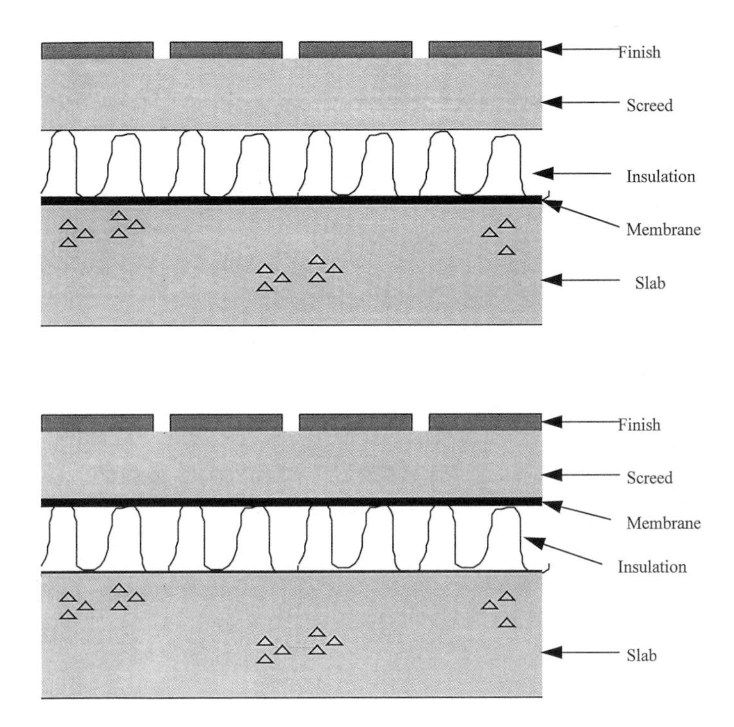

Figure 9.41. Top: inverted roof slab system; bottom: conventional roof slab system.

Figure 9.42. Priming of the deck to receive the waterproofing membrane.

Figure 9.43. Application of the rolled waterproofing membrane.

Figure 9.44. Ponding or water test on the deck.

Figure 9.45(a). BRC placed to avoid shrinkage cracks.

Figure 9.45(b). Screeding (cement mortar) is laid.

Figure 9.46. Ceramic floor tile finishes. Note the permanent gondola and its track for maintenance purposes.

9.11.2. *Drainage and Durability*

The biggest problem with flat roofs in tall buildings is associated with water leakages due to drainage problems and deterioration of roofing materials such as the waterproofing membrane [11–13]. Figure 9.47 shows an example of choking at the drainage outlet by the accumulation of debris, leaves etc. Figure 9.48 shows the "bubble" phenomenon on a waterproofing membrane, caused by the pressure of hot air evaporating from entrapped moisture introduced during construction or from elsewhere. Entrapped moisture before and after construction in concrete roofs can take several months to completely evaporate.

In addition, joints in exposed sheeting materials should be arranged to facilitate the flow of water. Special attention should be paid to the design of flashings at gutters and outlets.

9.12. Thermal Properties

Thermal design is concerned with the flow of heat through the roof construction and the effect of these on the performance of the roof and on the various components in the roofing system. All materials used in roof construction possess thermal insulating properties to varying degrees and for certain types such as woodwool slabs or aerated concrete units, the deck component alone can provide significant thermal insulation. Generally, a lighter and more cost-effective roof is obtained by adding a separate insulation layer, usually in the form of a rigid board insulation above the deck [14].

The rate of flow of heat through the roof is determined by the thermal conductivity of the elements making up the roof system. The Building Control Regulations of Singapore under Division 10 — Energy Conservation Requirements set out mandatory standards for the thermal transmittance or U-value of the roof. The "Handbook on Energy Conservation in Buildings and Building Services" published by the then Development & Building Control Division of PWD Singapore [15] provides detail in the assessment and calculation of the various required properties. The associated terms and methods of calculation are as follows:

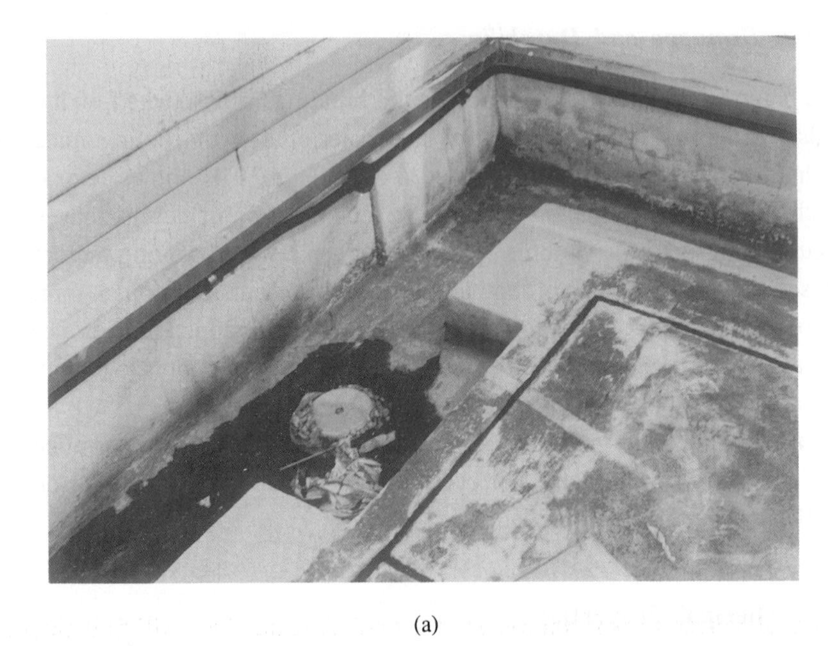

(a)

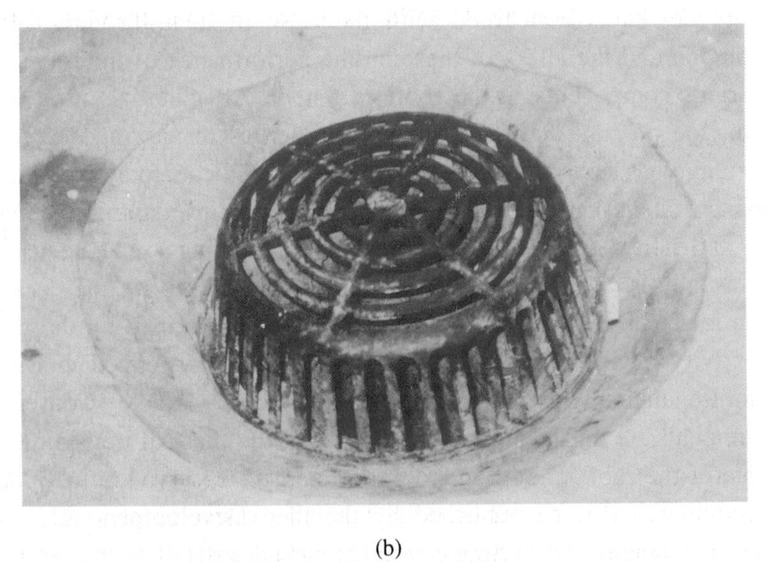

(b)

Figure 9.47. (a) and (b) Choking of the drainage outlet.

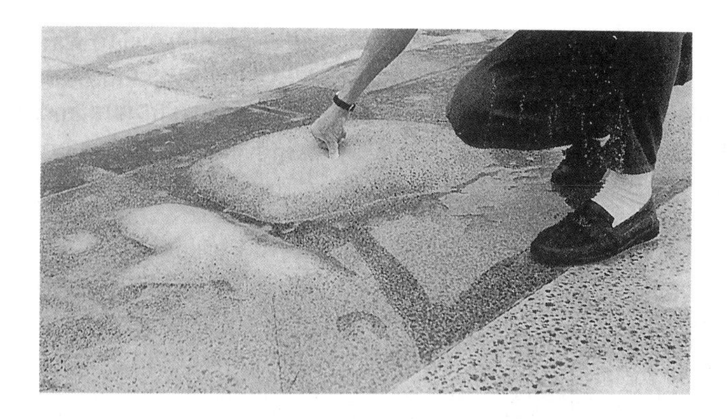

Figure 9.48. "Bubble" phenomenon or blistering caused by entrapped moisture.

9.12.1. *Thermal Conductivity* (*k*)

Thermal conductivity is a measure of the rate at which heat will flow through a material when a difference exists between the temperatures of its surfaces. It is expressed in W/m°C.

9.12.2. *Thermal Resistance* (*R*)

As the thickness of a material increases, its resistance to heat flow increases in direct proportion and can be calculated as:

$$\text{Thermal resistance } R = t/k$$

where t is the thickness in metres and k the thermal conductivity. The thermal resistance (R) is expressed in m^2°C/W.

At an exposed surface, the resistance to heat transfer by radiation and convection can also be regarded as thermal resistance, generally termed surface resistance. The surface resistance value depends on the emissivity of the surface, the direction of the flow of heat and additionally for the external surface, on the degree of exposure. The thermal resistance of airspaces depends on the size and ventilation of the cavity, the direction of the flow of heat, and on the emissivity of the surfaces of the cavity.

The total thermal resistance (R) of a roofing system is the summation of all the individual thermal resistance, taking into account the resistance of all the components of the roof including surface resistance and the resistance of cavities.

9.12.3. *Thermal Transmittance* (U)

The thermal transmittance of the roof is defined as the quantity of heat that flows through a unit area in unit time, per unit difference in temperature. It is expressed in $W/m^2{}^\circ C$ and is the reciprocal of the total thermal resistance of the roof:

$$U = 1/R$$

The value of U provides an easy method of assessment of the heat loss through the building structure and is not only required in calculations but also allows one to compare thermal performances of alternative roof constructions. The smaller the value of U, the better the insulation.

In the case where the roof deck consists of a number of individual components, the value of U is obtained from the total thermal resistance (R) of the roof structure which is calculated from the individual thermal resistance of each component of the roof.

$$U = 1/R_{si} + R_{so} + R_{cav} + R_1 + R_2 + R_3 + \dots$$

where R_{si} = internal surface resistance
R_{so} = external surface resistance
R_{cav} = resistance of any cavity

from standard thermal resistance values and R_1, R_2 and R_3 (and so on) are the thermal resistance of the materials, calculated from t/k where t is the thickness of material and k the thermal conductivity of the material.

Standard thermal properties of the materials are given in Table 9.1.

Table 9.1. Thermal properties of roof components [11].

	Thermal conductivity (k) W/m°C	**Thermal resistance (R)** m^{2}°C/W
Air surfaces		
Internal		0.105
External		0.045
Air cavities (unventilated)		
Low surface emissivity		0.32
High surface emissivity		0.18
Intermittent cavity		0.045
Decks		
Dense concrete	1.4	
Lightweight aerated concrete	0.16	
Plywood & chipboard	0.14	
Timber boarded deck	0.14	
Woodwool	0.093	
Metal decking	0	
Insulation		
Cellular glass slab	0.045	
Cork board	0.042	
Expanded polystyrene	0.034	
Extruded polystyrene	0.031	
Glass fibre roofboard	0.034	
Isocyanurate board	0.022	
Mineral wool slab	0.034	
Perlite board	0.05	
Perlite bitumen screed	0.076	
Polyurethane board	0.022	
Wood fibreboard	0.05	
Waterproofing		
Mastic asphalt		0.06
Bitumen felt		0.06

Table 9.2. An example of the calculation of the value of U [11].

Roof Element	Thickness t (m)	k (W/m°C)	R (m^2°C/W)
External surface resistance			0.045
Asphalt waterproofing			0.060
Insulation Board	0.050	0.042	1.190
Sand & cement screed	0.050	1.40	0.036
Concrete slab	0.125	1.40	0.089
Internal surface resistance			0.105

Total $R = 1.525$

Overall value of $U = 1/R = 0.65$ W/m^2°C.

9.12.4. *Example of the Calculation of the Value of U*

A roof deck consists of, from the top: 20 mm mastic asphalt, 50 mm insulation board ($k = 0.042$ W/m°C), 50 mm sand and cement screed ($k = 1.4$ W/m°C), and 125 mm concrete slab. The thermal resistance of each component from t/k are shown in Table 9.2. The value of U which is $1/R$ is thus 0.65 W/m^2°C.

References

[1] A. J. Elder and M. Vandenberg, *AJ Handbook of Building Enclosure*, The Architectural Press, London, 1974.

[2] R. C. Smith and C. K. Andres, *Principles and Practices of Heavy Construction*, Prentice-Hall, 1986.

[3] C. W. Griffin, *Manual of Low-Slope Roof Systems*, 3rd Edition, McGraw-Hill, New York, 1996.

[4] W. McElroy, *Roof Builder's Handbook*, PTR Prentice Hall, New Jersey, 1993.

[5] C. K. Andres, *Principles and Practices of Heavy Construction*, 5th Edition, Prentice Hall, New York, 1998.

[6] D. T. Coates, *Roofs and Roofing: Design and Specification Handbook*, Whittles, UK, 1993.

[7] S. Hardy, *Time-Saver Details for Roof Design*, McGraw-Hill, New York, 1997.

[8] H. W. Harrison, "Roofs and Roofing: Performance, Diagnosis, Maintenance, Repair and the Avoidance of Defects", Building Research Establishment, Watford, Herts, 1998.

[9] Council on Tall Buildings and Urban Habitat, Committee S37, "Cold-Formed Steel in Tall Buildings", McGraw-Hill, New York, 1993.

[10] D. A. Cuoco, "State of the Art of Space Frame Roof Structures", Proceedings for Annual Convention and Exhibit of American Society of Civil Engineers, New York, 26–30 October 1981, pp. 7–9.

[11] Bickerdike Allen Partners, "Flat Roof Manual — A Guide to the Repair and Replacement of Built-Up Felt Roofs", NHS Continuing Education Unit Publications, September 1985.

[12] R. Sperling, "Roofs for Warm Climates", Building Research Station, UK, 1984.

[13] "Technical Guide to Flat Roofing", PSA, UK, 1987.

[14] F. March, *Flat Roofing — A Guide to Good Practices*, Tarmac, 1983.

[15] The Development & Building Control Division (P.W.D.) Singapore, "Handbook on Energy Conservation in Buildings and Building Services", PWD, 1983.

INDEX